Housing Defects Reference Manual

OTHER TITLES FROM E. & F.N. SPON

Building Failures: Diagnosis and Avoidance
W.H. RANSOM

Builder's Reference Book
L. BLACK

Corrosion of Steel in Concrete
EDITED BY P. SCHIESSL

Defects and Deterioration in Building
B.A. RICHARDSON

Durability of Building Materials and Components
EDITED BY J.M. BAKER, P.J. NIXON, A.J. MAJUMDAR AND H. DAVIES

Handbook of Segmental Paving
A.A. LILLEY

Hazardous Building Materials
S.R. CURWELL AND C.G. MARCH

Maintenance of Brick and Stone Masonry Structures
EDITED BY A.M. SOWDEN

Rehabilitation and Re-use of Old Buildings
D. HIGHFIELD

Spon's Contractors' Handbooks
Minor Works, Alterations, Repairs and Maintenance
Roofing
Electrical Installation
Floor Wall and Ceiling Finishings
Painting and Decorating
Plumbing and Domestic Heating
EDITED BY SPAIN AND PARTNERS

Structural Integrity Monitoring
R.A. COLLACOTT

The Soiling and Cleaning of Building Facades
EDITED BY L.G.W. VERHOEF

For information on these and other titles please write to
The Promotion Department, E. & F.N. Spon, 2–6 Boundary Row,
London SE1 8HN.

Housing Defects Reference Manual

THE BUILDING RESEARCH ESTABLISHMENT
DEFECT ACTION SHEETS

E & FN SPON
An Imprint of Chapman & Hall
London · New York · Tokyo · Melbourne · Madras

UK	Chapman & Hall, 2–6 Boundary Row, London SE1 8HN
USA	Van Nostrand Reinhold, 115 5th Avenue, New York NY10003
JAPAN	Chapman & Hall Japan, Thomson Publishing Japan, Hirakawacho Nemoto Building, 7F, 1-7-11 Hirakawa-cho, Chiyoda-ku, Tokyo 102
AUSTRALIA	Chapman & Hall Australia, Thomas Nelson Australia, 102 Dodds Street, South Melbourne, Victoria 3205
INDIA	Chapman & Hall India, R. Seshadri, 32 Second Main Road, CIT East, Madras 600 035

First edition 1991

© British Crown copyright 1991 (year of first publication).
Published by permission of the Controller of
Her Britannic Majesty's Stationery Office.

Printed in Great Britain at the
The Bath Press, Avon

ISBN 0 419 16720 X 0 442 314914 (USA)

Apart from any fair dealing for the purposes of research or private study, or criticism or review, as permitted under the UK Copyright Designs and Patents Act, 1988, this publication may not be reproduced, stored, or transmitted, in any form or by any means, without the prior permission in writing of the publishers, or in the case of reprographic reproduction only in accordance with the terms of the licences issued by the Copyright Licensing Agency in the UK, or in accordance with the terms of licences issued by the appropriate Reproduction Rights Organization outside the UK. Enquiries concerning reproduction outside the terms stated here should be sent to the publishers at the UK address printed on this page.

The publisher makes no representation, express or implied, with regard to the accuracy of the information contained in this book and cannot accept any legal responsibility or liability for any errors or omissions that may be made.

British Library Cataloguing in Publication Data
Housing defects reference manual: The Building Research Establishment defect action sheets.
I. British Research Establishment
690

ISBN 0–419–16720–X

Library of Congress Cataloging-in-Publication Data
Available

Publishers Note

The Defect Action Sheets in this compilation were published between 1982 and 1990. New editions and revisions of many of the Sheets were issued by BRE, as indicated in the complete list on page xi. However, references in the Sheets to regulations, standards and other documents have not been revised at the time of publishing this compilation. Readers should therefore check that they are using the current editions of such documents.

Contents

The Defect Action Sheets have been collected together in this book into 14 subject groups for ease of use and reference to information on a particular topic. If you know the number of a DAS, you can find its position in the book by referring to the Numerical List of Defect Action Sheets on page xi. The Subject Index at the end of the book also enables information on a topic to be traced easily.

Numerical List of Defect Action Sheets

DAS number	*Title*	*First issued*	*Page*
DAS 1	Slated or tiled pitched roofs: ventilation to outside air (*Design*)	May 1982, minor revisions February 1985	156
DAS 2	Reinforced-concrete framed flats: repair of disrupted brick cladding (*Design*)	May 1982, minor revisions February 1985	62
DAS 3	Slated or tiled pitched roofs: restricting the entry of water vapour from the house (*Design*)	June 1982, minor revisions February 1985	158
DAS 4	Pitched roofs: thermal insulation near the eaves (*Site*)	June 1982	182
DAS 5	Pitched roofs: trussed rafters – site storage (*Site*)	July 1982, minor revisions February 1985	160
DAS 6	External walls: reducing the risk from interstitial condensation (*Design*)	July 1982, minor revisions February 1985	110
DAS 7	Pitched roofs: boxed eaves – preventing fire spread between dwellings (*Design*)	October 1982, minor amendments February 1985	166
DAS 8	Pitched roofs: separating wall/roof junction – preventing fire spread between dwellings (*Design*)	October 1982, New edition February 1985	168
DAS 9	Pitched roofs: sarking felt underlay – drainage from roof (*Design*)	November 1982, minor revisions February 1985	170
DAS 10	Pitched roofs: sarking felt underlay – watertightness (*Site*)	November 1982, minor revisions February 1985	172
DAS 11	Wood windows and door frames: care on site during storage and installation (*Site*)	December 1982, minor revisions February 1985	224
DAS 12	Cavity trays in external cavity walls: preventing water penetration (*Design*)	December 1982, minor revisions February 1985	122
DAS 13	Wood windows: arresting decay (*Design*)	January 1983, minor amendments February 1985	236
DAS 14	Wood windows: preventing decay (*Design*)	January 1983, minor revisions February 1985	238
DAS 15	Wood windows: resisting rain penetration at perimeter joints (*Design*)	January 1983, withdrawn April 1987, replaced by DAS 98	–
DAS 16	Walls and ceilings: remedying recurrent mould growth (*Design*)	January 1983, minor revisions February 1985	128
DAS 17	External masonry walls insulated with mineral fibre cavity-width batts: resisting rain penetration (*Site*)	February 1983, minor revisions April 1985	118
DAS 18	External masonry walls: vertical joints for thermal and moisture movements (*Design*)	February 1983, new edition February 1985	52
DAS 19	External masonry cavity walls: wall ties – selection and specification (*Design*)	February 1983, withdrawn June 1988, replaced by DAS 115	–
DAS 20	External masonry cavity walls: wall ties – installation (*Site*)	February 1983, withdrawn June 1988, replaced by DAS 116	–
DAS 21	External masonry cavity walls: wall tie replacement (*Design*)	March 1983	66

Numerical List of Defect Action Sheets

DAS number	*Title*	*First issued*	*Page*
DAS 22	Ground floors: replacing suspended timber with solid concrete – dpcs and dpms (*Design*)	March 1983, minor revisions February 1985	20
DAS 23	Pitched roofs: trussed rafters; bracing and binders – specification (*Design*)	March 1983, withdrawn August 1986, replaced by DAS 83	–
DAS 24	Pitched roofs: trussed rafters; bracing and binders – installation (*Site*)	March 1983, withdrawn August 1986, replaced by DAS 84	–
DAS 25	External and separating walls: lateral restraint at intermediate timber floors – specification (*Design*)	April 1983	76
DAS 26	External and separating walls: lateral restraint at intermediate timber floors – installation (*Site*)	April 1983, minor revisions March 1985	78
DAS 27	External and separating walls: lateral restraint at pitched roof level – specification (*Design*)	May 1983	80
DAS 28	External and separating walls: lateral restraint at pitched roof level – installation (*Site*)	May 1983	82
DAS 29	External and separating walls: cavity barriers against fire – location (*Design*)	June 1983, minor revisions February 1985	132
DAS 30	External and separating walls: cavity barriers against fire – installation (*Site*)	June 1983, minor revisions February 1985	134
DAS 31	Suspended timber floors: chipboard flooring – specification (*Design*)	July 1983	36
DAS 32	Suspended timber floors: chipboard flooring – storage and installation (*Site*)	July 1983	38
DAS 33	Flat roofs: built-up bitumen felt – remedying rain penetration (*Design*)	August 1983	146
DAS 34	Flat roofs: built-up bitumen felt – remedying rain penetration at abutments and upstands (*Design*)	August 1983	148
DAS 35	Substructure: dpcs and dpms – specification (*Design*)	September 1983, new edition March 1985	4
DAS 36	Substructure: dpcs and dpms – installation (*Site*)	September 1983, minor revisions March 1985	6
DAS 37	External walls: rendering – resisting rain penetration (*Design*)	October 1983, minor revisions March 1985	88
DAS 38	External walls: rendering – application (*Site*)	October 1983	90
DAS 39	Plastics drainage pipes: laying, jointing and backfilling (*Site*)	November 1983, minor revisions March 1985	256
DAS 40	Plastics drainage pipes: storage and handling (*Site*)	November 1983, minor revisions March 1985	254
DAS 41	Plastics sanitary pipework: jointing and support – specification (*Design*)	December 1983	262
DAS 42	Plastics sanitary pipework: jointing and support – installation (*Site*)	December 1983	264
DAS 43	Trussed rafter roofs: tank supports – specification (*Design*)	January 1984, minor revisions March 1985	184
DAS 44	Trussed rafter roofs: tank supports – installation (*Site*)	January 1984	186
DAS 45	Intermediate timber floors in converted dwellings – sound insulation (*Design*)	February 1984, minor revisions March 1985	40
DAS 46	Masonry walls: chasing (*Site*)	February 1984	56

DAS number	*Title*	*First issued*	*Page*
DAS 47	Suspended timber floors: notching and drilling of joists (*Design*)	March 1984, withdrawn April 1987, replaced by DAS 99	–
DAS 48	Electric cables – stripping plastics sheathing (*Site*)	March 1984	288
DAS 49	Clay-ware drainage pipes: storage and handling (*Site*)	April 1984	258
DAS 50	Flexibly-jointed clay-ware drainage pipes: jointing and backfilling (*Site*)	April 1984	260
DAS 51	Floors: cement-based screeds – specification (*Design*)	May 1984	12
DAS 52	Floors: cement-based screeds – mixing and laying (*Site*)	May 1984	14
DAS 53	Stairways: safety of users – specification (*Design*)	June 1984	46
DAS 54	Stairways: safety of users – installation (*Site*)	June 1984	48
DAS 55	Roofs: eaves, gutters and downpipes – specification (*Design*)	July 1984	270
DAS 56	Roofs: eaves, gutters and downpipes – installation (*Site*)	July 1984	272
DAS 57	Suspended timber floors: joist hangers in masonry walls – specification (*Design*)	August 1984	28
DAS 58	Suspended timber floors: joist hangers in masonry walls – installation (*Site*)	August 1984	30
DAS 59	Felted cold-deck flat roofs: remedying condensation by converting to warm-deck (*Design*)	September 1984	150
DAS 60	Single-wall flue pipes for class II appliances – location, fixing and shielding (*Design*)	September 1984	210
DAS 61	Cold water storage cisterns: overflow pipes (*Design*)	January 1985, withdrawn February 1990, replaced by DAS 141	–
DAS 62	Electrical services: avoiding cable overheating (*Design*)	February 1985, new edition August 1985	286
DAS 63	Flat or low-pitched roofs: laying flexible membranes when weather may be bad (*Site*)	August 1985	152
DAS 64	External walls – bricklaying and rendering when weather may be bad (*Site*)	August 1985	86
DAS 65	Oversites – concreting when weather may be bad (*Site*)	August 1985	8
DAS 66	Windows, doors and exterior joinery: applying putty, oil-based sealants and solvent-based paints when weather may be bad (*Site*)	August 1985	234
DAS 67	Inward-opening external doors: resistance to rain penetration (*Design*)	November 1985	226
DAS 68	External walls: joints with windows and doors – detailing for sealants (*Design*)	December 1985	228
DAS 69	External walls: joints with windows and doors – application of sealants (*Site*)	December 1985	230
DAS 70	External masonry walls: eroding mortars – repoint or rebuild? (*Design*)	February 1986	92
DAS 71	External masonry walls: repointing – specification (*Design*)	February 1986	94
DAS 72	External masonry walls: repointing (*Site*)	February 1986, new edition April 1986	96

Numerical List of Defect Action Sheets

DAS number	*Title*	*First issued*	*Page*
DAS 73	Suspended timber ground floor: remedying dampness due to inadequate ventilation (*Design*)	March 1986	22
DAS 74	Suspended timber ground floors: repairing rotted joists (*Design*)	March 1986	24
DAS 75	External walls: brick cladding to timber frame – the need to design for differential movement (*Design*)	April 1986	58
DAS 76	External walls: brick cladding to timber frame – how to allow for movement (*Site*)	April 1986	60
DAS 77	Cavity external walls: cold bridges around windows and doors (*Design*)	May 1986	112
DAS 78	External walls – dry lining: avoiding cold bridges (*Design*)	May 1986	114
DAS 79	External masonry walls: partial cavity fill insulation – resisting rain penetration (*Site*)	June 1986	120
DAS 80	Index	June 1986, withdrawn May 1987, replaced by DAS 100	–
DAS 81	Plasterboard ceilings for direct decoration: nogging and fixing – specification (*Design*)	July 1986	42
DAS 82	Plasterboard ceilings for direct decoration: nogging and fixing – site work (*Site*)	July 1986	44
DAS 83	Dual-pitched roofs: trussed rafters; bracing and binders – specification (*Design*)	August 1986, replaces DAS 23	190
DAS 84	Dual-pitched roofs: trussed rafters; bracing and binders – installation (*Site*)	August 1986, replaces DAS 24	192
DAS 85	Brick walls: injected dpcs (*Design*)	August 1986	100
DAS 86	Brick walls: replastering following dpc injection (*Design*)	August 1986	102
DAS 87	Wood entrance doors: discouraging illegal entry – specification (*Design*)	October 1986	244
DAS 88	Windows: discouraging illegal entry (*Design*)	October 1986	246
DAS 89	Domestic foul drainage systems: avoiding blockages – specification (*Design*)	November 1986	250
DAS 90	Domestic foul drainage systems: avoiding blockages – installation (*Site*)	November 1986	252
DAS 91	Domestic gas appliances: air requirements (*Design*)	December 1986	202
DAS 92	Balanced flue terminals: location and guarding (*Design*)	December 1986	206
DAS 93	Chimney stacks: taking out of service (*Design*)	February 1987	208
DAS 94	Masonry chimneys: dpcs and flashings – location (*Design*)	February 1987	218
DAS 95	Masonry chimneys: dpcs and flashings – installation (*Site*)	February 1987	220
DAS 96	Foundations on shrinkable clay: avoiding damage due to trees (*Design*)	March 1987	2
DAS 97	Large concrete panel external walls: re-sealing butt joints (*Design*)	March 1987	64
DAS 98	Windows: resisting rain penetration at perimeter joints (*Design*)	April 1987, replaces DAS 15	232

DAS number	*Title*	*First issued*	*Page*
DAS 99	Suspended timber floors: notching and drilling of joists (*Design*)	April 1987, replaces DAS 47	32
DAS 100	Index	May 1987, withdrawn June 1988, replaced by DAS 117	–
DAS 101	Plastics sanitary pipework: specifying for outdoor use (*Design*)	May 1987	266
DAS 102	External masonry walls: assessing whether cracks indicate progressive movement (*Design*)	June 1987	54
DAS 103	Wood floors: reducing risk of recurrent dry rot (*Design*)	June 1987	34
DAS 104	Masonry separating walls: airborne sound insulation in new-build housing (*Design*)	July 1987	140
DAS 105	Masonry separating walls: improving airborne sound insulation between existing dwellings (*Design*)	July 1987	142
DAS 106	Cavity parapets – avoiding rain penetration (*Design*)	August 1987	124
DAS 107	Cavity parapets – installation of copings, dpcs, trays and flashings (*Site*)	August 1987	126
DAS 108	Domestic hot water storage systems: electric heating – remedying deficiences (*Design*)	September 1987	280
DAS 109	Hot and cold water systems – protection against frost (*Design*)	November 1987	278
DAS 110	Dual-pitched roofs: trussed rafters – specification of remedial bracing (*Design*)	December 1987	194
DAS 111	Dual-pitched roofs: trussed rafters – installation of remedial bracing (*Site*)	December 1987	196
DAS 112	Dual-pitched roofs: trussed rafters – specification of remedial gussets (*Design*)	December 1987	198
DAS 113	Brickwork: prevention of sulphate attack (*Design*)	January 1988, withdrawn April 1989, replaced by DAS 128	–
DAS 114	Slated and tiled pitched roofs: flashings and cavity trays for step and stagger layouts – specification (*Design*)	May 1988	174
DAS 115	External masonry cavity walls: wall ties – selection and specification (*Design*)	June 1988, replaces DAS 19	68
DAS 116	External masonry cavity walls: wall ties – installation (*Site*)	June 1988, replaces DAS 20	70
DAS 117	Index	June 1988, withdrawn June 1989, replaced by DAS 134	–
DAS 118	Slated or tiled pitched roofs – conversion to accommodate rooms: ventilation of voids to the outside air (*Design*)	August 1988	162
DAS 119	Slated or tiled pitched roofs – conversion to accommodate rooms: installing quilted insulation at rafter level (*Site*)	August 1988	164
DAS 120	Solid floors: water and heating pipes in screeds (*Design*)	October 1988	16
DAS 121	Solid floors: water and heating pipes in screeds (*Site*)	October 1988	18
DAS 122	Windows and doors: reconstituted stone non-structural components; 'plastic' repair using Portland cement mortar – specification (*Design*)	November 1988	240

Numerical List of Defect Action Sheets

DAS number	Title	First issued	Page
DAS 123	Windows and doors: reconstituted stone non-structural components; 'plastic' repair using Portland cement mortar on site (*Site*)	November 1988	242
DAS 124	Pitched roofs: renovation of older-type timber roofs – re-tiling or re-slating (*Design*)	December 1988	176
DAS 125	Pitched roofs: re-tiling or re-slating of older-type timber roofs (*Site*)	December 1988	178
DAS 126	Domestic chimneys: solid fuel – flue design (*Design*)	February 1989	212
DAS 127	Domestic chimneys: solid fuel – flue installation (*Site*)	February 1989	214
DAS 128	Brickwork: prevention of sulphate attack (*Design*)	April 1989, replaces DAS 113	98
DAS 129	Freestanding masonry boundary walls: stability and movement (*Design*)	April 1989	72
DAS 130	Freestanding masonry boundary walls: materials and construction (*Design*)	April 1989	74
DAS 131	External walls: combustible external plastics insulation – horizontal fire barriers (*Design*)	May 1989	136
DAS 132	External walls: external combustible plastics insulation – fixings (*Site*)	May 1989	138
DAS 133	Solid external walls: internal dry-lining – preventing summer condensation (*Design*)	June 1989	116
DAS 134	Index	June 1989, withdrawn March 1990, replaced by DAS 144	–
DAS 135	External masonry painting (*Design*)	July 1989	104
DAS 136	Domestic draughtproofing: balancing ventilation needs against heat losses (*Design*)	September 1989	204
DAS 137	Internal walls: ceramic wall tiles – loss of adhesion (*Site*)	November 1989	106
DAS 138	Domestic chimneys: re-building or lining existing chimneys (*Site*)	November 1989	216
DAS 139	Unvented domestic hot water storage systems – safety (*Design*)	December 1989	282
DAS 140	Unvented domestic hot water storage systems – installation and inspection (*Site*)	December 1989	284
DAS 141	Water storage cisterns: warning pipes (*Design*)	February 1990, replaces DAS 61	276
DAS 142	Slate-clad roofs: fixing of slates and battens (*Design*)	February 1990	180
DAS 143	Drainage stacks – avoiding roof penetration (*Design*)	March 1990	268
DAS 144	Index	March 1990, replaces DAS 134	290

Introduction

Defect Action Sheets were produced by the Housing Defects Prevention Unit, a small group of staff at BRE brought together in 1981 by the then Minister of Housing, Mr John Stanley, to help Local Authorities identify the reasons for defective technical performance in the housing stock, to provide advice on how to avoid the recurrence of defects, and to offer effective remedial measures. The Minister had become concerned at the comparatively large numbers of dwellings standing empty because of defects.

Each Local Authority in Britain nominated a contact point to provide information to the unit, and to be a distribution point for information being provided to the staff of that authority. A series of symposia was held each year for the first five years of the life of the unit, during which valuable feedback was obtained, and which gave direct insight into the problems being encountered in the field. A Technical Committee was set up, with representatives from Local Authority Associations, and on which The National Federation of Housing Associations and the National House Building Council were also represented, to provide further advice to the unit.

Defect Action Sheets began to be issued in May 1982. As the main output from the unit, they were short, sharp and well-illustrated, and addressed specifically to designers or to site staff, and written specifically for each audience. Evidence from research pointed to the fact that mistakes were being made in spite of the existence of relevant information, so the need was to provide a brief reminder rather than a technical treatise. This led to the standard format, of a single sheet, with a concise statement of the problem on the first page, together with a memorable photograph or drawing that was more likely to be recalled in time of need. On the second page were reminders about the principles and practices of suitable solutions.

The feedback that BRE has obtained indicates that the sheets have been a popular and effective source of technical guidance. They have been favourably reviewed in the technical press, and indeed have provided a source of basic material for other publications on housing defects.

The BRE staff involved are now turning their attention more widely to defects in all building types, and the Technical Committee has been disbanded. The set of Defect Action Sheets now covers most of the common defects to be found in housing, and new sheets will only be produced as particular problems arise, or updating becomes necessary.

This book provides a convenient compendium of all the sheets current when the regular publication ceased in March 1990. Practices in the industry change relatively slowly, and principles rarely, and it is therefore to be expected that most of the information contained in the sheets will remain current for some time to come.

H.W. Harrison

PART ONE

Foundations and Substructure

Foundations on shrinkable clay: avoiding damage due to trees

FAILURE: **Cracks in foundations and superstructure**

DEFECT: **Foundations not deep enough where close to trees or to locations of felled trees**

Figure 1 Trees can shrink clay under foundations.

Cracking is fairly common in dwellings on shrinkable clay soils and near to trees. Foundations are affected if trees are close enough for their roots to dry and shrink clay soil below foundation level. This can happen even where foundations are deep enough not to be susceptible to normal seasonal movements. Where trees have been removed, swelling of clay soil, as it slowly wets up, can cause damage: for large trees this swelling may well continue for some years.

If ***foundations*** are ***not below the depth at which clay soils may shrink or swell*** due to the proximity or removal of trees, both ***foundations and superstructure*** may be ***damaged***.

Foundations on shrinkable clay: avoiding damage due to trees

DAS 96
March 1987

PREVENTION

Principle — Foundations on shrinkable clay soils must be designed so that the superstructure will not be damaged by differential shrinkage or swelling.

Practice

- Establish whether soil is shrinkable clay[1]:
  - — any location in the shaded areas of Figure 2 is suspect; elsewhere, local knowledge will be a guide.

If so:

- Identify the species of any trees on or adjacent to the site and establish their likely maximum height. Table 1 gives maximum height of some common species:
  - — include any mature trees which have been removed in, say, the last 10 years.
- Use Table 1 to establish the rule of thumb for the 'safe' distance that a house with normal foundations should be from any tree (or site of tree) Figures 3a and 3b:
  - — if species is above the line, the 'safe' distance is the mature tree height, H, but see footnote to Table 1;
  - — if species is below the line, the 'safe' distance is 0.5H.

Table 1

Tree species	Maximum tree height H (m)	Safe distance from building
Poplar	24	
Oak	23	
Willow	15	
White Beam/Rowan	12	H (see footnote)
Flowering Cherry/Plum	8	
Plane	30	
Elm	25	
Cypress	25	
Lime	24	
Maple/Sycamore	24	0.5 H
Common Ash	23	
Beech	20	
Birch	14	
Pear	12	
Apple	10	

Footnote H may be reduced to 0.5 H where the 'shrinkage potential'[1] of the clay has been established as medium or low. Tests to determine the properties of clay soil are specified in BS 1377 and are readily available from soil testing laboratories at little cost.

Where tree (or site of tree) is closer to the proposed house than the 'safe' distance:

- Consider bored piles and ground beams[2,3,4] where trees are to remain, or any trees larger than saplings are to be removed, Figure 4.
- Consider, where trees have been or might be removed, providing in addition a suspended ground floor, and:
  - — clearance not less than 100 mm below pc ground beams;
  - — compressible filling not less than 150 mm thick below beams cast in trenches.
- Design drainage, external services and external works to accommodate differential movements, particularly at points of entry to inspection chambers and to buildings.
- Check that drains, boundary walls and other external works will not be damaged by the action of trees that are to remain.

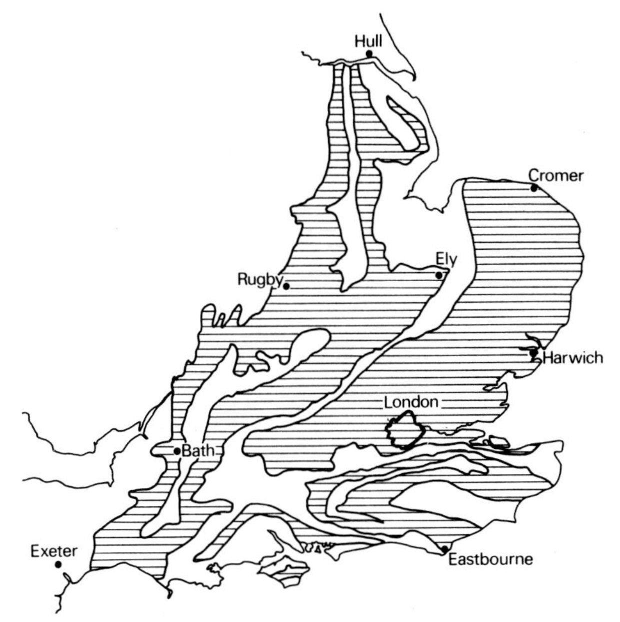

Figure 2 General distribution of firm shrinkable clays — based on BRE records and Institute of Geological Sciences maps.

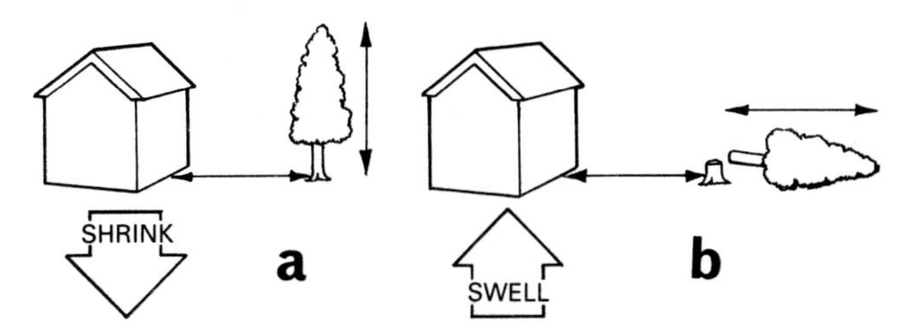

Figure 3 Compare tree height with distance from foundations.

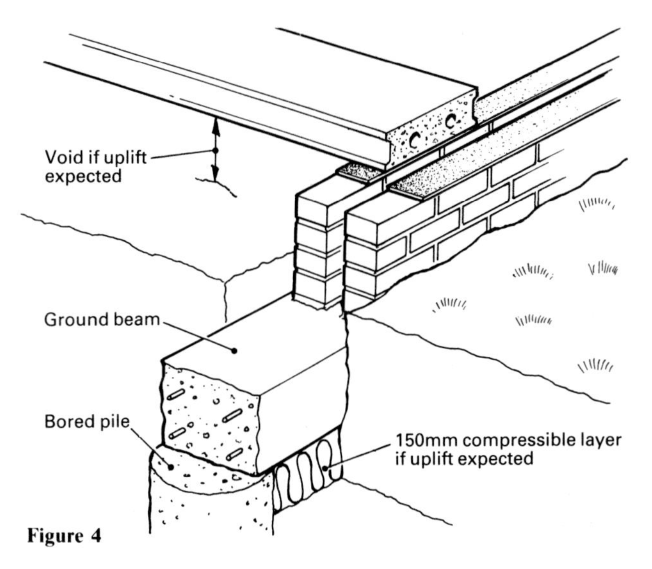

Figure 4

REFERENCES AND FURTHER READING

1 *BRE Digest* 240, 'Low-rise buildings on shrinkable clay soils'. Part 1.

2 *BRE Digest* 241. 'Low-rise buildings on shrinkable clay soils'. Part 2.

3 *BRE Digest* 242. 'Low-rise buildings on shrinkable clay soils'. Part 3.

4 *BRE Digest* 298, 'The influence of trees on house foundations in clay soils'.

□ *BRE Digests* 63, 64, 67 'Soils and foundations, Parts 1, 2 and 3'.

□ *British Standard* BS 8004: 1986 'Code of Practice for foundations'.

□ *British Standard* BS 5837: 1980: 'Code of Practice for trees in relation to construction'.

DAS 35

September 1983 New edition March 1985

Design

CI/SfB 8 (21.9)(13.9)

Substructure: dpcs and dpms — specification

FAILURES: **Rising damp; consequential damage to wall and flooring materials and finishes**

DEFECTS: **Dpcs less than 150 mm above finished ground level or paving; specified dpcs not wide enough to cover full wall or leaf thickness; dpcs bridged by pointing, rendering (or by debris in the cavity); no specified lap between dpc and dpm; dpms not specified to be lapped or sealed; wrong choice of dpm**

Figure 1 Sloped paving bridges DPC

This *Defect Action Sheet* does not deal with resisting groundwater under pressure, nor with DPMs in heated floors.

All UK Building Regulations require protection of walls and floors against moisture from the ground. In a BRE survey of housing under construction, defective DPCs and DPMs were common. (Figure 1.)

If ***DPCs*** are ***too narrow***, pointed over or bridged, or ***too low*** in relation to finished ground level, or if ***laps*** are ***insufficient***, buildings will be at risk from ***rising damp***.

All the materials commonly used for DPCs can be satisfactory: their effectiveness is determined by design and site practices; but not all of the materials available for DPMs are satisfactory for all flooring finishes and their adhesives. Otherwise, for the majority of new housing, the risk of rising damp through minor imperfections in DPMs is probably low. However, rectifying failure can be very costly.

PREVENTION

Principle — Moisture from the ground must not reach vulnerable materials or the inside of the building.

Practice

- Specify DPCs wide enough to project slightly on the external face of the wall[2]; specify that they must not be pointed over (see Figure 2).
- Specify laps in flexible DPCs to be not less than 100 mm.
- Specify DPCs to BS 743[3], or choose a material with an acceptable third party certificate.
- Specify all horizontal and stepped DPCs protecting walls from moisture from the ground to be not less than 150 mm above all finished ground levels (including paving).
- Specify external render coats to be stopped above DPC level.
- Specify DPMs in solid floors to be for preference above the floor slab and under a screed of 50 mm minimum thickness (Figure 2):
  - 'Construction water' in slabs, which can take many months to dry out, can behave like rising damp.
- Specify DPCs under partitions.
- Specify that DPMs and DPCs must lap (see Figure 2):
  - If a power-floated slab is used instead of a screed, a polythene DPM can only be within or below the slab: instruct site staff to check that DPM is not punctured and to ensure that power-floating does not cut off the projecting DPM needed to form the lap with the DPC — if it is cut off there is no easy remedy.
- Specify how to achieve continuity between DPCs and DPMs at changes in levels, for example at steps in terraces (Figure 3).
- Specify a minimum thickness of 0.2 mm (200 microns) (in practice 1200 g [300 microns] will be needed) if a polythene sheet DPM is to be used.
- Specify laps in polythene sheet to be double welted:
  - Although, strictly to comply with the Code[1], laps are to be at least 150 mm, sealed with adhesive tape, this may not be practicable in wet weather.
- Specify how continuity is to be achieved where services puncture the DPM. Taping on a polythene sheet collar is considered to be sufficient to comply with the Code[1] (see DAS 36 Figure 5).
- Do not specify polythene sheet as a DPM for moisture-sensitive finishes[1] such as magnesite, cork, wood, or chipboard, nor for certain adhesives such as PVA and PVC emulsions used with sheet floor finishes:
  - Hot-applied bitumen about 3 mm thick can be regarded as completely impermeable, or alternatively a similar thickness formed for example by several coats of bitumen/rubber emulsion.

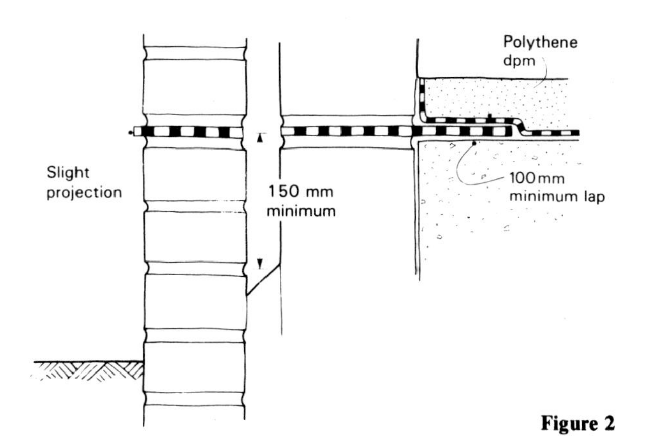

Figure 2

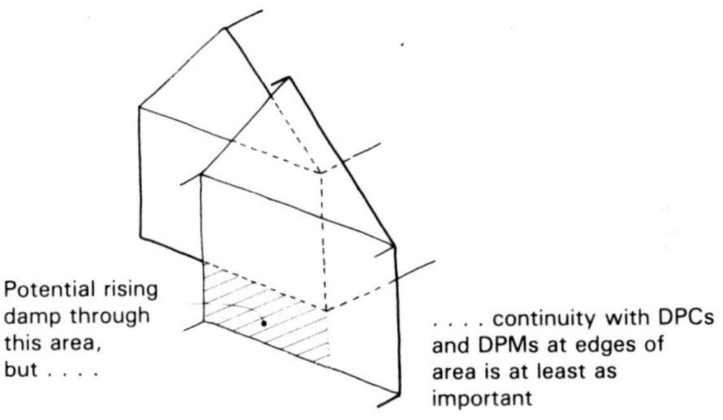

Figure 3

REFERENCES AND FURTHER READING

1 *British Standard Code of Practice* CP102: 1973, 'Protection of buildings against water from the ground'.

2 *British Standard Code of Practice* CP121: Part 1: 1973, 'Walling; Brick and block masonry'.

3 *British Standard* BS 743: 1970, 'Materials for damp-proof courses'.

BRE Digest 77, Damp-proof courses.

Defect Action Sheet DAS 36 'Substructure: DPCs and DPMs — installation'

Substructure: dpcs and dpms — installation

FAILURES: **Rising damp; consequential damage to wall and flooring materials and finishes**

DEFECTS: **Dpcs pointed or rendered over; dpcs bridged by mortar droppings; dpms punctured; dpcs and dpms not lapped; fill and paving not kept at least 150 mm below dpcs**

Figure 1 DPC less than 150 mm above paving

This *Defect Action Sheet* does not deal with resisting groundwater under pressure.

In a BRE survey of housing under construction, many instances of DPCs not covering the full wall or leaf thickness were seen. For example, sometimes the DPCs specified were not wide enough, sometimes they were badly positioned. DPCs were not always high enough (Figure 1). Sometimes lengths of DPC were not properly lapped, or the DPM was not lapped with the DPC. Often, DPCs and DPMs were damaged during construction operations.

If ***DPCs and DPMs*** are ***not*** carefully laid to achieve ***an uninterrupted barrier***, and if care is not taken to avoid puncturing or otherwise damaging them, there is a risk that ***rising damp*** will occur in the house; walls, floors and their finishes will be permanently damaged.

PREVENTION

Principle — Moisture from the ground must not reach vulnerable materials or the inside of the building.

Practice

(Note: if water table apparent on excavation is near lowest DPM, tell the designer — and this DAS does not then apply.)

- Check that edges of sheet DPMs project enough to lap later with the DPC. — Note that a power-float can easily cut off the projecting DPM (Figure 2).
- Check that polythene sheets in a DPM are joined with a double welt (Figure 3).
  — Strict compliance with the Code[1] requires the sheets to be sealed with adhesive tape (Figure 4); it also requires all punctures in the membrane to be sealed by patching — the patches to overlap at least 150 mm all round, taped as above.
- Check how continuity is to be achieved where service pipes perforate the DPM (Figure 5 shows one method).
- Ensure that horizontal DPMs are continuous with vertical DPMs at changes of level.
- Ensure that all DPCs are of the specified width, are laid on a full mortar bed, with a full mortar bed laid over.
- Ensure that DPCs are not laid until ready to continue brickwork above, otherwise they may get damaged.
- Ensure that DPCs that will remain exposed for some time — eg at thresholds — are protected.
- Ensure that laps in all flexible DPCs are at least 100 mm.
- Check that DPCs are not pointed over, and not bridged by rendering or by mortar droppings within the cavity.

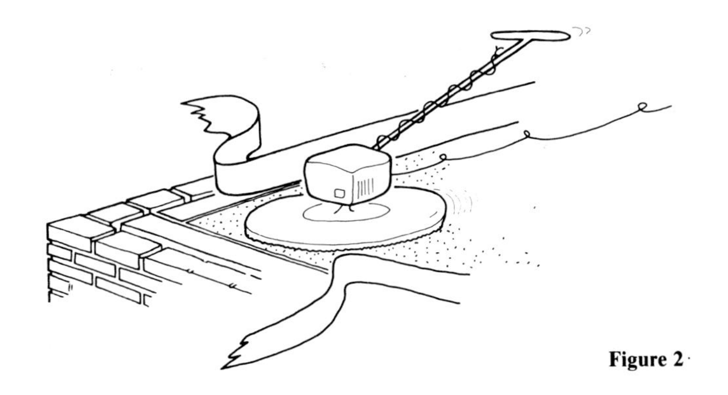

Figure 2

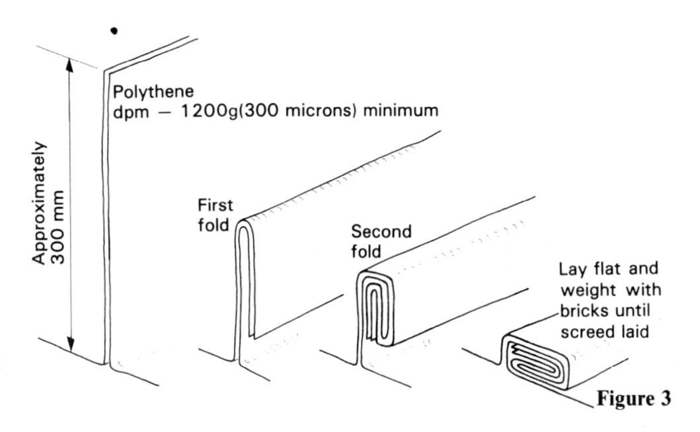

Figure 3

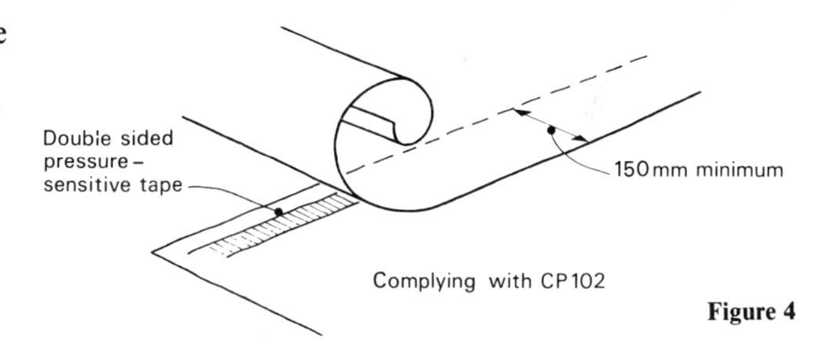

Figure 4

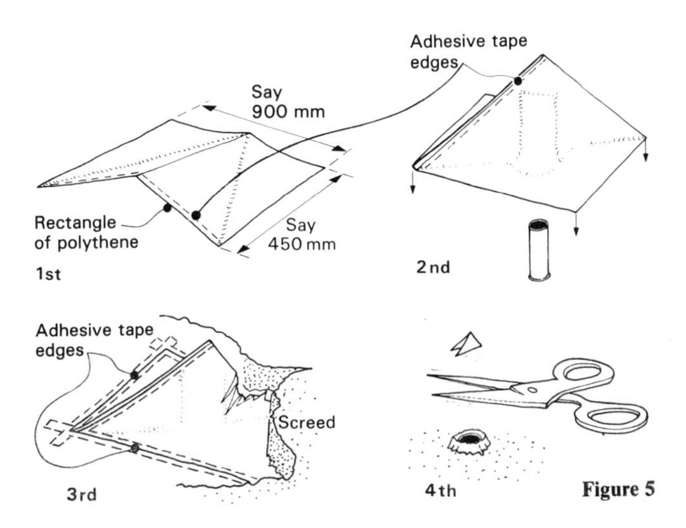

Figure 5

REFERENCES AND FURTHER READING

1 *British Standard* CP102: 1973: 'Protection of buildings against water from the ground'.

Defect Action Sheet DAS 35 'Substructure: DPC's and DPMs—specification'.

Oversites — concreting when weather may be bad

FAILURES: **Frost damage; low strengths or crumbly concrete surfaces caused by rain washing out fines**

DEFECTS: **Concrete laid when temperature too low or during rainfall; inadequate protection from weather after laying**

Figure 1 Concreting in bad weather

Provided sufficient precautions are taken it is possible to lay and cure concrete successfully in any weather conditions likely to be encountered in the UK. Elaborate precautions such as heated working enclosures, however, are rarely economically justified in housing. Care must therefore be taken to lay concrete only when weather conditions are suitable, both at the time of laying and throughout the subsequent curing period. Weather protection may be needed in addition to other measures taken (such as a covering of damp sand to promote curing).

If air ***temperature*** is ***low*** enough to delay initial set, ***concrete*** oversites will ***not*** develop their ***full strength.*** ***If heavy rain*** falls on a freshly laid slab, the ***fines*** may be ***washed out,*** and the ***concrete*** surface will be ***weak*** and ***crumbly.***

PREVENTION

Principle — Concrete should not be laid during bad weather, or when bad weather is likely, unless protection can be given.

Practice

- Do not permit mixing or laying of concrete when the air temperature is below 2°C (Reference 1) nor when it is expected to fall below 2°C within the next few hours:
  - local conditions may increase risk of low temperatures (eg frost hollows – seek local knowledge);
  - temperatures are lower on higher ground (by about 0.6°C for every 100 m height above sea level.
- Check that aggregates are not frost-bound, Figure 2;
  - provided air temperature is above 2°C, and if concreting cannot be delayed, frost-bound aggregates must be thoroughly heated.
- Do not permit unprotected concrete slabs to be laid in more than the lightest rain or snow showers.
- Protect fresh concrete if rain or snow is likely.
- Remember that pre-recorded forecasts are available — see 'Weather line' numbers in your dialling code booklet. Forecasts specific to contractors' needs can be obtained, for a fee, from your local Weather Centre (see directory).
- For your locality, the Meteorological Office can compute the proportion of working hours in which conditions will be unsuitable for the operations covered by this DAS – a charge is made – England and Wales, telephone 0344 420242, Extension 2278, Scotland, 031 334 9721, Extension 524, Northern Ireland, 0232 228457. A list of threshold conditions for many building operations (together with local data for Plymouth) is given in Reference 2.

Figure 2

REFERENCE AND FURTHER READING

1 *Cement and Concrete Association,* 'Winter concreting': 3rd edition 1978.

2 *BRE Report,* 'Climate and Construction Operations in the Plymouth area; Keeble and Prior. (In preparation.)

PART TWO

Ground Floors

Floors: cement-based screeds — specification

FAILURES: **Shrinkage-cracking and curling; crushing of screeds under flooring; reduced impact sound resistance; damp-staining of wall plaster**

DEFECTS: **Incorrect specification of mix, screed thickness, curing and drying; lack of design provision to maintain isolation of floating screeds; wet screed in contact with wall plaster**

Figure 1 Indentation of thin flooring over weak screed

In housing, reported screed problems mostly concern breakdown and crushing of floating screeds or damp-staining of adjacent wall plaster. Crushing of non-floating screeds is less often reported: floor loads are usually too low. However, such screeds sometimes crack locally — eg where wall cavities at thresholds become filled with builder's rubbish prior to screeding. Occasionally, screed areas are large enough for curling and subsequent cracking to occur.

If ***floating screeds*** are ***too thin or weak*** they may ***crush*** under point or impact loads; if the ***resilient layer*** beneath is ***bridged*** the ***screed may crack.*** If ***non-floating screeds*** are ***too weak*** flooring may be ***indented;*** if they are ***too thin over buried services*** they may ***break down.*** If ***drying time*** of any type of screed is ***insufficient,*** adjacent wall ***plaster*** may become ***damp-stained*** or ***floorings*** may become ***detached*** — and subsequently both may be wrongly diagnosed as rising damp. If ***curing*** of any type of screed is ***inadequate*** it will probably ***break down.***

PREVENTION

Principle — Screeds must neither break down nor permit indentation of flooring; floating screeds must not bridge resilient layer below.

Practice

All screeds:

- Specify a 1:4 cement:fine aggregate[1] mix by weight with lowest water content compatible with full compaction:
  - — screeds of 50 mm thickness or more may be of fine concrete, 1:1½:3, with 10 mm maximum aggregate.
- Specify that the required minimum thickness is to be maintained at all points including over any embedded services, Figure 2;
  - — central heating pipes should preferably not be buried[2] except over short distances, without joints, and suitably wrapped.
- Specify movement joints only where such joints occur in the base;
  - — screeds rarely need subdivision into bays in housing but consider subdividing if either dimension exceeds about 5 m, particularly if ratio of sides is significantly greater than 1½:1.
- Specify 7 days curing (eg under polythene sheet) and then adequate drying time before finishes are laid.
- Instruct site supervisors to ensure that wall plaster does not reach down to screed surface, Figure 3.

Integral ('monolithic') screeds:

- Specify that base concrete is to be brushed with a stiff broom and that the screed must be laid within 3 hours of casting base.
- Specify a 10 mm minimum achieved screed thickness (integral screeds should not be thicker than 25 mm to avoid shrinkage stresses).

Screeds laid directly on set concrete bases:

- Specify that base is properly prepared:
  - — the best bond is achieved by mechanically hacking, cleaning off loose material, damping and brushing with a cement slurry (or using a bonding agent) immediately prior to screeding.
- Specify that the screed thickness achieved is to be between 25 and 40 mm.

Unbonded screeds (eg laid on dpm):

- Specify a 50 mm minimum screed thickness: if the screed contains heating elements specify a 65 mm minimum achieved screed thickness.

Floating screeds (eg on resilient, sound insulating, materials):

- Specify a 65 mm minimum achieved screed thickness; if the screed contains heating elements specify a 75 mm minimum achieved screed thickness.
- Specify that resilient layer is to be carried up at perimeter and protected from penetration of the screed into any joints or from seepage into the resilient layer:
  - — eg, specify polythene sheet and wire mesh over resilient layer.

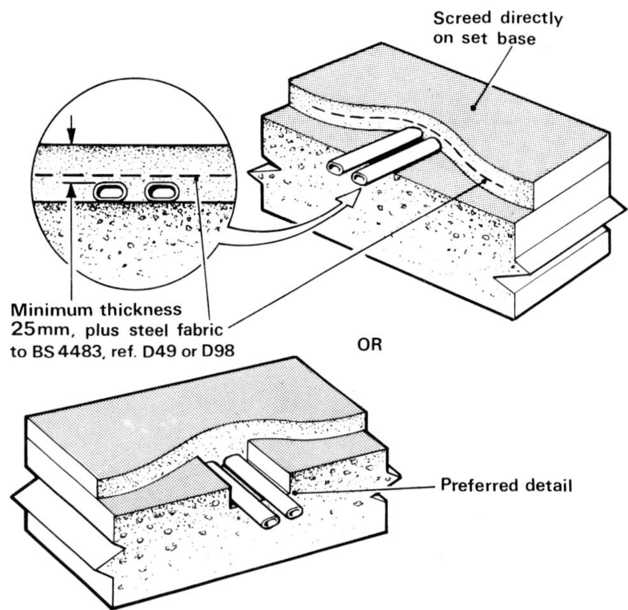

Figure 2

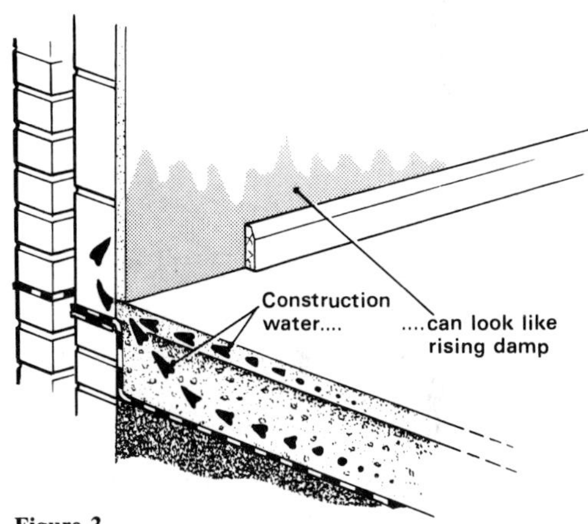

Figure 3

REFERENCES AND FURTHER READING

1 British Standard BS 882:1201:Part 2:1973 'Aggregate from natural sources for cements'

2 British Standards BS 5276:Part 2:1976 'Code of Practice for selection and installation of gas space heating'.

Defect Action Sheets DAS 35 and 36 'Substructures: DPCs and DPMs'

BRE *Digest* 104, 1979 'Floor screeds'

British Standards *Code of Practice* CP204:Part 2:1970 'In-situ floor finishes'

Defect Action Sheet DAS 52 'Cement-based screeds — mixing and laying'

Floors: cement-based screeds — mixing and laying

FAILURES: **Shrinkage-cracking and curling; crushing of screeds under flooring; reduced impact sound resistance; damp-staining of wall plaster**

DEFECTS: **Poor batching, mixing, compaction, curing; screed too thin or weak; bridging of floating screeds; insufficient drying out time; wet screed in contact with wall plaster**

Figure 1 Poorly compacted screed

Poor compaction, often accompanied by others of the defects listed above, is a major cause of breakdown in screeds. If screed mixes are ***not batched accurately, not mixed thoroughly*** without segregation, and particularly if they are ***not compacted fully*** throughout the screed thickness, ***screeds break down.*** If screeds are ***too thin, or not carefully cured,*** they ***curl and crack.*** If screed mixes contain ***too much fine sand or water,*** shrinkage will cause ***curling and cracking.*** If screeds are given too ***little drying time,*** or are in contact with wall plaster, construction water will cause ***problems with most floorings,*** and can ***stain wall plaster.***

PREVENTION

Principle — Screeds must neither break down nor permit indentation of flooring; floating screeds must not bridge resilient layer below.

Practice

All screeds:

- Check that fine aggregate is clean and well graded (too much 'fines' will give too much shrinkage).
- Check that mix proportions are as specified.
- Check that mixing is thorough and do not permit unspecified additives;
  — screed mixes tend to segregate in drum mixers.
- Check that water content is right:
  — as a guide, squeeze a handful of the mix tightly; it should hold together but not exude water.
- Ensure that screeds are laid immediately after mixing.
- Ensure that screeds are thoroughly compacted:
  — banging the float down achieves almost nothing; use a heavy tamper or, better, a mechanical compactor; Figure 2;
  — thicker screeds are best compacted in two layers (one immediately after the other), particularly if only hand tamping if possible.
- Check that the specified minimum thickness is maintained everywhere, particularly over services buried within screed thickness.
- Ensure that screeds are cured (eg under polythene sheet) for at least 7 days.
- Plan work to allow, ideally, one month for drying for every 25 mm of thickness, including any part of the base above the dpm, before laying flooring.
- Before laying moisture-sensitive flooring, ensure screed is dry by testing[1].
- Check that wall plaster is not taken down into contact with wet screeds.

Integral ('monolithic') screeds:

- Check that integral screeds are laid within 3 hours of casting base and after brushing base with a stiff broom just before the set.
- Check that screed thickness is not less than 10 mm or more than 25 mm.

Screeds laid directly on set concrete bases:

- Check that base is prepared as specified:
  — mechanical hacking, cleaning, damping, or overnight soaking may be needed; if cement grout is specified, screed must be laid before grout sets; keep muddy boots off the slab; do not permit hacking of pre-cast units.

Unbonded screeds:

- Check that surface is clean and dry and, if a dpm, undamaged.

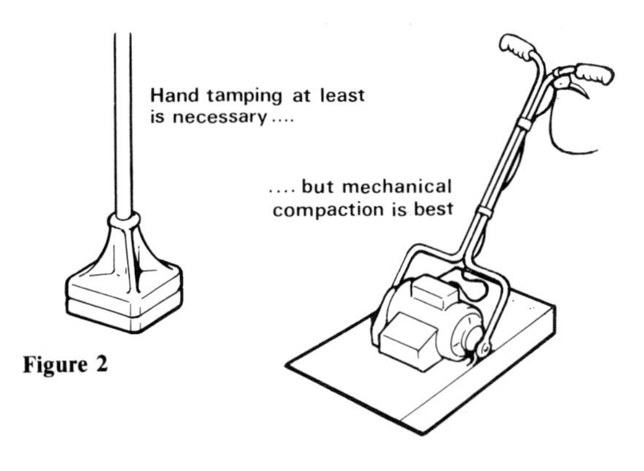

Figure 2

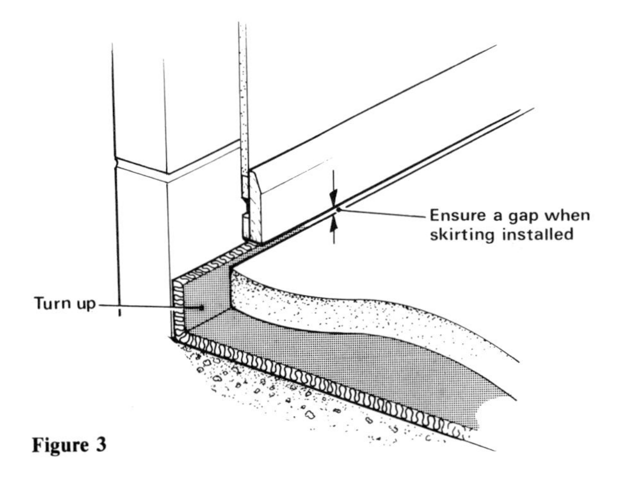

Figure 3

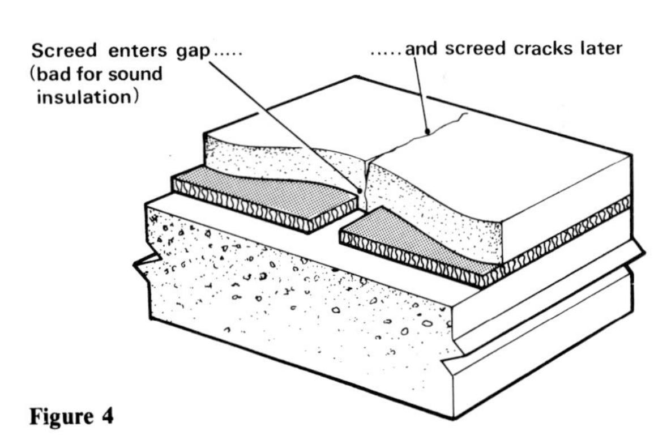

Figure 4

Floating screeds:

- Ensure that floating floors are not bridged:
  — check that resilient layer turned up at perimeter, Figure 3; no gaps causing bridges in resilient layer, Figure 4; sheet (eg polythene) laid over resilient layer plus wire mesh if specified.

REFERENCES AND FURTHER READING

1 *BRE Digest* 163, 1974 'Drying out buildings'
BRE Digest 104, 1979, 'Floor screeds'
Cement and Concrete Association, 1980, 'Laying floor screeds'.
British Standard *Code of Practice* CP204:Part 2:1970 'In-situ floor finishes'.

Solid floors: Water and heating pipes in screeds

FAILURE: **Wetting of floors, screeds, flooring and finishings**

DEFECTS: **Pipes buried directly in screeds; inadequate access for inspection and repair of ducted piped services.**

Figure 1

Water heating pipes buried in a floor can leak for a long period before the failure is observed. Many BRE investigations of dampness and consequential damage have revealed that leakage from buried pipes has occurred undetected over long periods; the appreciable cost of locating and rectifying leaks in buried pipes is further increased by the costs of reinstatement of floors disrupted by repair work, Figure 1.

If ***buried*** supply and heating ***pipes,*** or their joints, ***leak*** during service, then such ***leakage*** is likely to be ***undetected until*** consequential ***damage*** has ***occurred; repairs*** are ***costly*** and ***disruptive.***

PREVENTION
Principle — Pipework in floors screeds should be laid in purpose made ducts and provision made for inspection so that leaks and defective joints can be readily located and repaired.

Practice

- Specify, wherever practicable, locations for pipework other than within solid floors and screeds, Figure 2.
- Specify, where pipework within a solid floor is unavoidable, a pipe duct, with a continuous removable cover, Figure 3.
- Specify that the duct is to be laid so that the top will be flush with the finished floor level.
- Specify that the ductwork is fixed by screwing or nailing to the subfloor so that subsequent screeding does not disturb it.
- Specify that all joints in the ductwork are to be sealed with waterproof tape.
- Advise building owner of the locations and content of ducts.

NOTE: Where ducts cross fire 'boundaries' eg in district heating schemes fire stopping will be needed.

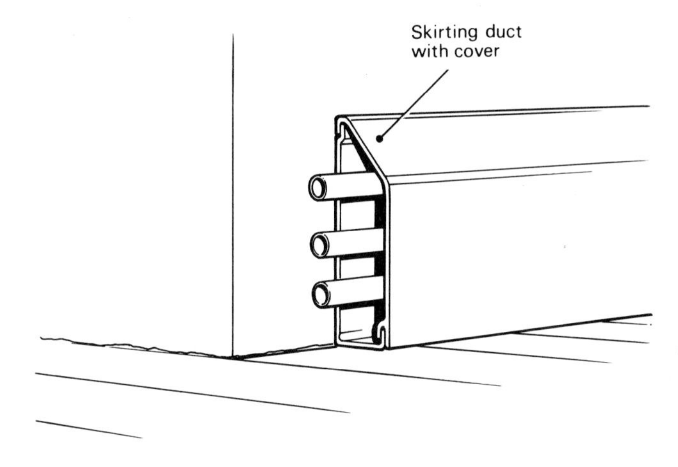

Figure 2

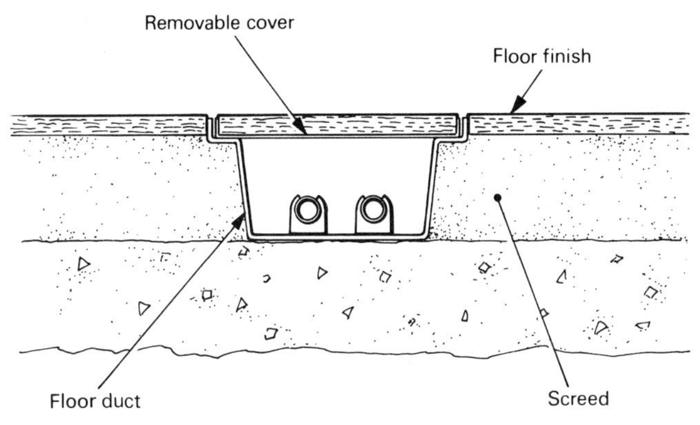

Figure 3

FURTHER READING

BSI *British Standard* BS 6700:1987.
Specification for design, installation, testing and maintenance of services supplying water for domestic use within buildings and their curtilages.

Water Research Centre. *Water Supply Byelaws Guide.* 1986 ed. S F White and G D Mays.

DAS 121
October 1988

Site
CI/SfB 8(23)(5-)P(D6)

Solid floors: Water and heating pipes in screeds

FAILURE: Leakage of water wetting floors, screeds and finishes.

DEFECTS: Pipework buried directly in floors, inaccessible for inspection and repair.

Figure 1

Leakage from pipes buried in the floor can result in considerable damage; disruption to locate the fault may be extensive, Figure 1.

If ***service or heating pipes*** are ***buried in floor screeds*** then ***failure*** of a pipe or a joint may go ***undetected*** with resulting ***damage*** to ***floors, screeds*** and ***coverings.***

PREVENTION

Principle — Services carrying water within a floor must be laid in purpose made ducts and thus be able to be inspected so that any point of leakage can be readily located and repaired.

Practice

- Check whether ducts have been specified for any hot or cold water pipes or heating services run in a solid floor.

- Ensure that ducts are installed at the right level in relation to finished floor level;
  — ducts should not be concealed by any permanent floor finish.

- Ensure that the ductwork is securely fixed to prevent movement during screeding or other work, Figure 2.

- Check that joints in the ductwork are sealed with waterproof tape, Figure 3.

- Ensure that covers to ductwork are screwed down after pipework is commissioned, and also that these screws are easily located if required.

- Check that no leaks are evident when the pipework is filled and the system is commissioned.

- Ensure that any fire stopping specified completely fills the duct at the required locations.

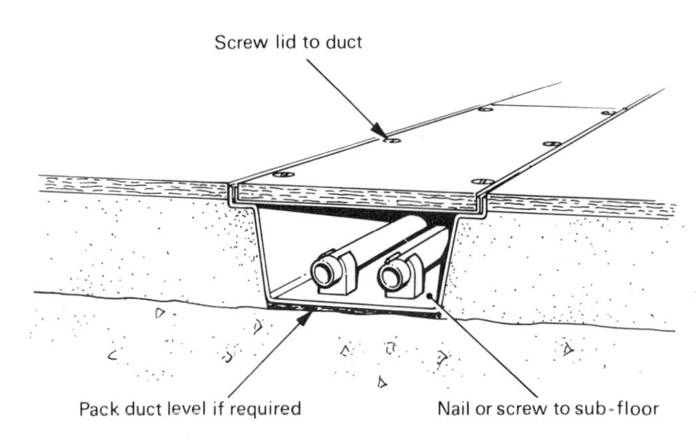

Figure 2

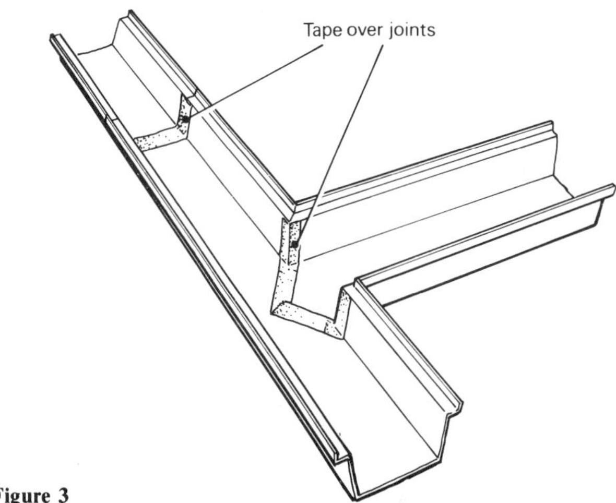

Figure 3

Ground floors: replacing suspended timber with solid concrete — dpcs and dpms

FAILURE: **Rising damp**

DEFECT: **Original wall damp-proof course bridged by new floor construction**

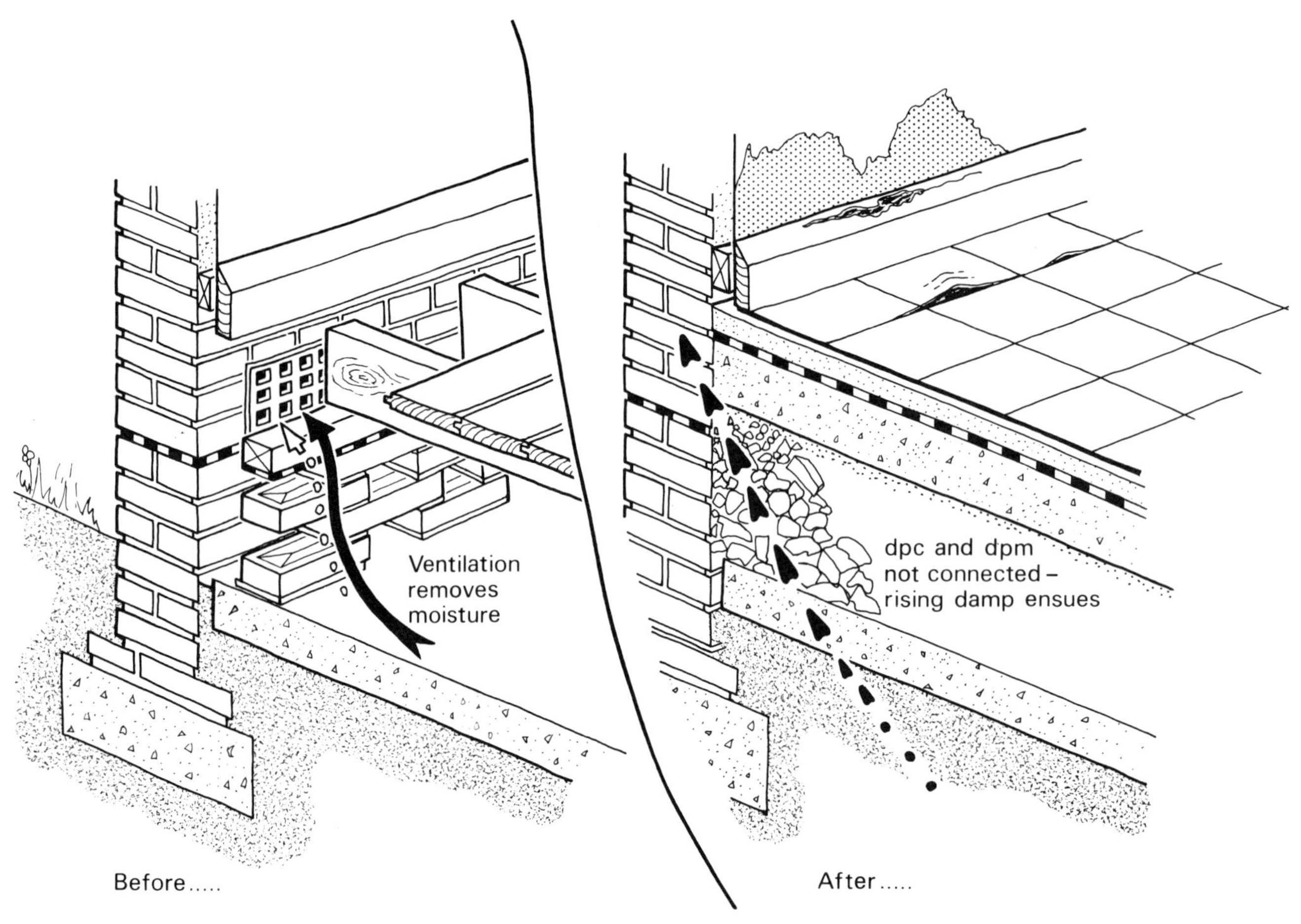

Figure 1

In rehabilitation work a suspended timber ground floor is sometimes replaced by a solid concrete floor. Subsequently, the occurrence of rising damp is sometimes reported to BRE and investigation shows that the new construction enables rising damp to by-pass the original dpc in the wall (Figure 1).

If a ***solid floor replaces a suspended timber floor,*** even if any original wall dpc is still effective, that original ***dpc is too low. If an additional dpc is not installed at the level of the dpm*** in the new slab, or ***if the original dpc*** is ***not linked*** by vertical dpm ***to the dpm in the slab, rising damp can occur.*** Effective resistance to rising damp is crucial to the soundness of investment in rehabilitation work.

PREVENTION

Principle — Rising damp must be resisted by an uninterrupted barrier formed by dpc and dpm.

Practice

If wall dpc is absent or ineffective, or if a vertical membrane is impracticable:

- Specify a new dpc of inserted sheet material (eg copper) at the same level as the new dpm; check detailing at the junction of dpc and dpm to ensure that there is no path for rising damp (for example, detail as indicated in Figure 2a).

If existing dpc is effective:

- Specify a vertical membrane (for example, not less than 3 coats of bitumen/latex emulsion or bitumen emulsion) to link existing dpc to dpm (Figure 2b). Note that if bricks or mortar are powdery it may not be possible to achieve a good film of emulsion; in that case, use a vertical dpm of 1200 gauge polythene sheet.

Alternatively:

- Consider using chemical injection dpc (Figure 2c).

Then:

- Consider specifying the new dpm to be placed immediately below the screed (with screed not less than 50 mm thick); construction water in the screed will usually dry out in 6 – 8 weeks. If the dpm is placed below the slab, construction water from the slab will take several months to dry out before floor finishes can safely be laid, and this delay may be unacceptable in rehabilitation work.

— Note the opportunity to improve thermal insulation of the floor[1].

- Take account of the structural aspects of replacement.

REFERENCE AND FURTHER READING

1 *BRE Digest* 145 'Heat losses through ground floors'.

British Standard Code of Practice CP 102: 1973 'Protection of buildings from water from the ground'.

BRE Digest 54, 1971 edition, 'Damp-proofing solid floors'.

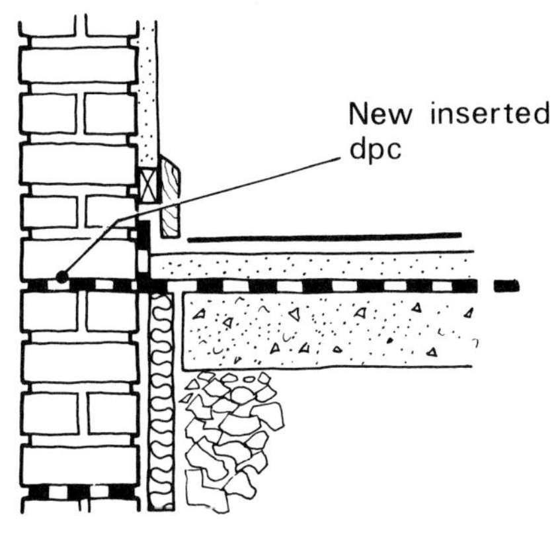

Figure 2a

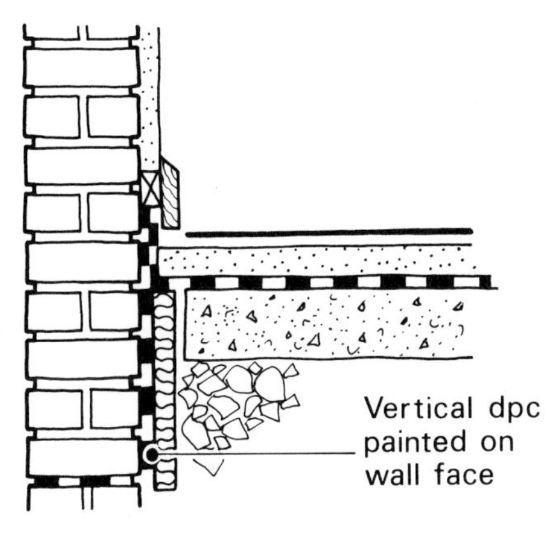

Figure 2b

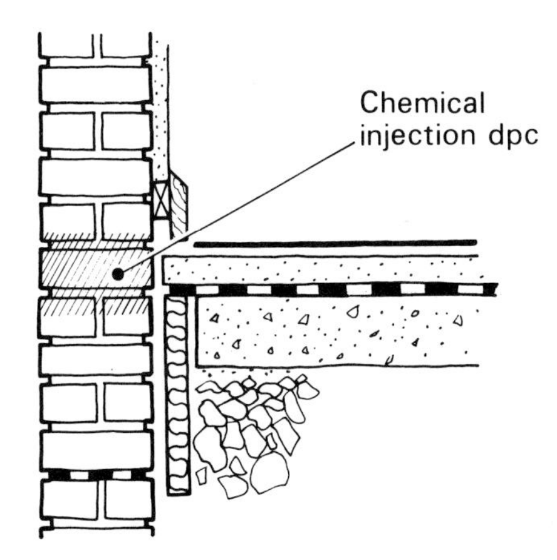

Figure 2c

DAS 73
March 1986

Design
CI/SfB 8 (13.1)(L31)(W7)

Suspended timber ground floor: remedying dampness due to inadequate ventilation

FAILURE: **Rot in joists, wall plates, floor boards**

DEFECTS: **Insufficient or obstructed air-bricks; cross-flow ventilation obstructed**

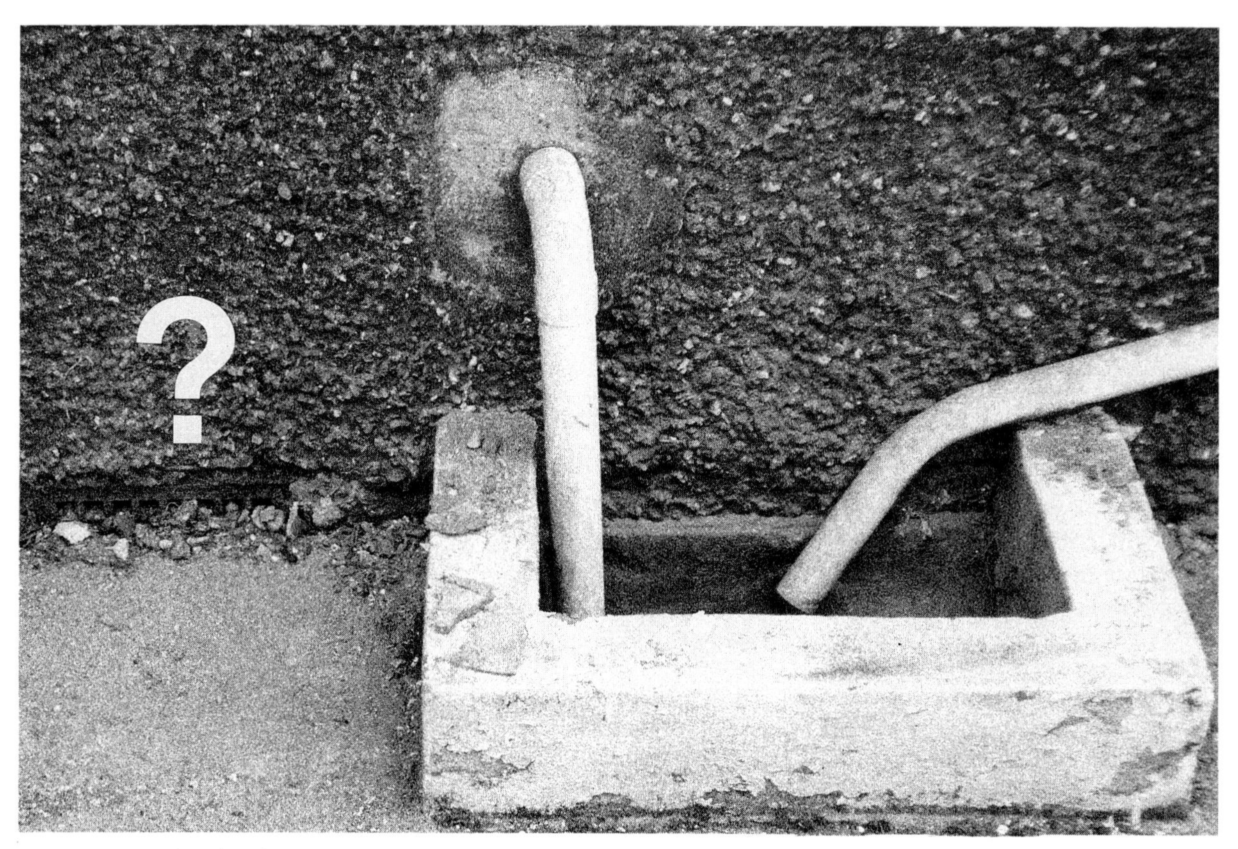

Figure 1 Not the best location . . . . . .

Ventilation cannot protect timber in contact with, say, persistently wet brickwork; however, in normal circumstances it will help to keep timber adequately dry.

In older property, voids below suspended timber ground floors are often inadequately ventilated by current standards. Air-bricks may be absent, insufficient, or obstructed by raised soil or paving levels, or cross-ventilation may be obstructed. Joist ends in solid external walls are often the first parts affected by rot, and local lack of rigidity in the floor may be the first indication of a problem. When dry rot occurs, special measures are needed[1]; if some wet rot has occurred, repair is often possible, see DAS 74.

If remedial ***work does not include*** sufficient ***improvement to ventilation, rot can recur.***

REMEDIAL WORK

Principle — Risk of rot must be reduced by improving ventilation of sub-floor void.

Practice

- Ensure that provision for ventilation via air-bricks is adequate:
  - — air bricks not blocked;
  - — arranged on opposite sides of void;
  - — where practicable not less than 3200 mm² open area per metre run of wall[2]; (strict compliance with the Code[2] will require eg one 200 mm × 200 mm air-brick to BS 493 per metre run; use of smaller more closely-spaced air-bricks may reduce wall strength unacceptably).
- Ensure that there are no 'dead air' pockets in void;
  - — local areas of solid floor may need ducts through them to permit cross-flow, Figure 2;
  - — solid sleeper walls need perforation (areas as given above) if solid blocking present between joists, Figure 3;
  - — where adequate sub-floor cross ventilation cannot be provided, eg in back-to-back terraces, consider using solid or suspended concrete floor.
- Consider extra action needed if rot has begun:
  - — identify sources of water — eg rising damp, penetrating rain, leaks from water supply or drainage pipes, and correct; check that DPC is present, not bridged, and at least 150 mm above ground or paving level, and that sources of persistent wetting of wall above DPC are corrected — eg blocked gullies, broken or blocked down-pipes and gutters, leaking plumbing; check that DPC is present under wall plates on sleeper walls.
  - — consider whether timbers suffering localised wet rot can be repaired — see DAS 74.

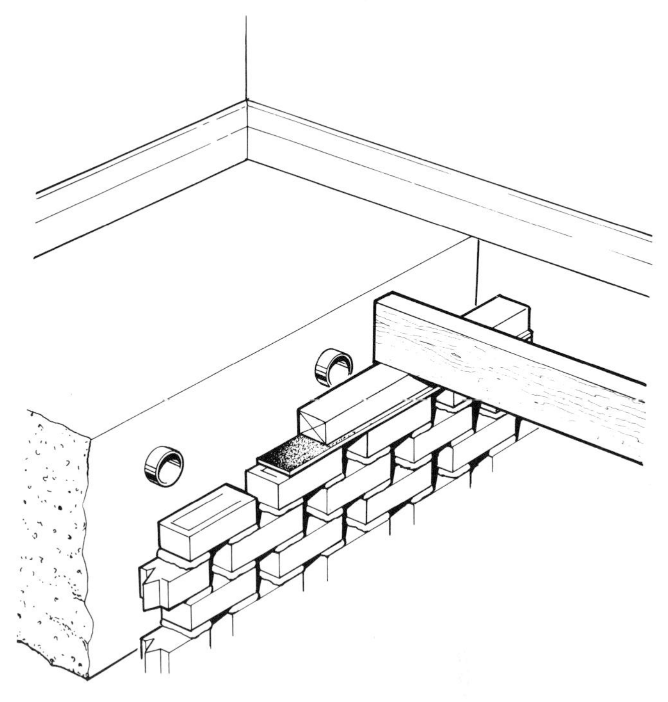

Figure 2

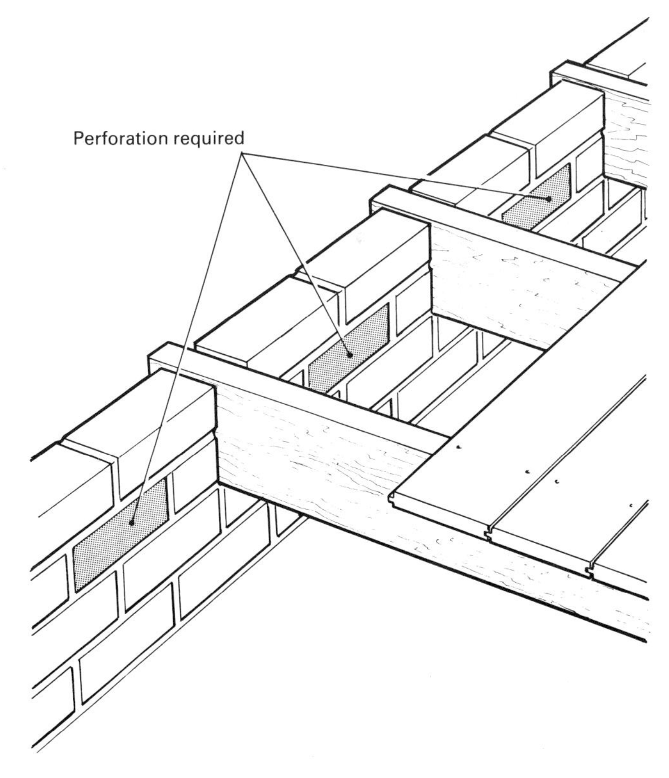

Figure 3

REFERENCES AND FURTHER READING

1 *BRE Digest* 299.

2 *British Standard Code of Practice:* CP102:1973 'Protection of buildings against water from the ground'.
PRL Technical Note 44; 'Decay in buildings, recognition, prevention and cure'.

Suspended timber ground floors: repairing rotted joists

FAILURES: Floor uneven or locally unsupported; rot in replacement timber

DEFECTS: Inadequate connection in repaired joists; dampness or poor ventilation not corrected and replacement timbers not adequately preservative-treated

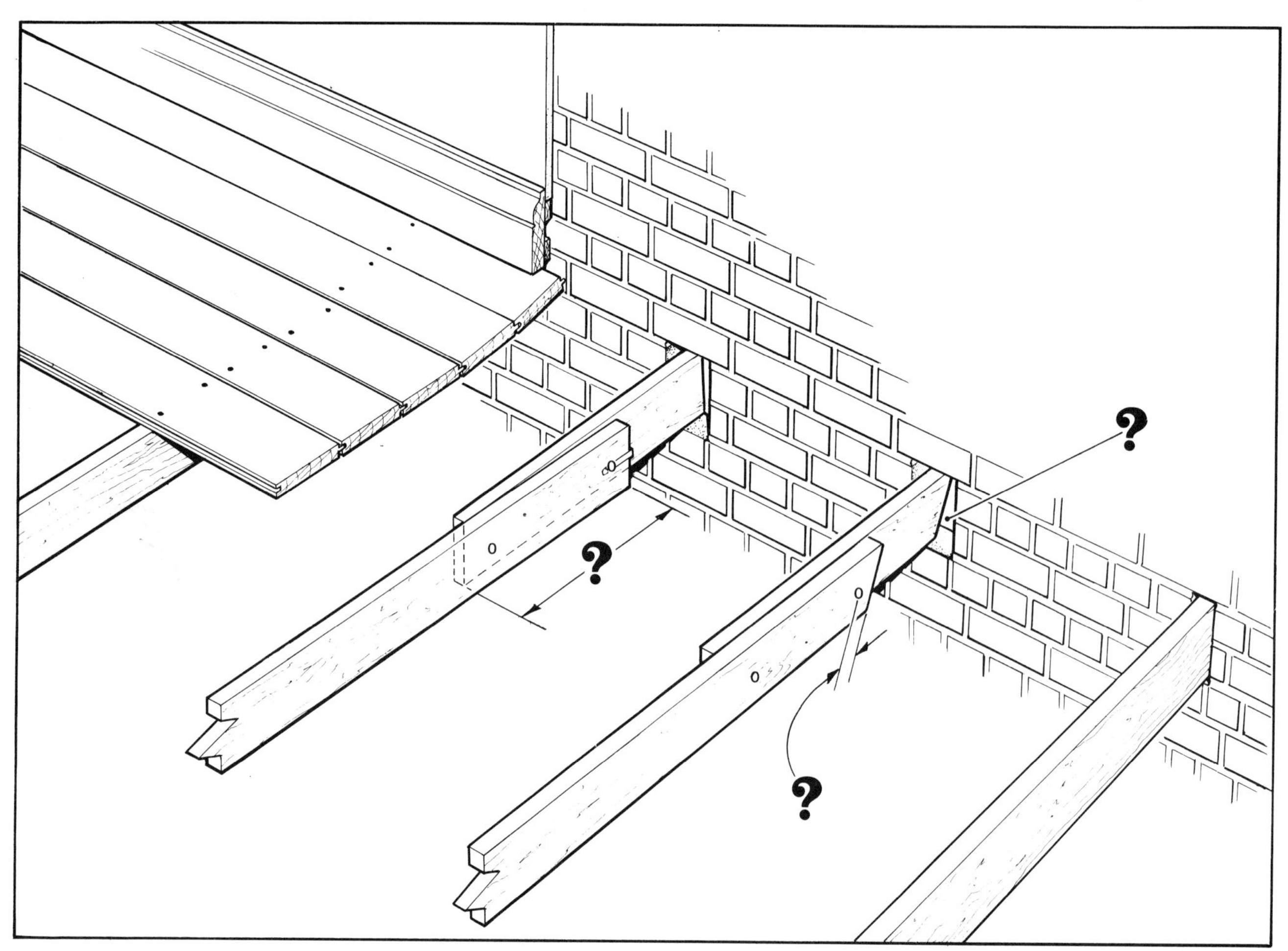

Figure 1

Where ventilation of voids below suspended timber floor is inadequate, or DPC's are defective or by-passed, rot can occur in untreated timbers, particularly at joist ends if joists are housed in an external wall. Such joists can often be repaired.

If ***timber floors*** are ***repaired without protection from dampness*** and without correction of inadequate ventilation, ***or without using treated timber, rot can recur;*** if new ***ends*** of joists are ***not securely connected*** or ***not restrained*** against twisting, ***sag or collapse*** may occur.

REMEDIAL WORK

Principle — repaired joists must be structurally adequate and durable; causes of rot must be corrected.

Practice:

- Correct causes of inadequate sub-floor ventilation and dampness (see DAS 73)
- Eradicate any dry rot[1]:
  - — depending on extent of rot, it may be advisable to renew entire floor, otherwise proceed as follows.
- Check extent of any wet rot:
  - — if long span joists have rot at one end only, not more than 2 or 3 joists being so affected, adopt following 'rule of thumb' repair:

    Rotted joist end to be cut back to sound new wood;

    New end to overlap old joist by not less than 4 times joist depth;

    Ends to be clamped for drilling in-situ (so that holes align) and drilled a close fit for not less than four M12 (12 mm dia) black (mild steel) bolts;

    Round toothed connectors and 3 mm minimum thickness plain washers to be used, Figure 2;

    New end to be supported clear of external wall and not in the pockets in which original joists ends rotted, Figure 3;

    Old pockets to be filled and joists to be blocked to prevent rotation (or herringbone strutted if blocking would impede ventilation).
  - — if long span joists have rot at one end only but more than 2 or 3 are to be repaired, confirm by calculation[2,3] that 'rule of thumb' repair will be satisfactory.
  - — if long span joists have rot elsewhere than at one end only, replace with similar new joists;
  - — if short span joists (carried on intermediate sleeper walls, Figure 4) have rot at *any* point, replace with similar new joists.
- Specify double vacuum, or pressure CCA, preservative treated replacement timber and three full brush coats of preservative on all cut ends of both new and old timber.
- Specify that, in all cases, the ends of replacement joists must be kept out of contact with external walls.
- Consider, where entire floor must be replaced, specifying solid concrete — see DAS 22.

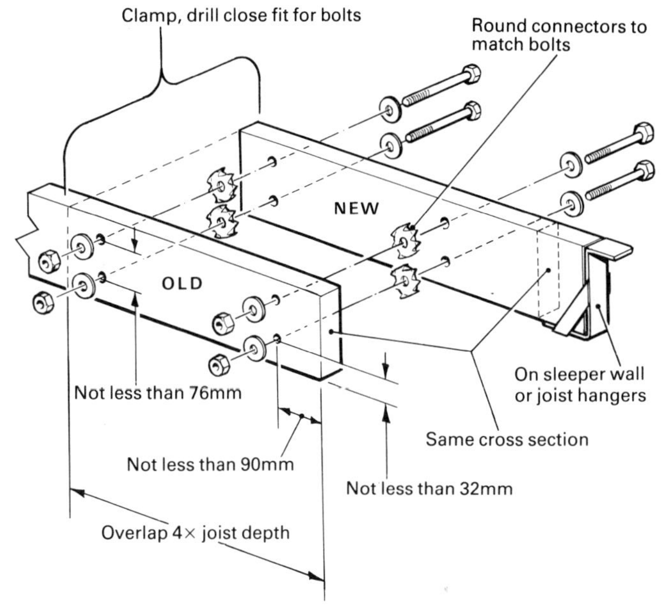

Figure 2

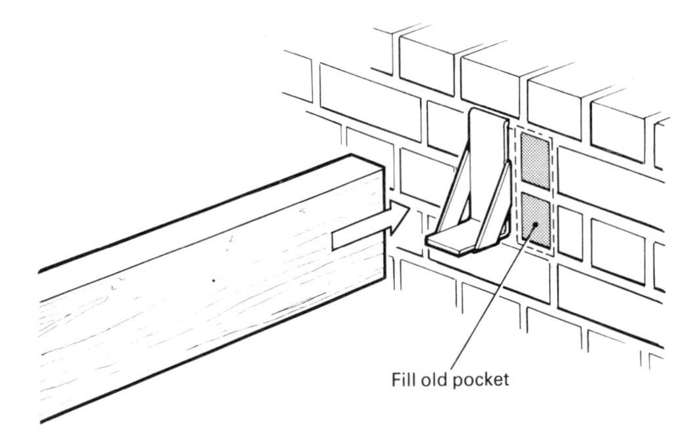

Figure 3

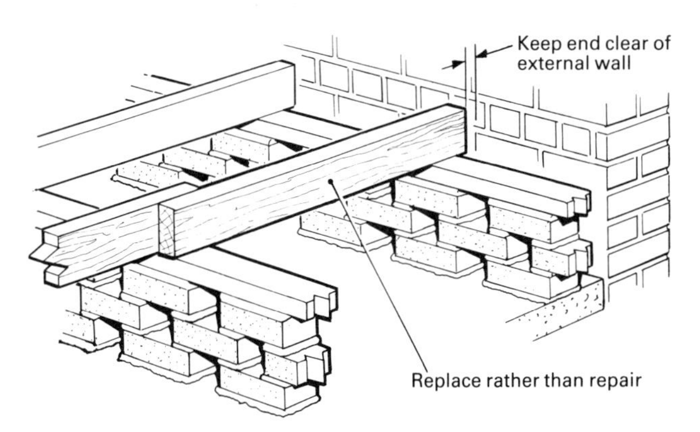

Figure 4

REFERENCES AND FURTHER READING

1 *BRS Digest* 299.

2 *BS* 5268 Code for the use of structural timber.

3 *BS* 1579 Specification for connectors for timber.

PART THREE

Suspended Floors Ceilings and Stairways

Suspended timber floors: joist hangers in masonry walls — specification

FAILURES: **Masonry over-stressed locally; uneven and springy floors, uneven ceilings; cracked plaster at ceiling perimeter; cold bridging**

DEFECTS: **Incorrect choice of hanger; uneven coursing; incorrect joist length; perimeter nogging omitted; brick course in thermal block wall**

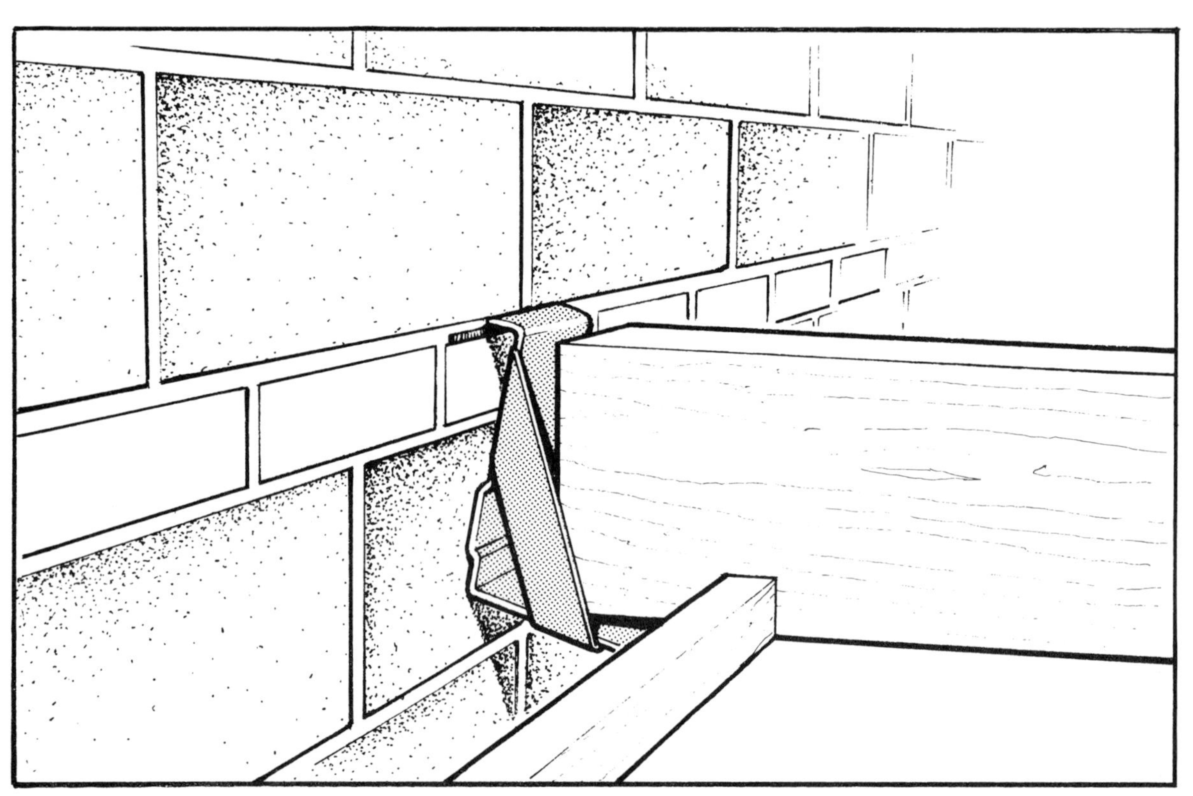

Figure 1 Hanger not tight to wall, joist cut short, hanger on packing, cold bridge at brick course used to level, nogging too far from wall

In a BRE survey, joist hangers were often inadequately fixed and supported, or joists were too short to bear adequately on hangers.

If ***hangers do not bear directly,*** and without packing, on level masonry, if backs of hangers are ***not tight against masonry face,*** if gaps between joist ends and hangers are too large, or if an inappropriate grade of hanger is specified, ***masonry or hangers*** may be ***over-stressed*** and ***floors*** may be ***unevenly supported.***

PREVENTION

Principle — Joist hangers must provide firm, secure and level support for floors.

Practice

- Identify what crushing strengths have been specified for the masonry at locations where joist hangers are to be used;
  - — note that, in any floor, joists may span on to walls of different masonry strengths.

Suspended timber floors: joist hangers in masonry walls — specification

DAS 57
August 1984

- Specify hangers of appropriate grade for use with masonry of those strengths*:
  — hangers complying fully with BS 6178[1] are made (and individually marked) for minimum masonry crushing strengths of 2.8, 3.5 and 7.0 N/mm²; it is essential that hangers are not used with masonry of lower strength than the hanger grade marking — instruct site staff to check, particularly if more than one grade is to be used on site. Alternatively, consider specifying one grade that will suit all strengths of masonry used.

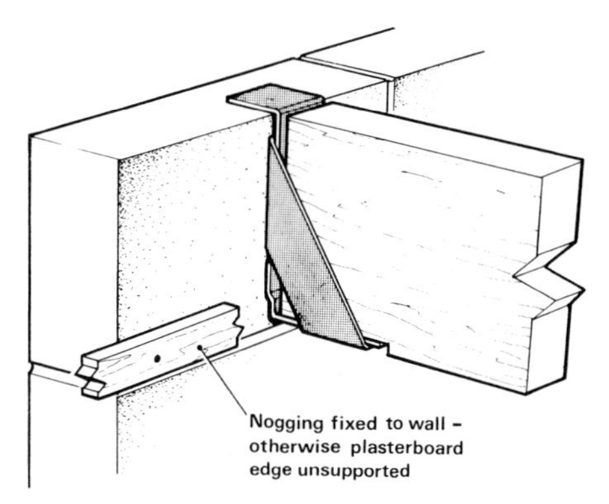

Figure 2

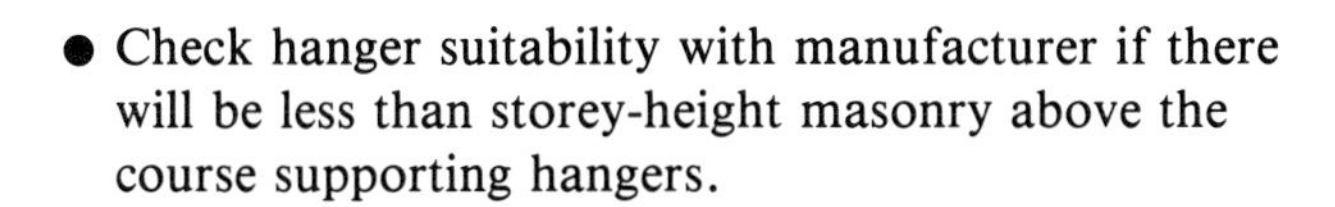

- Check hanger suitability with manufacturer if there will be less than storey-height masonry above the course supporting hangers.

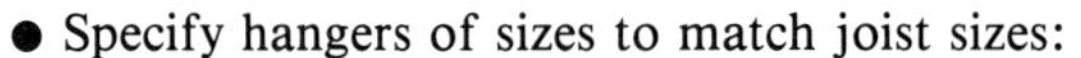

- Specify hangers of sizes to match joist sizes:
  — check the numbers of joists of each size. (Note: hangers complying fully with BS 6178[1] are designed to carry the maximum loads for which joists of corresponding size are suitable).

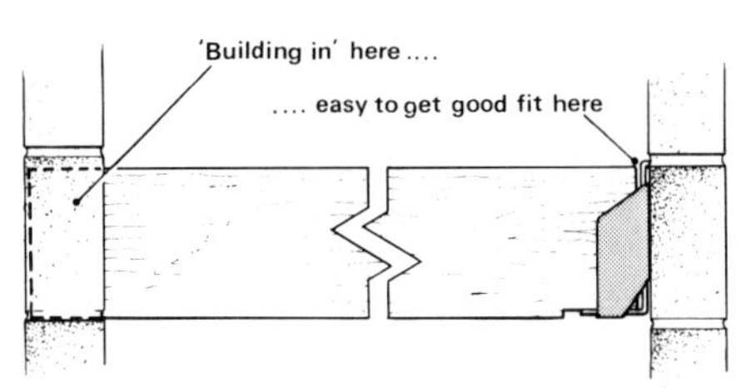

Figure 3

- Specify that hangers are to be bedded directly on masonry.

- Check that coursing can achieve correct height above datum; minor adjustments to level should be made in the bed joint below the course on which hangers bed;
  — if a brick-size course will be needed to achieve coursing, specify 'brick sized blocks' of same material as blocks so that intended strength and thermal insulation are maintained. (Check thermal insulation — a course of clay bricks in a 'thermal block' wall will be a substantial cold bridge.)

- Specify that joists are to be notched just sufficiently to lie flush with underside of hangers so that ceiling plaster board can 'run through';
  — note that perimeter noggings fixed to wall may be needed for plasterboard, Figure 2.

- Check that provision is made for lateral restraint of walls where needed[2].
  — either use the special 'restraint hangers' (see Reference 2 and BS 5628) or incorporate, in addition, lateral restraint straps.

- Specify the number, size and type of fixings to be used to connect joists to restraint hangers.

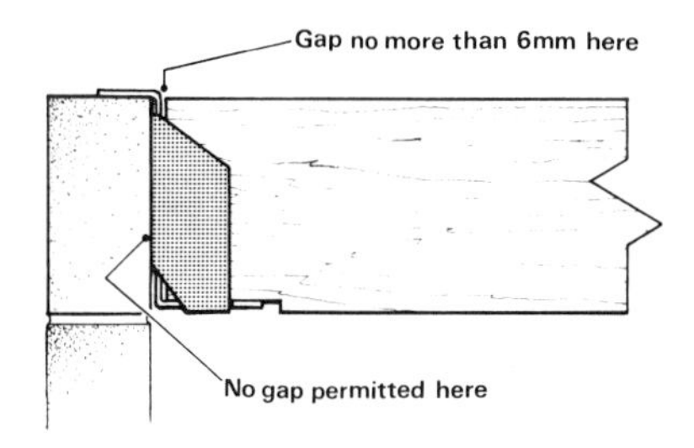

Figure 4

- Instruct site staff to check that hanger back-plates are tight against masonry face and that gaps between joist ends and back plates do not exceed 6 mm, Figure 3.
  — note that, if joists are on hangers at both ends, variation in wall to wall distances may require joists to be individually cut to appropriate length; consider alternative of 'building in' joists at one end, Figure 4.

*This must be obtained from manufacturers' literature.

REFERENCES AND FURTHER READING

1 ***British Standard*** BS 6178 : Part 1 : 1982 'Specification for joist hangers for building into masonry walls of domestic dwellings'.
2 ***Defect Action Sheet*** DAS 25 (Design) 'External and separating walls: lateral restraint at intermediate timber floors — specification'.
British Standard BS 5628 Code of practice for the structural use of masonry : Part 1 : 1978 'Unreinforced masonry'.

DAS 58
August 1984

Site
CI/SfB 8 (23.9)(D6)

Suspended timber floors: joist hangers in masonry walls — installation

FAILURES: Uneven and springy floors; ceilings uneven or cracked at edges

DEFECTS: Coursing not level; packing directly under hangers; wrong grade of hanger used; hangers not tight to wall; gaps too great between joist ends and back plate; brick course used in thermal block wall

Figure 1 Poor fit of joist hanger to brick separating wall

BRE surveys have found instances of hangers bearing on packing instead of directly on masonry, hangers not tight to the wall, gaps too large between joist ends and back plates, joists inadequately fixed to 'restraint hangers' and badly twisted joists pulling hangers out of plumb.

If ***hangers*** are ***not bedded*** directly on masonry and installed ***tight to the wall,*** or if ***joists*** are ***cut short, hangers*** may ***move*** under load and the floor will 'spring' or settle. If ***hangers designed for*** use on ***high strength blockwork*** are ***used*** on ***low strength blockwork,*** the ***blockwork*** may be ***crushed*** locally under the masonry flange and heel of hanger, causing settlement and ***disruption*** of the ***floor.*** If joists ***hangers*** are ***out of plumb*** they may ***move*** under floor loads. ***If bricks levelling course are used*** in block walls ***thermal insulation*** will be ***impaired.***

Suspended timber floors: joist hangers in masonry walls — installation

DAS 58
August 1984

PREVENTION
Principle — Joist hangers must provide firm, secure and level support for floors.

Practice

- Check that no damaged or corroded hangers are used.
- Ensure that masonry course to carry hangers has been brought to correct height and is level:
  - minor adjustments to level should be made in the bed-joint **below** the top course, so that hangers can bear directly on masonry; any 'make-up' course of smaller units to be of same material as remainder of wall.
- Check that masonry units supporting hangers are of specified strength.
- Ensure that correct grade and type of hanger is used:
  - hangers complying fully with BS 6178[1] are made (and individually marked) for minimum masonry crushing strengths of 2.8, 3.5 and 7.0 N/mm^2: it is essential that hangers are not used with masonry of lower strength* than the hanger grade marking. Beware — often a floor spans on to walls of differing strengths, and various grades of hanger may have been specified for use on the site. Check also that special 'restraint hangers' are used where specified.
- Ensure that all hanger back-plates are tight against masonry face — no gap is permissible here, Figure 2.
- Ensure that gap between joist end and back plate does not exceed 6 mm, Figure 3.
- Check that undersides of joists are notched out for the hanger flanges just sufficiently for the plasterboard ceiling to 'run through'.
- Check how perimeter noggings are specified to be fixed to give support to the edges of plaster-board[2].
- Check that all hangers are plumb, and restraint hangers well-connected to joists with the specified number, type and size of fixings.

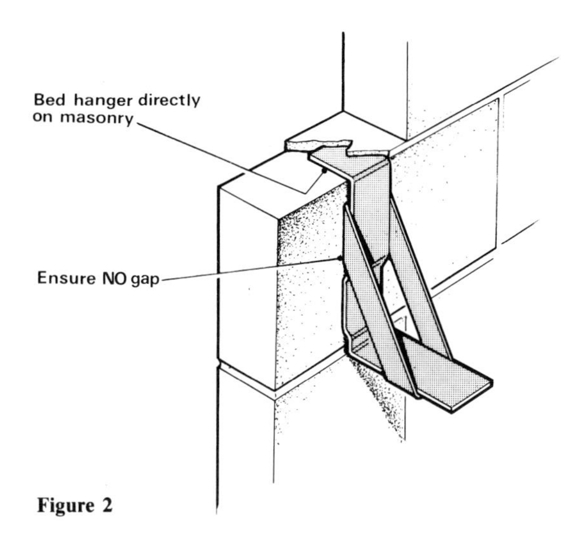

Figure 2

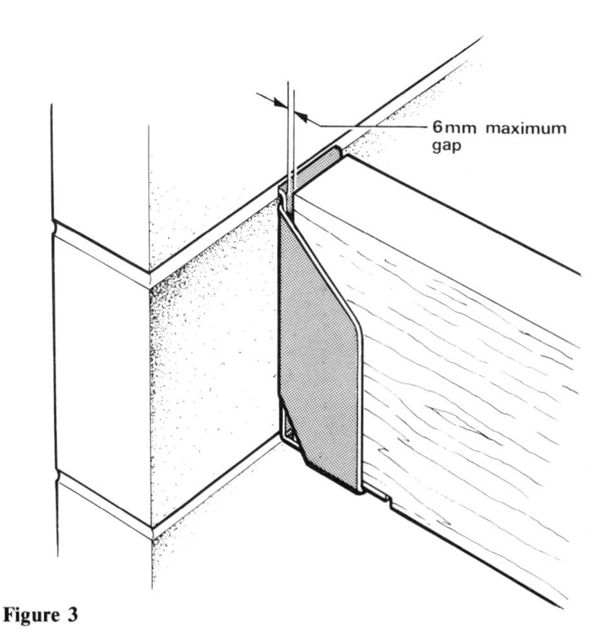

Figure 3

REFERENCES AND FURTHER READING

1 *British Standard* BS 6178 : Part 1 : 1982 'Specification for joist hangers for building into masonry walls of domestic dwellings'.

2 *Defect Action Sheet* DAS 57 (Design) 'Suspended timber floors : joist hangers in masonry walls — specification'.

*This must be obtained from manufacturers' literature

Suspended timber floors: notching and drilling of joists

FAILURE: **Reduction of designed strength of joists**

DEFECTS: **Holes drilled, or notches cut, outside permitted zones; holes or notches too close together; notches wrongly made.**

Figure 1 Notches too close, cut too deep, outside permitted zones

A BRE survey observed that frequently notches cut in the tops of joists were too near the centre of the span; holes drilled through joists were too close to joist ends or not on the neutral axis (centre line). This DAS gives an accepted 'rule of thumb' approach which avoids problems with domestic floors; any notching or drilling outside the limits shown here must be subject to calculation.

Holes, or notches, too close together, ***holes*** drilled ***near joist ends*** or ***off the neutral axis,*** and ***notches*** badly made or ***near the centre of the span,*** can ***weaken joists unacceptable*** unless specifically taken into account in design.

PREVENTION

Principle — The design strength and stiffness of joists must not be endangered.

Practice

- Design services layout so that notches and holes will be in permitted zones:
  - notches should only be made on the top edge of joists in the zones indicated in Figure 2;
  - holes may only be made on the neutral axis of joists and only in the zones indicated in Figure 3;
  - in rehabilitation work, check first that joists are not undersized.

- Specify the maximum permissible depth of notches, Figure 2.

- Specify the maximum permissible size of holes*, Figure 3.

- Specify that a hole is not to be within 100 mm of a notch. Holes within the permitted zones must be at least 3 diameters (centre to centre) apart, Figure 4.

Note: where joists are supported in hangers, joist ends may be notched underneath by no more than is necessary to accommodate the thickness of the hanger. Trimming and trimmer joists may be notched outside the permitted zones, along their top edges, to a depth no greater than is necessary to accommodate the top flanges of hangers carrying trimmer and common joists.

If design requirements for notches or holes cannot be met within these limitations:

- Increase the calculated minimum required joist depth by the depth of notches or diameter of holes[1].

* Joists in timber frame construction may be subject to specific requirements. Refer to design manual or specification, or to the NBA/TRADA manual.

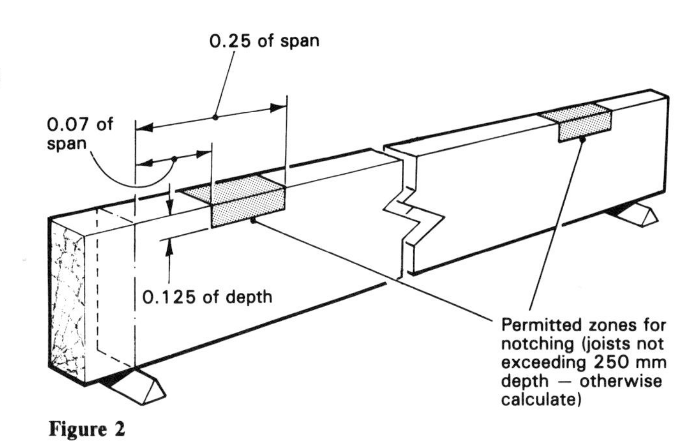

Figure 2

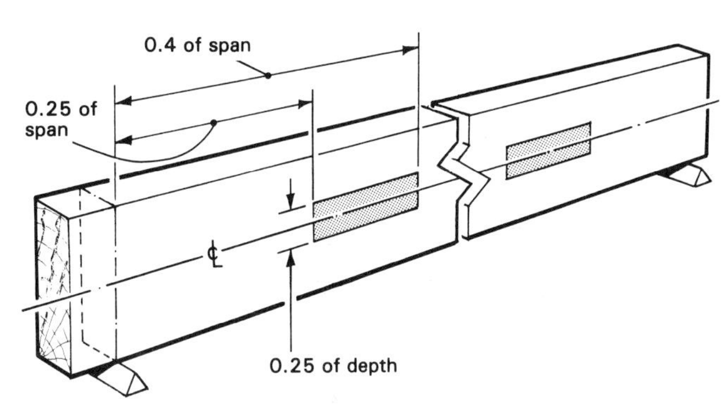

Figure 3

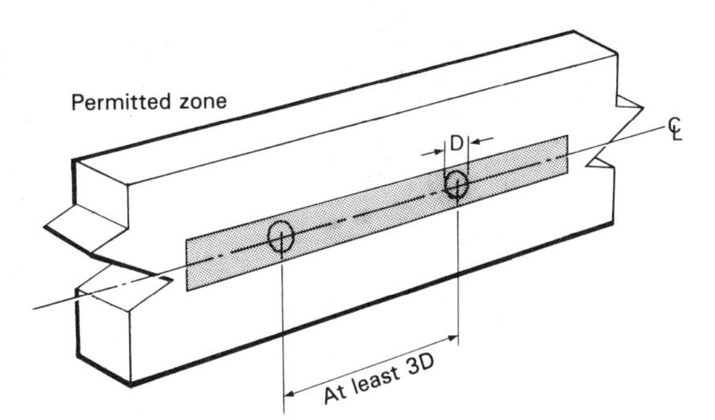

Figure 4

REFERENCES AND FURTHER READING

1 *British Standard* BS 5268 : Part 2 : 1984 The structural use of timber.

British Standard Code of Practice BS 5449: Part 1: 1977: Code of practice for central heating for domestic premises.

DAS 103
June 1987

Design
CI/SfB 8(23.9)(W7)(L64)

Wood floors: reducing risk of recurrent dry rot

FAILURE: Recurrence of dry rot after treatment

DEFECTS: Sources of dampness incorrectly diagnosed; dampness inadequately cured; drying out too slow; fungus not fully eradicated; replacement timber inadequately treated with preservative

Figure 1 Sheets of dry rot mycelium

Dry rot can recur unless correct remedial measures are specified and executed thoroughly. Priority should be given to correcting defects in the construction which encourage dry rot. ***If conditions*** are ***not*** sufficiently ***improved*** by eliminating all sources of dampness and by increasing ventilation, ***if infected timber*** has ***not*** been fully ***removed, if fungus*** within the masonry is ***not*** effectively ***contained*** and if ***replacement timbers*** are ***not*** adequately ***treated*** with preservative, ***dry rot can recur.***

Wood floors: reducing risk of recurrent dry rot

DAS 103
June 1987

PREVENTION

Principle — Change environment to one which prevents regrowth; remove visible fungus growths; contain any fungus remaining within masonry.

Practice

- Check that the outbreak is indeed dry rot[1,2].
- Identify and eliminate all sources of dampness:
  - — leaks and overflows from services, rainwater pipes, gutters and gulleys; rising damp; rain penetration; condensation etc;
  - — locally high moisture content in timber (say, greater than 20%) will be a guide to problem areas.
- Check that ventilation of floor voids is adequate;
  - — air bricks may be blocked, insufficient in number, or not located to achieve thorough cross ventilation; if necessary specify measures to improve ventilation, see DAS 73 and Figure 2.
- Specify measures to promote rapid drying of the structure eg, by extra heating and ventilation;
  - — remove floorboards adjacent to wet walls to increase air flow.
- Determine the full extent of the outbreak;
  - — inspect all woodwork in the vicinity, removing skirtings and floorboards to enable inspection of joists etc;
  - — look out for timber embedded in the walls, eg bond timbers, wall plates, disused lintels;
  - — note that fungus may have spread from or to adjacent rooms or property;
  - — strip plaster (approximately 300 mm) around adjacent woodwork to ensure the infection has not spread. Recoat with a fungicidal plaster specifically formulated to inhibit dry rot (check suitability with manufacturer).
- Specify that fungal growths and rotted wood are carefully removed and burned;
  - — it is usually adequate to cut away timber 300-450 mm beyond visible rot.
- Specify that all masonry within, say, 450 mm of the limits of the outbreak shall be treated with an effective fungicide by brushing or coarse spraying;
  - — flame treatment is ineffective.
- Redesign the bearing for replaced joists, eg to avoid rehousing new ends in external walls;
  - — support joist ends clear of walls on suitably protected RSJs, hangers or, at ground level on new sleeper walls, Figure 2, (note: new dpcs will be needed).
- Specify* protection for all existing timbers at continuing risk of decay (ie potentially remaining above 20% moisture content) for example those which must remain embedded in walls;
  - — suitable treatments include localised wall irrigation with preservative; preservative plugs or pastes; pressure injection. See DAS 13.

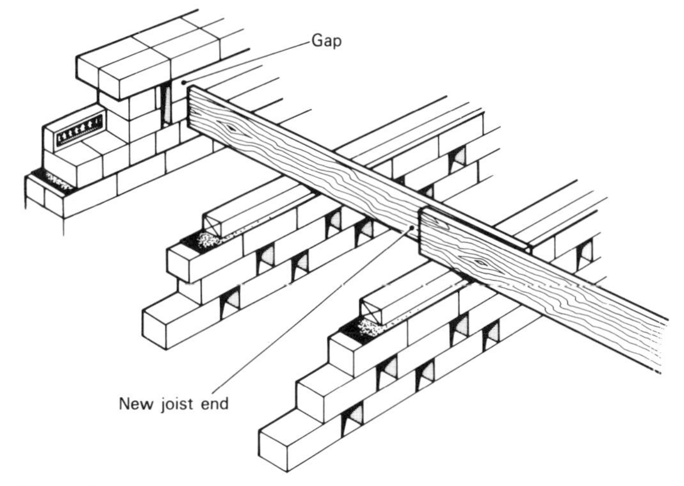

Figure 2

- Specify* timber pressure impregnated with preservative for all replacement work[3,4].
- Specify* application of three full brush coats of an effective preservative (eg type F in BS 5707) to all ends exposed by cutting.
- Check, on completion of works that moisture content of all timbers does not exceed 20% (after further drying if necessary).
- Check periodically that improved conditions are maintained.

---

*Specify only preservatives and methods of application cleared as safe under the Control of Pesticides Regulations, 1986.

REFERENCES AND FURTHER READING

1 **BRE (PRL) Technical Note** 44: 'Decay: recognition, prevention and cure'.
2 **BRE Digest** 299. 'Dry rot: its recognition and control'.
3 **British Standard** BS 5268: Part 5: 1977: Preservative treatments for constructional timber.
4 **British Standard** BS 5589: 1978: Code of Practice for preservation of timber.
□ **British Standard** BS 5707: Solution of wood preservatives in organic solvents.
□ **BRE Report** BR98 'Recognising wood rot and insect damage in timber'.

Suspended timber floors: chipboard flooring — specification

FAILURES: **Chipboard sagging, buckling, loose or squeaking; loss of strength and stiffness due to wetting or high humidity**

DEFECTS: **Flooring grade board not specified; perimeter gaps not specified; nogging, nailing and glueing specification inadequate; access traps not specified; moisture resistant board not specified for potentially wet or humid areas**

Figure 1 Failure from water percolating through tile joints

This *Defect Action Sheet* does not deal with floating floors constructed for sound insulation purposes.

Unsatisfactory chipboard floors were a source of complaint in an AMA survey of housing defects. In fact BRE surveys have found several different kinds of defects in chipboard floor decks, including the use of incorrect grade, joints neither glued nor nogged where one or other should have been specified, failure to use moisture-resistant board or to protect by 'tanking' in potentially wet areas, failure to provide perimeter gaps for movement, and decks cantilevered at perimeters by about 150 mm.

Chipboard can provide a satisfactory floor deck, but if requirements for ***grade, thickness*** in relation to span, ***fixings*** and ***edge support*** are ***not correctly specified,*** the ***floor deck*** may be ***unsound.*** If boards are ***not moisture resistant grade*** or ***not protected from moisture*** in potentially wet areas (such as kitchens and bathrooms) significant ***permanent loss of strength*** may occur. Similar permanent loss of strength may also occur if boards are not kept dry in storage or are not protected during wet trades' operations such as plastering. ***Chipboard should never get wet.***

PREVENTION

Principle—Chipboard decks must be well supported, firmly fixed, and protected from wetting at any stage.

Practice

- Choose a flooring grade of chipboard[1] and specify that boards are marked to BS 5669 type II or, for improved moisture resistance, type II/III, and instruct site supervisors to check board marking.
- Specify 18 mm board for joist spacings up to 450 mm and 22 mm board for joist spacings up to 600 mm (Figure 2).
- Specify laying T&G boards with long edge at right angles to the direction of joist span; square edge boards can be laid with long edge parallel to or at right angles to joists: (the former uses less nogging).
- Specify that all edges of square-edge or loose tongue boards are to be supported on joists or on noggings at least 38 mm wide (whether loose tongues are to be glued or not) (Figure 2).
- Specify that boards with integral tongues are laid so that every short edge is supported on a joist, with staggered joints.
  — tongues, whether integral or not, are recommended to be glued.
- Specify a gap for movement at the perimeter of each floor deck (Figure 3). The gap should be 2 mm per metre run and never less than 10 mm.
  — a floor in a room 3 metres across can expand about 7 mm with a change in moisture content from 9% to 16%. Larger floors therefore need bigger gaps.
- Specify that all boards are to be fixed with 10 gauge (3.35 mm) ring-shank nails of length at least 2½ times board thickness.
- Specify that chipboard must not cantilever.
- Provide access traps for services, fixed with 50 mm × 8 gauge screws, and supported at all edges (Figure 4) (nailed chipboard does not withstand removal like ordinary floorboards).
- Protect type II chipboard floors in potentially wet areas like bathrooms with for example an unjointed plastics sheet turned up behind skirtings; tiles will not do.
  — specify all service pipes to the protected areas to be taken through walls, or fitted with a sealed waterproof sleeve where unavoidably penetrating the waterproof layer.
- Specify that all chipboard decks are to be protected while wet trades (eg plastering) are in progress.

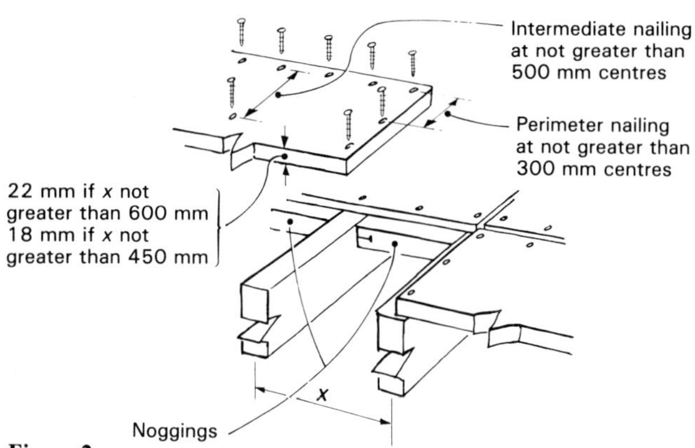

Figure 2

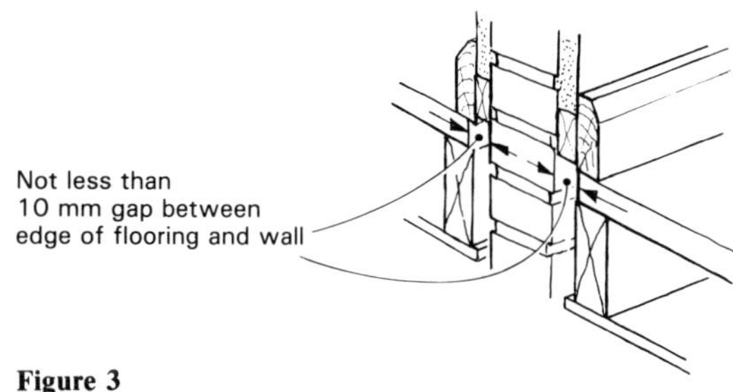

Figure 3

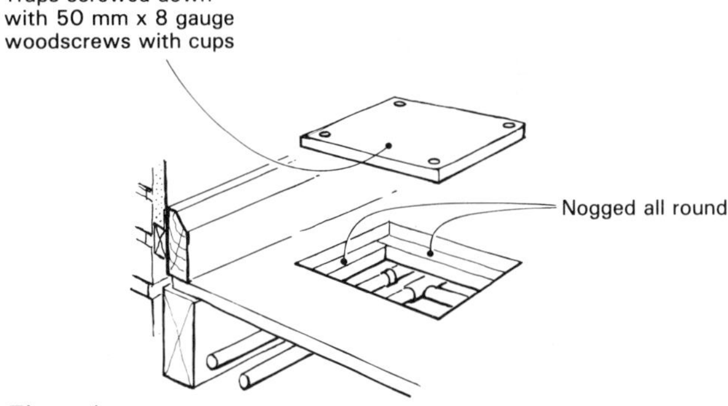

Figure 4

REFERENCES AND FURTHER READING

1 *British Standard* 5669:1979 'Specification for wood chipboard and methods of test for particle board'.

BRE Digest 239 'The use of chipboard'.

BRE Information Paper IP 31/79 'Chipboard — guidance on the selection of the appropriate type'.

Defect Action Sheet 32 'Suspended timber floors: chipboard flooring — storage and installation'.

DAS 32
July 1983

Site
CI/SfB 8 (23)j7

Suspended timber floors: chipboard flooring — storage and installation

FAILURES: **Chipboard sagging, buckling, loose or squeaking; loss of strength and stiffness due to wetting or high humidity**

DEFECTS: **Flooring grade board not used; boards not stored under correct conditions; no perimeter gaps for expansion; nogging, nailing and glueing inadequate; moisture resistant board not used in potentially wet or humid areas**

Figure 1 Chipboard must not get wet

This *Defect Action Sheet* does not deal with floating floors constructed for sound insulation purposes.

If the correct grade and thickness of chipboard is used it can provide a satisfactory floor deck, but BRE surveys have found many examples of unsatisfactory installations. ***If joints*** between boards with plain edges or loose tongues ***are not nogged,*** or if boards with tongues and grooves formed in them are ***not glued***, or if boards are not supported at floor perimeters, ***chipboard floors can sag and break.*** If plain nails are used instead of ring-shank nails, boards can spring. ***If boards become wet*** before installation, during construction or when in service, ***they can fail or suffer*** a ***permanent loss in strength.***

Suspended timber floors: chipboard flooring — storage and installation

DAS 32
July 1983

PREVENTION

Principle—Chipboard decks must be well supported, firmly fixed, and protected from wetting at any stage.

Practice

- Check that boards are to BS 5669, flooring grade, type II (red stripe) or II/III (red/green stripe) depending on design specifications.
- Check that thickness is as specified (18 mm for joists at 450 mm maximum centres, 22 mm for joists at 600 mm maximum centres).
- Ensure that boards in site storage are stored flat and kept dry, and that stacks are not higher than 15 boards (Figure 2a). Alternatively store on pallets or on level bearers at not more than 600 mm centres, with additional bearers (vertically above lower bearers) every 10 to 15 boards.
- Ensure that boards are given at least 24 hours to become stable in atmospheric conditions; they should be stacked on joisted floors immediately prior to laying.
- Check that stacks do not exceed about 6 boards deep on any one suspended floor to avoid overloading joists (Figure 2b).
- Boards with integral tongues should be fixed with their lengths across joists; short edges should be located over joists and staggered.
- Check that every edge of square-edge or loose tongue boards is supported by a joist or a nogging, whether loose tongues are glued or not.
- Check that boards are fixed with ring-shank nails of 10 gauge (3.35 mm) and length not less than 2½ times board thickness; nailing to be at not more than 300 mm centres at board perimeters and not more than 500 mm centres elsewhere; nail heads should be punched 2 to 3 mm below surface (Figure 3).
- Check that a gap for movement is present at the perimeter of each floor deck (Figure 4). The gap should be 2 mm per metre run, and never less than 10 mm.
- Ensure that boards are fully supported at perimeter walls (Figure 4).
- Check that any access traps are supported by noggings at all edges and are fixed with 50 mm × 8 gauge screws, not nails.
- Ensure that type II/III moisture resistant boards (red/green stripe) are used where specified (eg bathrooms, kitchens). Otherwise, if plastics sheet 'tanking' is specified, ensure that it is turned up at edges and sealed where services penetrate — check particularly where imperfections will be hidden from view, eg under bath. (Figure 5).
- Check that all installed floors are protected when wet trades are operating.

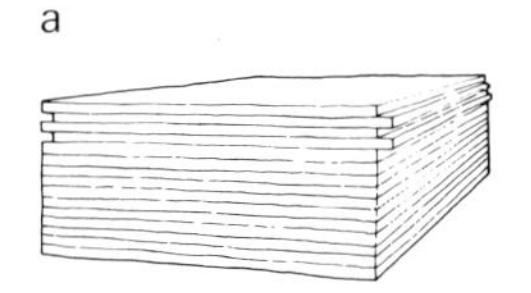

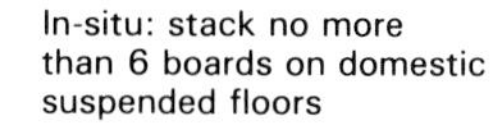

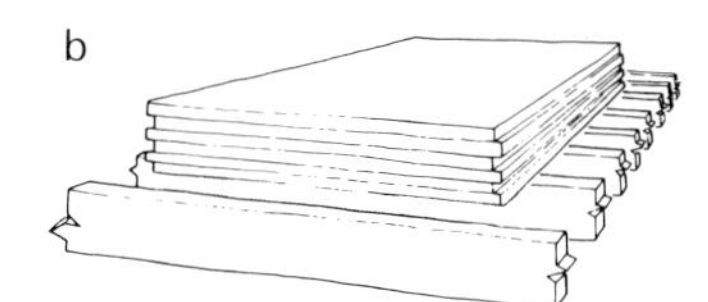

Figure 2a and 2b

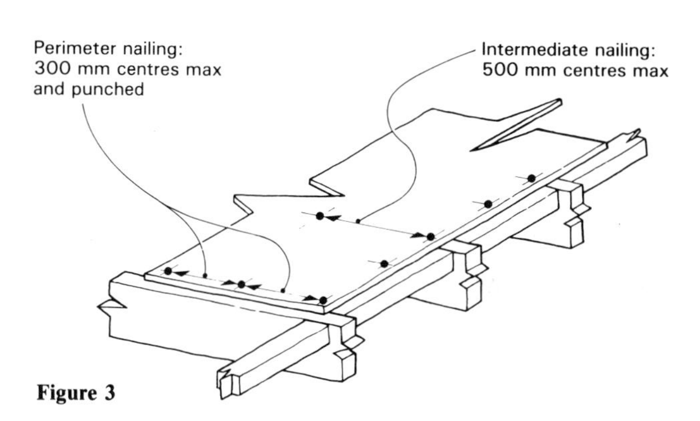

Figure 3

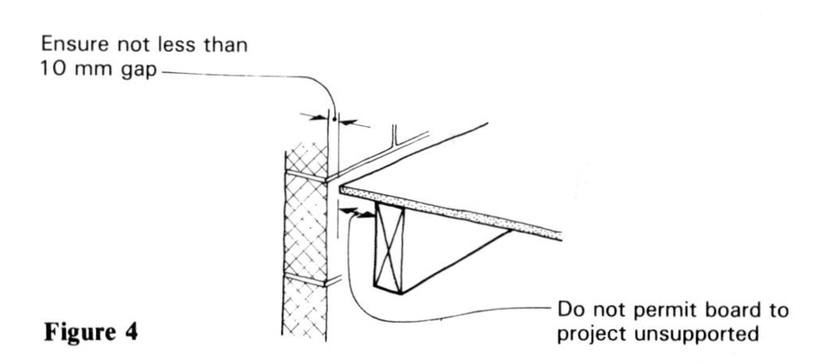

Figure 4

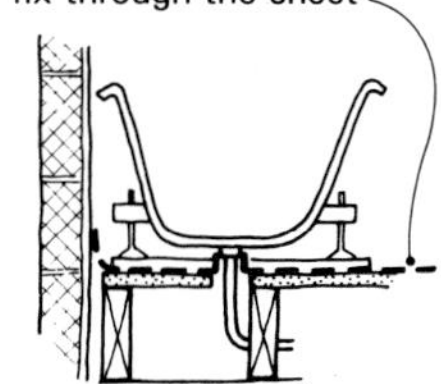

Figure 5

REFERENCES AND FURTHER READING

Defect Action Sheet 31 'Suspended timber floors: chipboard flooring — specification'.

Intermediate timber floors in converted dwellings — sound insulation

FAILURE: **Unacceptable sound transmission through floors which are required to act as separating floors in dwellings converted to multiple occupancy**

DEFECTS: **Insufficient mass; inadequate isolation/resilience against impact sound; airborne sound paths**

Figure 1 A common problem in converted dwellings

In the conversion of a house into flats, the original intermediate floors need to be upgraded to act as separating floors. The transmission of airborne and impact sounds must be reduced.

Airborne sounds — radio, TV, speech etc — travel through the floor, along flanking walls and through air paths such as cracks. They can be reduced by increasing mass or by structural isolation, sealing air paths at the same time, as described in this *Defect Action Sheet.*

Impact sounds — mostly footsteps — travel mainly through the floor. They can be reduced by structural isolation and further reduced by resilient floor coverings.

If the modifications to a floor ***add insufficient mass, do not provide a resilient or isolating layer,*** or ***do not close airborne sound paths,*** sound insulation is likely to prove ***unsatisfactory.*** As the sound insulation of a floor improves, the passage of sound by other routes becomes the dominant factor. Further improvement to the floor becomes unproductive unless the flanking walls are also improved.

REMEDIAL WORK

Principle — A 'converted' floor should have sound resistance comparable with that of separating floors in new construction.

Practice

If available headroom allows and existing ceiling is sound:

- Specify, in the lower room, an additional ceiling (which incorporates a rock wool or glass fibre quilt, minimum thickness 25 mm, density not critical) to be supported only by the walls and not connected to the floor above (Figure 2):
  - — a coving may be needed to eliminate airborne sound paths at the perimeter of the added ceiling;
  - — if floor boards in the upper room are plain-edged use screws where necessary to avoid squeaking boards and cover with hardboard. Note that skirting boards may need to be removed and re-fixed.

If available headroom precludes the above, the alternative is a floating floor to the upper room, raising floor levels by about 65 mm:

- Check that suitable details can be devised for door frames, services, sanitary fittings, stair flights etc.
- Check that joists will carry the increased dead loads satisfactorily with added members if necessary, then specify construction typified in Figure 3:
  - — decide whether to remove floor boards or ceiling; floor boarding can be retained if sound and sufficiently level; ceiling can be retained if in good condition and of equivalent mass to 31 mm of plasterboard — 2 coat (sanded undercoat) work is likely to comply; skirting boards will need to be removed in either case;
  - — insert rock wool or glass fibre quilt, not less than 100 mm thick, of density not more than 36 kg/m³, between joists. (If the decision above was to remove the ceiling, quilt cut slightly wider than joist spacing will retain itself in place pending plasterboarding);
  - — if decision was to remove ceiling, replace with 2 layers of plasterboard, with staggered joints, not less than 31 mm total thickness;
  - — if decision was to remove floor boarding, re-fix boarding if sound and reasonably plane using screws where necessary to avoid squeaking boards; otherwise substitute plywood, not less than 12 mm thick, nailed to joists;
  - — lay 25 mm of rock wool or glass fibre, of density 60 to 80 Kg/m³, followed by not less than 19 mm of plasterboard laid loose and with a gap at perimeter walls;
  - — lay T & G flooring grade chipboard, not less than 18 mm thick, with all joints glued but boards **not nailed down,** leaving a perimeter gap (not less than 10 mm — see DAS 31);
  - — fix ceiling coves to eliminate airborne sound paths at perimeter, or fill gaps with flexible sealant;
  - — re-fix skirting boards to cover gap at chipboard perimeter but just clear of the chipboard;
- Specify that all potential airborne sound paths — eg where services penetrate — are sealed.
- Check that the overall design meets fire requirements.
- Note that alternative solutions that require sand pugging* on the existing ceiling pose practical difficulties in construction. Also it is difficult to assess the ability of an existing ceiling to support the sand, and to predict the performance of such solutions in fire.

*The term 'deafening' is used in Scotland.

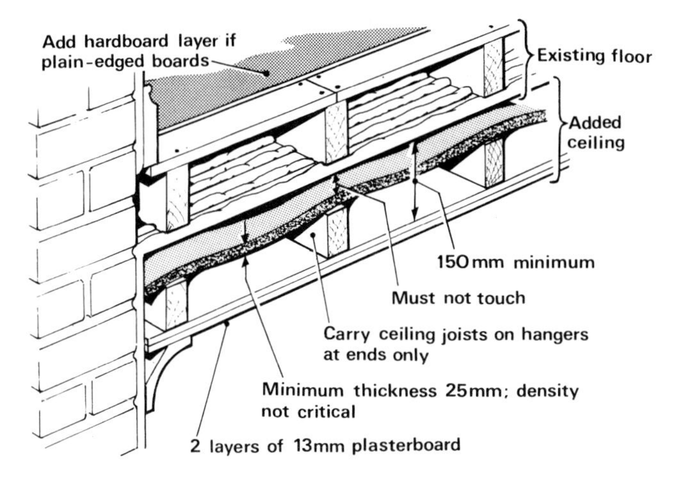

Figure 2

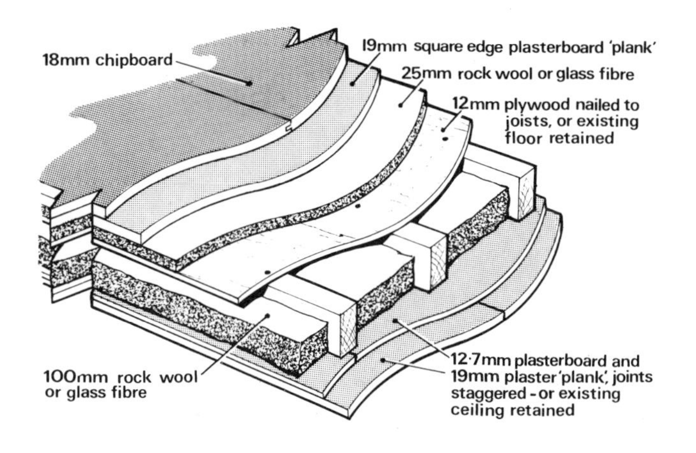

Figure 3

REFERENCES AND FURTHER READING

Defect Action Sheet DAS 31 'Suspended timber floors: chipboard flooring — specification'.

Plasterboard ceilings for direct decoration: nogging and fixing — specification

FAILURES: Ceilings uneven, sagging; cracking at board joints and perimeter

DEFECTS: Incorrect support, insufficient noggings, perimeter noggings omitted; insufficient nails, nails incorrect type

Figure 1 Board edges not supported .... cracked ceiling

Plasterboard ceilings are often inadequately nogged, both generally and at ceiling perimeters — the latter particularly when joist hangers are used (see DAS 57).

If board ends and edges, particularly at ceiling perimeters, are ***not adequately supported,*** boards may sag and finishes may crack; ***if nails are incorrect type and size,*** or their ***spacing excessive, boards*** may be ***in-secure.***

PREVENTION

Principle — Plasterboard for ceilings must be securely fixed to correctly sized and spaced supporting members.

Practice

- Specify appropriate board thickness for joist or truss tie spacing; the recommended spacing for board thickness of 12.7 mm is 450 mm, maximum spacing 600 mm; the recommended spacing for 9.5 mm board is 400 mm, maximum spacing 450 mm.

Note: 12.7 mm thick wallboard is commonly specified for 600 mm centres; since this is the ***maximum*** recommended spacing it is particularly important to ensure that boards will be well supported and remain dry in service.

- Specify that boards are fixed with bound edges at right angles to joists or ties.
- Specify that all board ends and edges are adequately supported:
  — manufacturer's recommend that all edges are supported;
  — NHBC do not require support to bound edges except at perimeter of ceilings under trussed rafters, or to cut edges already within 50 mm of a support.
- Specify 38 × 38 mm noggings for general use, Figure 2, and 25 × 38 mm battens for perimeter fixings, Figure 3;
  — unless joist or tie spacing and board positioning and cutting are accurate, adequate support of end joints under joists or ties may be difficult to achieve (particularly with 35 mm wide joists or ties); consider specifying a 19 x 38 mm batten fixed to the side of the joist or tie, Figure 4.
- Specify a butted joint for adjacent bound edges and a 3 mm gap when either edge is cut.
- Specify the appropriate length of zinc coated plasterboard nails to BS 1202 Part 1:
  — a nail length of 40 mm is recommended for 12.7 mm boards, 30 mm for 9.5 mm boards.
- Specify that nails are to be at not more than 150 mm centres.
- Specify, for the most commonly used 'wallboard' type of plasterboard, that it is to be fixed with the decoration (white) face exposed.

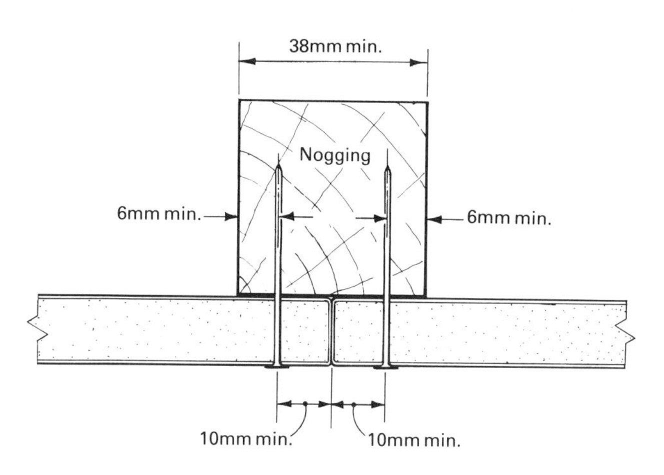

Figure 2 Edge joint

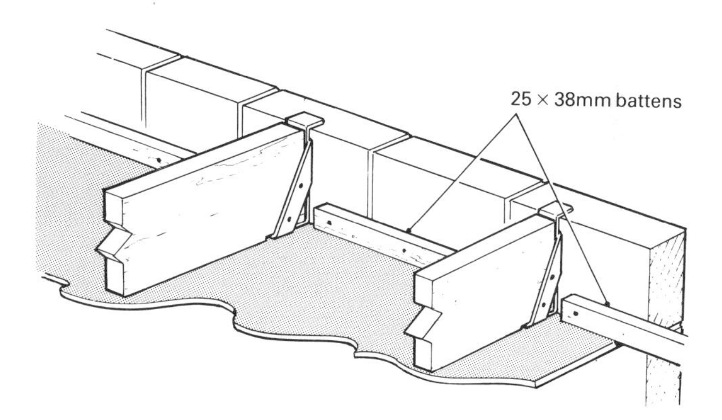

Figure 3

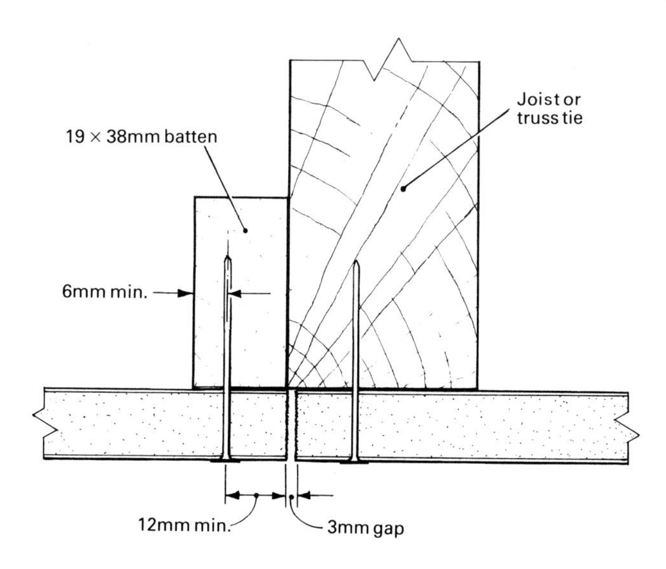

Figure 4

Plasterboard ceilings for direct decoration: nogging and fixing — site work

FAILURES: **Ceilings uneven, sagging; finishes cracking at board joints and perimeter**

DEFECTS: **Nogging omitted at joints or perimeter; nails incorrect size or too widely spaced; boards incorrect thickness for support centres**

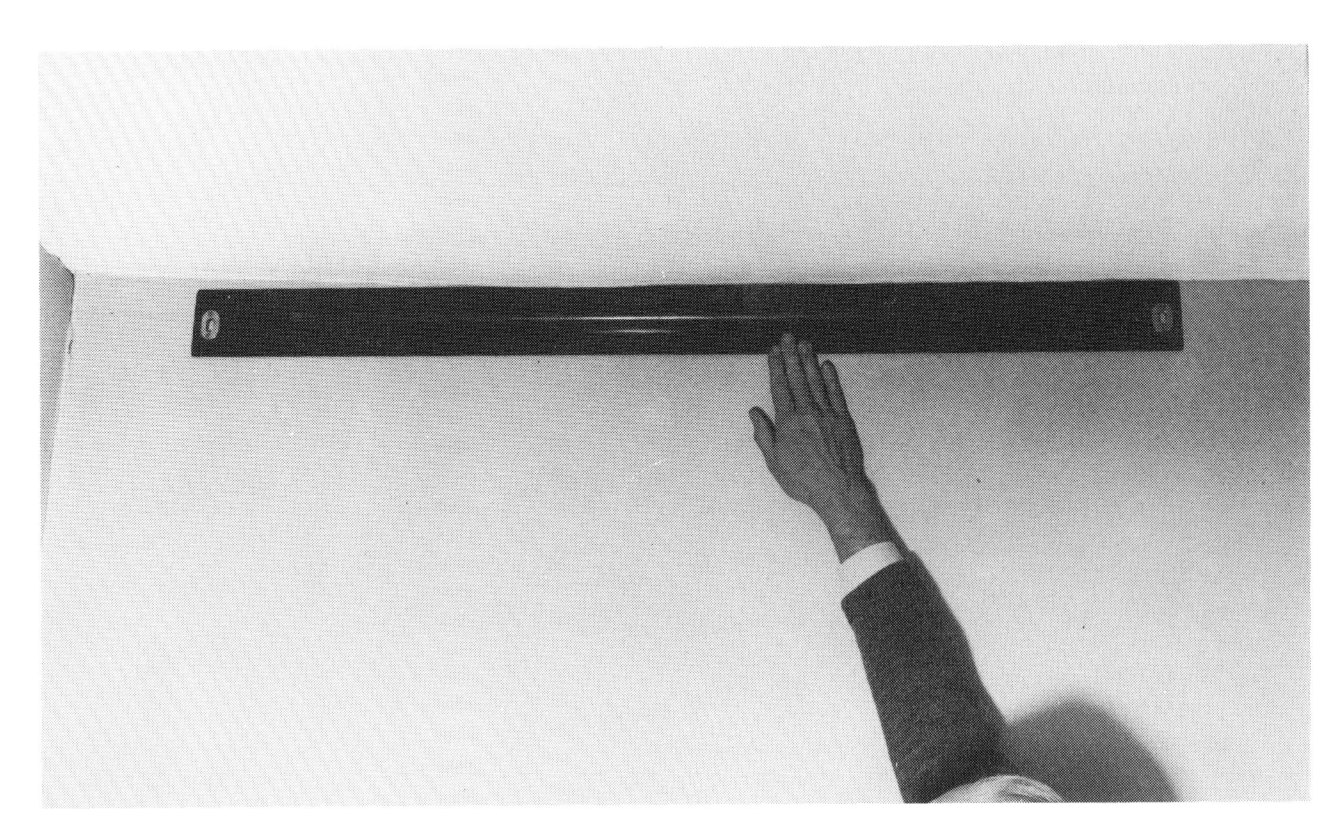

Figure 1 Perimeter not nogged .... ceiling sags

Plasterboard, of the type known as 'wallboard' and 12.7 mm or 9.5 mm thick, is commonly used for domestic ceilings. Plasterboard ceilings are often inadequately nogged and noggings are omitted at ceiling perimeters — particularly where joist hangers are used to support joists or the heels of mono-pitch trusses.

If ***boards*** are ***not supported at all joints, and at ceiling perimeter***, boards may sag, finishes may crack; if plasterboard ***nails*** are of ***incorrect size or at excessive spacing***, ***boards*** may be ***insecure.***

PREVENTION

Principle — Boards must be securely fixed to correctly sized, and correctly located, supporting members.

Practice

- Check that board thickness is as specified:
  — the recommended spacing for 12.7 mm board is 450 mm, maximum spacing 600 mm; the recommended spacing for 9.5 mm board is 400 mm, maximum spacing 450 mm.

Note: 12.7 mm wallboard is commonly specified for use at maximum spacing and when so used it is important to ensure that boards are well supported and kept dry at all stages.

- Check that boards are fixed with decoration (white) face exposed.
- Ensure that boards are fixed with their bound edges at right angles to joists or truss ties.
- Check that end joints are staggered, Figure 2.
- Check that adjacent bound edges are butted.
- Check that boards are fixed with a 3 mm gap when either edge is cut.
- Ensure that board ends are securely located under joists or truss ties:
  — where inaccuracies, for example, in joist or truss tie positioning make secure fixing impossible to achieve, a 19 mm wide × 38 mm batten should be fixed to the side of the joist or tie.
- Check whether specification requires that joints made at right angles to truss ties or joists are to be nogged:
  — if so noggings should be at least 38 × 38 mm.
- Ensure that plasterboard is supported at perimeters of ceilings under trussed rafters by battens not less than 25 mm wide.
- Check whether specification requires perimeter battens at ceilings under intermediate floors:
  — if so, these battens should be not less than 25 mm wide.
- Check that nails are correct length (40 mm for 12.7 mm board, 30 mm for 9.5 mm board).
- Ensure that nails are at not more than 150 mm centres:
  — nails should not be closer than 10 mm to bound edges and 12 mm to cut edges, Figure 3;
  — nails should be driven until just below the surface but without breaking the paper (if paper is broken renail 50-60 mm away).

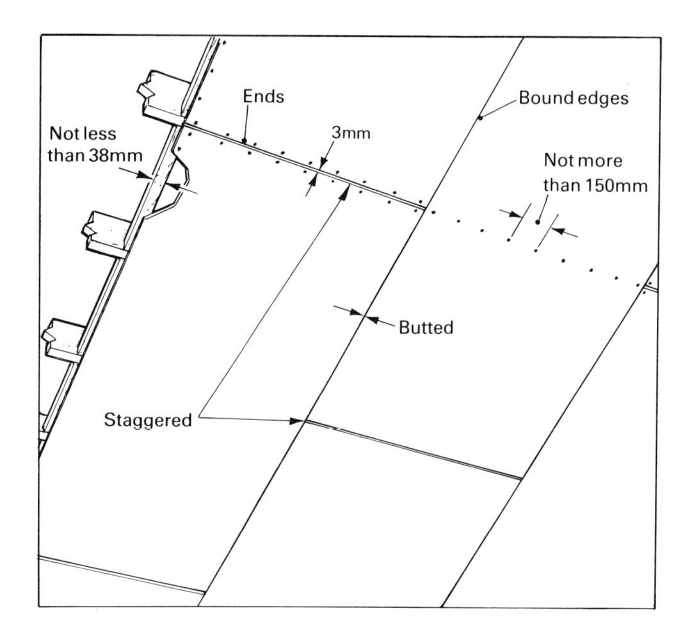

Figure 2

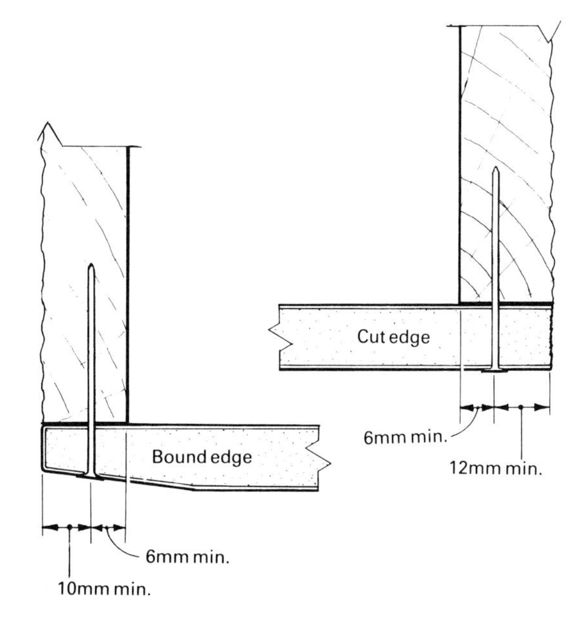

Figure 3

DAS 53
June 1984

Design
CI/SfB 8 (24)(U47)(A3u)

Stairways: safety of users — specification

FAILURE: **Unsafe stairways**

DEFECTS: **Top nosing, newels, flights and handrails not firmly fixed; widths, headroom, handrail height incorrect; handrails trap or injure hand, or trap clothing; lighting inadequate; glazing dangerous**

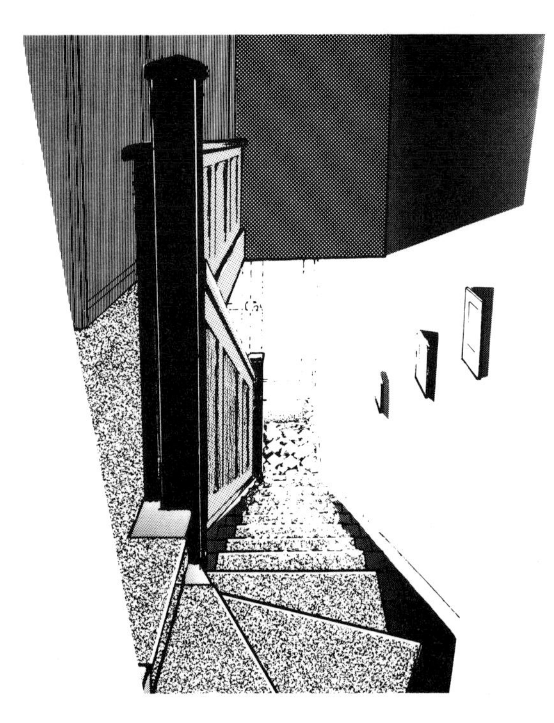

Figure 1 Glare on stair with unprotected glazing at foot

Several different faults can occur in stairways — particularly in rehabilitated construction, but also in new-build. Some of these have safety implications. Stairs are used by people of all ages and all degrees of fitness. A BRE study has shown that there are in the UK over 200,000 accidents on stairways in housing every year. About 600 are fatal. Only a proportion of these accidents is likely to be due to inadequate design. However, BRE site surveys have revealed many defects in design and construction of stairways, even to the extent of total removal of stairway balustrades in rehabilitated housing.

Stairways in new construction must comply with Building Regulations. Stairways in rehabilitation work need not comply (unless the stair is entirely renewed) but consideration should be given to upgrading where necessary[1].

If ***stairs or nosings*** give ***insecure footing,*** if handrails give an ***insecure hand-hold*** or catch clothing, or if ***lighting*** is ***poor or badly sited,*** there is ***risk of accident;*** if there is ***unprotected*** or non-safety ***glazing*** near the foot of the stairway, the risk of ***serious injury*** resulting is increased.

PREVENTION

Principle — Stairways must be safe for all likely users.

Practice

- Consider insertion of extra newels where existing balustrades will not meet the current strength recommendation[1]:
  — specify how such newels are to be fixed — eg through-bolted to joist — to give permanent rigidity.
- Specify how flights are to be fixed, eg at string and at foot, to provide rigidity.
- Specify how nosings or top treads are to be fixed to achieve permanent rigidity (or avoid using flights with separate top nosings or half treads):
  — note that trimmer joist hangers, if not let into joist, may produce unstable floor boarding at top of flight (see DAS 54); instruct site supervisors to check that floor boards and top treads or half treads do not rock;
  — check that design will provide continuity of level between top tread or nosing and flooring.
- Check that headroom will not be less than 2 m anywhere:
  — note: measured from pitch line, Figure 2.
- Check whether the appropriate minimum widths recommended by the Code of Practice[1] can be achieved:
  — 800 mm if serving a single dwelling;
  — note that width is measured to the **handrail,** or between handrails, Figure 3, and that length of tread must not be less than this;
  — going of landings must not be less than width of stairway.
- Ensure that maximum gap in balustrades, or open risers, would not pass a 100 mm sphere.
- Specify that handrail is to be securely fixed at recommended[1] height (not less than 840 mm nor more than 1 m vertically above the pitch line).
- Check that continuous passage of hand on handrails will not be obstructed eg by apron linings or handrail fixings, Figure 4:
  — instruct site supervisors to check on completion that handrails are smooth and that there are no points at which a hand may be injured — eg by sharp-edged fixings.
- Check all stairways for conformance, where possible, with the Code of Practice[1]. Particularly:
  — avoid winders at the top of stairs;
  — locate glazing so that someone falling will not hit it — otherwise guard glazing or specify a suitable type of safety glass or plastics[2].

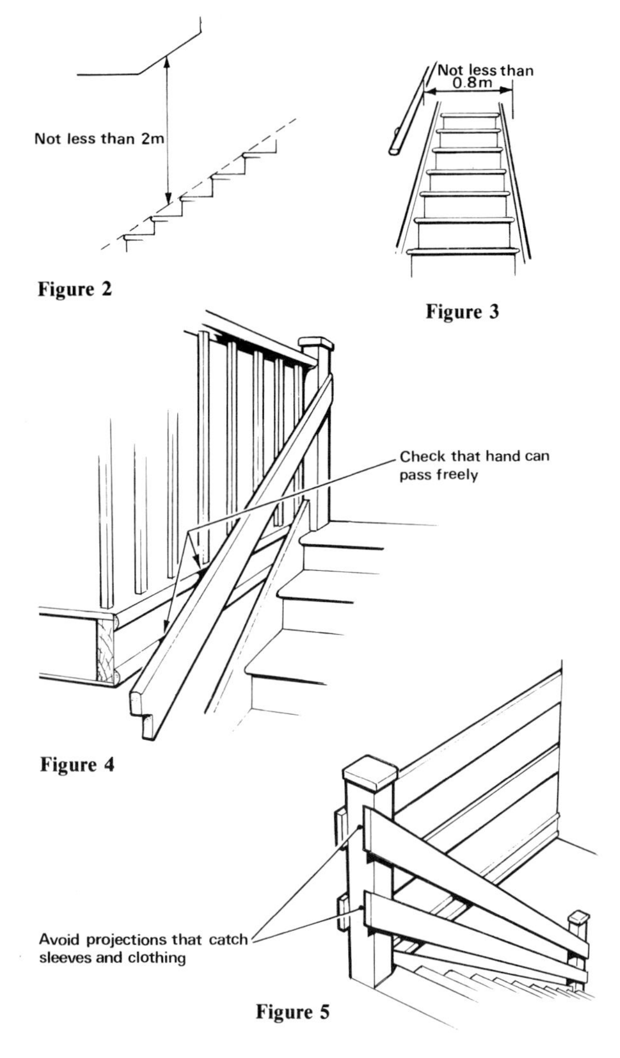

Figure 2

Figure 3

Figure 4

Figure 5

  — locate natural and artificial light sources to enhance visibility of nosings, particularly at top and bottom of stair and at changes of direction, but avoid glare.
  — ensure that no door or pivotted window could obstruct the stairway.
- Check overall design in new-build to ensure conformance with Building Regulations.

REFERENCES AND FURTHER READING

1 *British Standard* BS 5395; 1977 'Stairs Ladders and Walkways : Part 1 : Code of Practice for straight stairs'.

2 *British Standard* BS6262 : 1982 'Glazing for Buildings'.
British Standard BS 6180 : 'Code of Practice for protective barriers'.
BRE Information Paper IP 18/81, 'Accidents involving glass'.

DAS 54
June 1984

Site
CI/SfB 8 (24)(D6)(U47)

Stairways: safety of users — installation

FAILURE: Unsafe stairways

DEFECTS: Newels, flights, top nosings or floor boarding, handrails, balustrades, not firmly fixed; handrails rough or splintered; handrail brackets or fixing screws projecting; handrails fixed at wrong height; stairs not protected from damage by following trades

Figure 1 200 000 accidents on stairs in the home every year

Every year there are about 200,000 accidents on stairs in the home. BRE site surveys have found many defects: newels and top nosings were dangerously insecure, flights were not rigidly fixed, handrails were not sanded smooth, handrail brackets presented sharp obstructions, wooden spacers between handrail and apron lining obstructed passage of the hand, and mouldings on the apron lining were fixed so that they wedged the hand between moulding and handrail. On over a quarter of the sites inspected, unprotected stairs were damaged after installation: loosened nosings etc may not be re-fixed.

If ***flights, newels, nosings, handrails and balustrades*** are ***not firmly fixed,*** or ***if handrails*** either trap or ***injure the hand*** or force the user to let go, the ***risk of accident,*** already high, is increased.

PREVENTION

Principle — Stairways must be built so that they are safe for all users.

Practice

- Check how newel is to be fixed:
  — if unspecified, consider through-bolting to joists.

- Ensure that trimmer joist hangers are let into joists so that top nosing, or floor board at top of flight, will seat without rocking, Figure 2.

- Check how flight is to be fixed:
  — if unspecified, consider using for example metal dowels; plugging and screwing string to wall; screwing through bottom riser into batten plugged and screwed to floor, Figure 3.

- Ensure that flight is fixed at correct pitch angle.

- Check that balustrades are securely fixed at specified centres.

- Ensure that any separate top nosing or half tread is immovable and has been securely fixed, Figure 4.

- Check that handrail is securely fixed at specified height.

- Check that no handrail brackets, fixing screws or spacers, present sharp edges and that handrail is smooth and also unobstructed throughout its length — particularly at the apron lining.

- Check that flights are protected from damage both before and after installation.

- Check at handover that any damage has been made good.

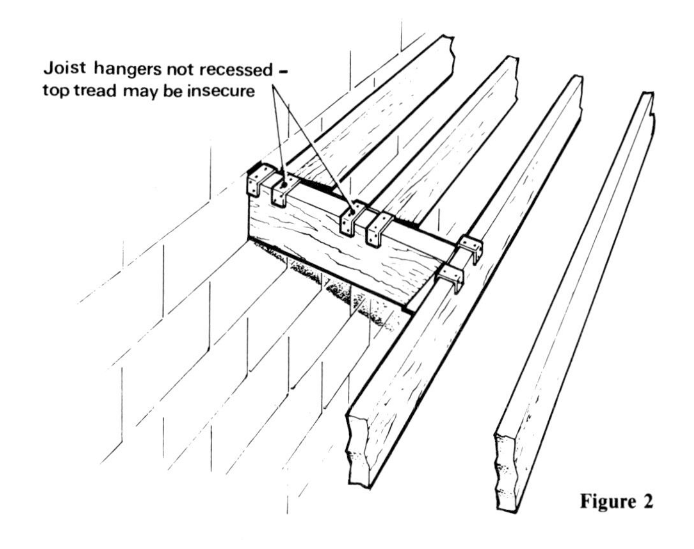

Figure 2

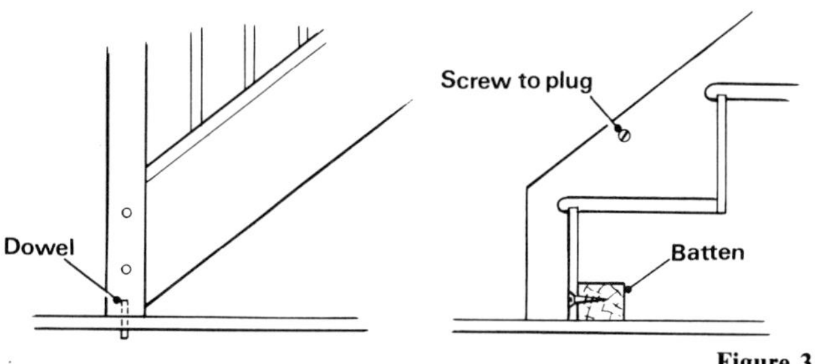

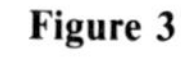

Figure 3

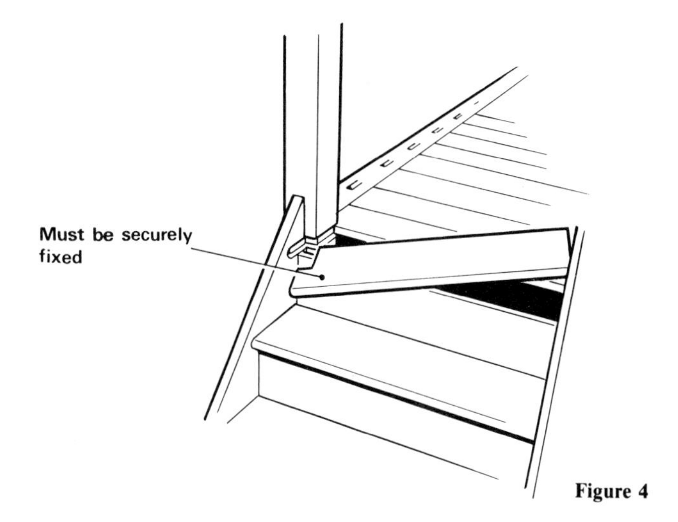

Figure 4

PART FOUR

Walls 1: Movement and Structural Aspects

DAS 18
February 1983 New edition February 1985

Design
CI/SfB 8 (21)(F47)

External masonry walls: vertical joints for thermal and moisture movements

FAILURE: **Significant cracking — generally vertical but sometimes stepped — of the external leaf of masonry walls**

DEFECT: **Inadequate provision for horizontal movements of masonry**

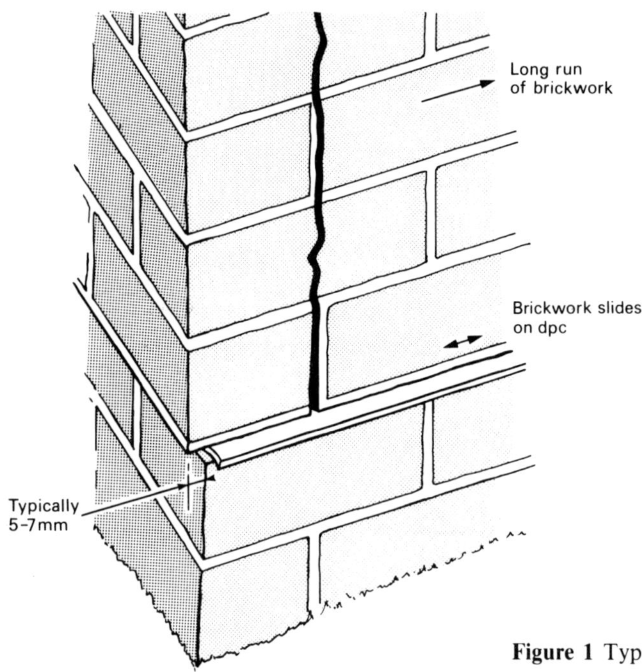

Figure 1 Typical crack caused by horizontal movements

All building materials, and the elements made from them, 'move' as a result of the expansion and contraction caused by temperature or moisture content changes. The amount of movement depends on materials and conditions. Some kinds of movement cease early in a building's life, others continue indefinitely.

Many long terraces in a BRE survey of housing under construction had no provision for movements. If no design provision is made, for both initial and long-term movements, masonry walls may crack. A masonry wall can 'absorb' a limited amount of expansion or contraction without distress. But, ***if a wall is long enough*** to produce more ***expansion*** than it can absorb, ***displacement and consequential cracking*** will occur. And, ***if it is long enough*** to produce more ***contraction*** than it can absorb — by distribution among fine insignificant cracks — ***major cracks*** will occur. If ***appropriate movement joints*** are provided the risk of ***major cracking*** can be ***minimised.*** Whilst this DAS provides basic rules, careful design consideration is needed, consulting References 1, 2 and 3.

PREVENTION

Principle Movements resulting from the response of masonry to temperature or moisture changes, must not produce significant cracking.

Practice

- Provide long runs of fired clay brickwork (say, more than 3 houses) with a joint capable of accommodating 10 mm of movement at about every 12 metres (Reference 2 should be consulted), see Figure 2. Note: this is not a 10 mm joint, it is a joint capable of accepting 10 mm of movement — its designed width must be related to the jointing materials used[1] and may need to be of the order of 15 to 25 mm. If narrower joints are required they must be spaced correspondingly closer.
- Provide calcium silicate brickwork with a movement joint at about every 7.5 metres[2,3]. In this case the movement is predominantly shrinkage and jointing materials must be capable of continued functioning after joints have reached their widest condition.
- Provide concrete brick- or blockwork with a movement joint at about every 6 metres.
- Calculate in accordance with Reference 1 when in doubt about providing a movement joint:
  - in housing, cracks in inner leaf blockwork are usually small and made good with plaster or concealed by dry lining;
  - movement joints (for thermal and moisture changes) are unnecessary in the footings of small houses, even in long terraces: conditions below dpc are relatively stable. But joints to accommodate differential soil movements may be needed.
- Locate movement joints where there is lateral support (eg at separating walls) Figure 4. If joints are located in spandrels ensure that stability is adequate and that they will not produce cracking elsewhere (eg in tiled sills bridging the joint, Figure 5).
- Instruct site staff to put in extra ties, at 300 mm vertical spacing, at each side of the joint. Wire ties, if the structural design permits, are preferable: they can accommodate movement by flexing, Figure 4.
- Protect short returns (say, less than 800 mm) by movement joints, Figure 3; suitable fillers for movement joints are given in Reference 1, Table 5.
- Instruct site staff not to allow the use of clay bricks warm from the kiln: keep them for about a week, by which time most of any initial irreversible expansion will have occurred.
- Avoid specifying unnecessarily strong mortars: they cannot absorb stresses so easily[2].

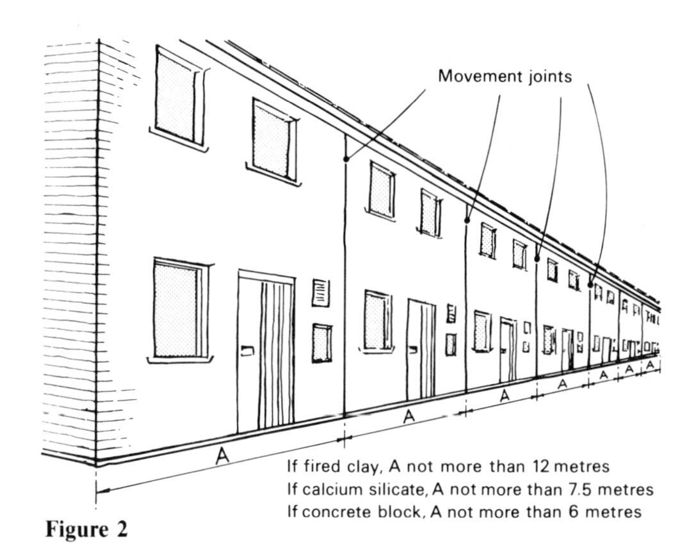

Figure 2

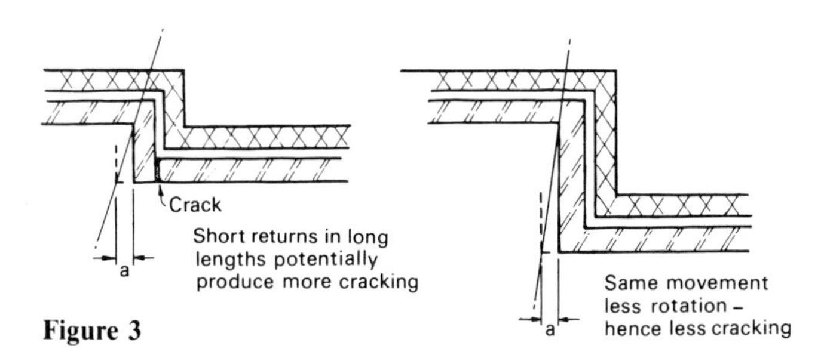

Figure 3

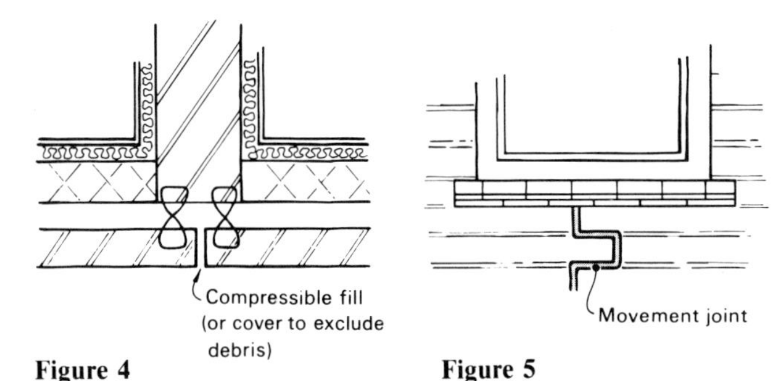

Figure 4

Figure 5

REFERENCES AND FURTHER READING

1 *British Standard* 6093:1981 'Joints and jointing in building construction'.

2 *British Standard Code of Practice* CP 121, Walling Part 1:1973 Brick and block masonry'.

3 *BRE Digest* 157 'Calcium silicate brickwork'.

BRE Digests 227, 228, 229 'Estimation of thermal and moisture movements (Parts 1, 2 and 3 respectively)'.

CIRIA Technical Note 107 'Design for movement in buildings'.

External masonry walls: assessing whether cracks indicate progressive movement

FAILURE: Loss of structural integrity

DEFECT: Cracks caused by progressive movement

Figure 1

In rehabilitation work it is often necessary to decide what, if anything, needs to be done about cracks in masonry walls.

Sometimes quite minor cracks need action to prevent consequential damage — eg where cracks in brickwork have produced corresponding cracks in rendering, potentially leading to saturation and sulphate attack. Often cracks in masonry are long-standing and of little significance. Very occasionally cracks are indicative of progressive movement and it is with these that this DAS is concerned.

All buildings move and most crack. If cracks are caused by progressive movement, action is needed. If ***cracks*** are ***wrongly diagnosed as progressive, incorrect*** and unnecessary ***action*** may be ***taken*** — possibly ***at considerable cost***.

External masonry walls: assessing whether cracks indicate progressive movement

DAS 102
June 1987

PREVENTION

Principle — Movement must be monitored if it is to be reliably diagnosed as progressive.

Practice

- Check that cracks are not merely superficial — eg confined to rendering or, in cavity walls, to one leaf.
- Record width, taper, location, direction and distribution of cracks;
  — record both exterior and interior cracks on elevation sketches;
  — distinguish between cracks indicative of shear movement, Figure 2, and tension cracks;
  — note particularly the direction of taper in tapered cracks (which will often be stepped, following mortar joints), Figure 3.
- Deduce from the above the probable directions of movements in the structure;
  — repeated readings taken with a surveyor's level on a course of brickwork near ground level may help to confirm directions of movement and the locations where movement is greatest.
- Seek confirmation in terms of the presence or absence of potential causes;
  — eg volume changes in clay soils, ground subsidence, consolidation, nearby excavations, mining, unequal loading etc.
- Monitor movement at least 4 times over a minimum period of 6 months to establish whether movement has ceased or is progressive;
  — cyclic movement may well be found to be superimposed on progressive movement and a longer period of monitoring may be needed to distinguish between 'seasonal' and 'progressive';
  — the traditional glass 'tell-tales' do not tell a tale — they do not distinguish between movements of progressive or other kinds. Better methods are available, eg Figure 4.

If movement is proved to be progressive

- Consult references 1 and 2 or seek engineering advice.

If movement is proved NOT to be progressive

- Repair crack (only if necessary, eg, for appearance or weather exclusion).

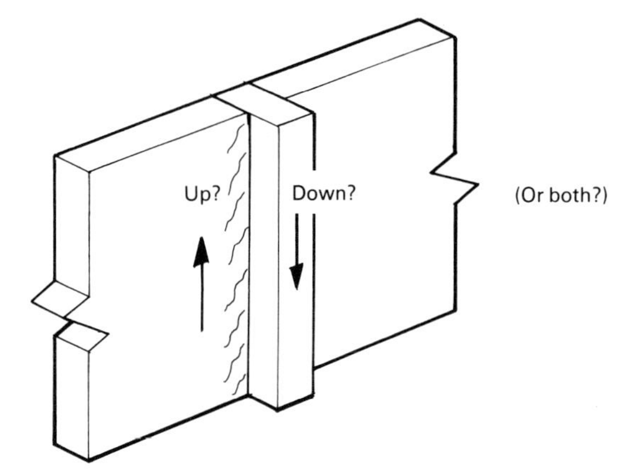

Figure 2

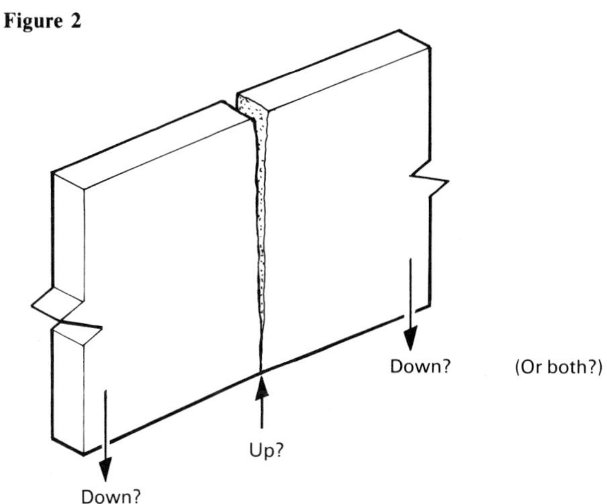

Figure 3

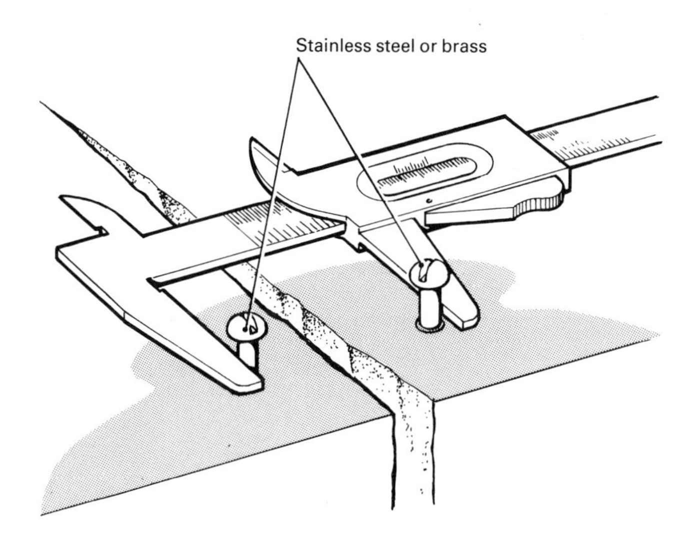

Figure 4

REFERENCES AND FURTHER READING

1 BRE Digest 298 'The influence of trees on house foundations in clay soils'.
2 BRE Digest 313 'Mini-piling for low-rise buildings'.
□ BRE Digest 75 'Cracking in buildings'.
□ BRE Digest 251 'Assessment of damage in low-rise buildings'.

Masonry walls: chasing

FAILURES: **Cracking, reduced strength, instability; impaired sound insulation in separating walls**

DEFECTS: **Chases greater than permitted depth, masonry broken away at chase intersections or where a vertical chase reaches to the bearing area at the top of a wall; chases and outlet boxes back-to-back on separating walls**

Figure 1 Back-to-back socket outlet boxes in a separating wall

Unsatisfactory chasing was seen in many dwellings in a BRE survey. If ***chases*** are ***too deep***, edges and intersections ***not cleanly cut***, chases ***close (or coincident) on opposite sides*** of the wall, ***or if a chase breaks out the bearing*** for supported construction, ***stability*** of the wall or the construction it supports can be ***endangered***.

Many masonry separating walls, as constructed, only marginally provide the required level of sound insulation. Their sound insulation may be reduced, unacceptably, if chases or recesses (eg for electrical cables or outlet boxes) are back-to-back or too deep (Figure 1).

PREVENTION

Principle — The number, size or position of chases in a wall must neither impair its stability nor unacceptably reduce its sound insulation.

Practice

- Ensure that chases and socket box recesses are not back-to-back:
  - — where shown on drawings nominally back-to-back, consult the designer before re-siting (Figure 2);
  - — note that back-to-back cooker socket outlets in separating walls are a common cause of poor sound insulation.
- Do not permit walls of hollow blocks to be chased; shallow chases may be permissible in multi-perforated bricks.
- Do not allow horizontal or diagonal chases to be cut unless specifically approved by the designer; horizontal and diagonal chases seriously weaken a wall. All such chasing must be approved and carried out under close supervision[1].
- Calculate the maximum permitted depth of any chase (from Figure 3) and tell the operative — he cannot see how thick a completed wall is:
  - — if chasing already plastered walls, ignore plaster thickness.
- Do not allow chases to exceed the permitted depth (Figure 3).
- Ensure that chases are not cut with impact tools (either mechanical or hand)[1].
- Check that chases are cut cleanly, without spalling or other damage, particularly near bearing surfaces (Figure 4).

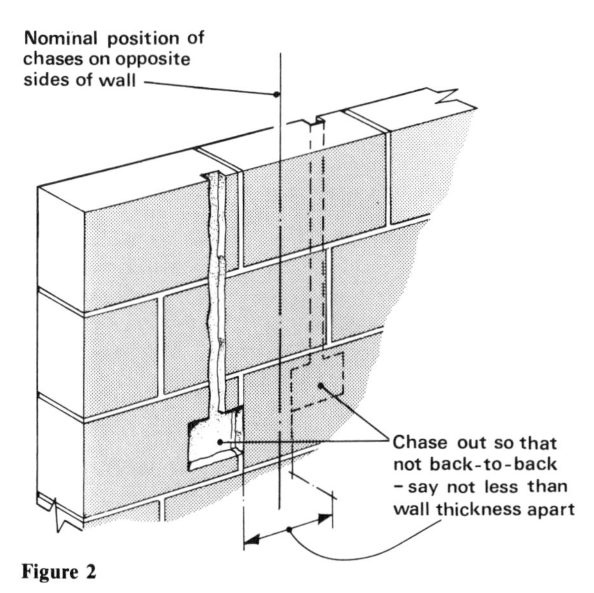

Figure 2

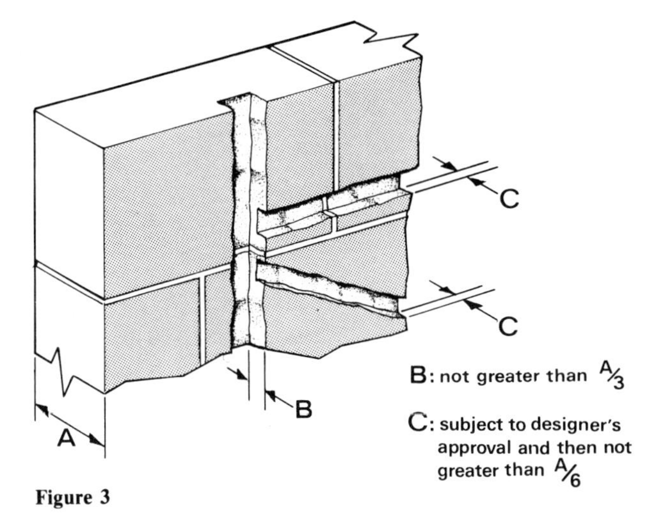

Figure 3

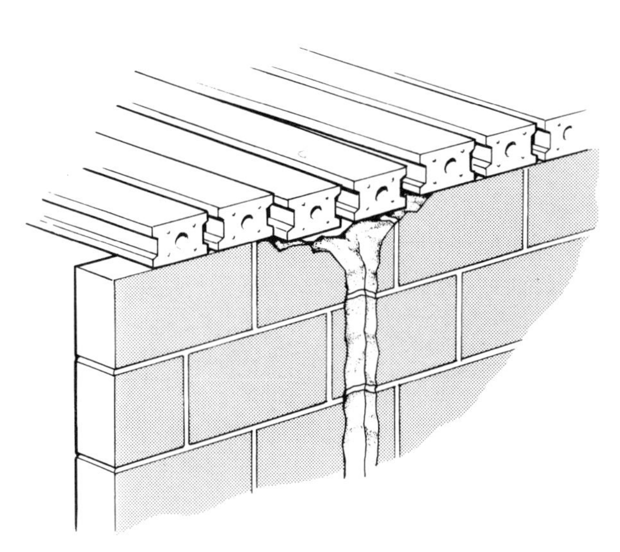

Figure 4

REFERENCES AND FURTHER READING

1 British Standard *Code of Practice* CP 121: Part 1: 1973: Code of practice for walling — brick and block masonry.

External walls: brick cladding to timber frame — the need to design for differential movement

FAILURE: **Disruption at eaves, verges, sills or porches, following vertical shrinkage of frame relative to brickwork**

DEFECT: **Adequate shrinkage joints not clearly specified at all necessary points**

Figure 1 Some key points for shrinkage

BRE surveys and site investigations of failures show that there is not always adequate provision for shrinkage of timber frame shells relative to brick cladding. Design provision needs not only to be indicated on drawings but means must be found (eg training) to ensure that site staff understand its importance.

If ***gaps or soft joints*** are ***not specified*** for locations where relative vertical movement occurs, ***rafters may bear on cladding, verges*** may be ***disrupted, porch roofs*** may be ***disturbed, sills*** below windows or below lightweight cladding may be ***tilted to a back-fall; instability, detachment or rain penetration*** may follow. Brickwork, saturated following disruption of weathering details, may suffer.

PREVENTION

Principle — A timber frame shell must be free to shrink vertically without unintended loading of any parts of the dwelling.

Practice

- Identify all points at which frame shrinkage could produce contact (and thus unintended loading) between brickwork and parts attached to, or supported by, the frame:
  — eg sills below windows or cladding attached to the frame Figure 2, porch trusses, canopy brackets, eaves Figure 3, protruding rafters, verges, beams over bays, floors oversailing brick cladding to lower storey, penetrating services, balanced flues.

- Make design provision[1,2] for movement at all such points, noting particularly where appreciable thicknesses of cross-grain timber potentially produce marked shrinkage.

- Do not design so that parts such as vertical battens for tile hanging, porch roof trusses or gallows brackets, are fixed both above and below a floor zone unless the fixings will accommodate shrinkage of cross-grain timber in the floor zone:
  — porch trusses or gallows brackets should be fixed to studs of either upper or lower storey, not to both, Figure 4;
  — vertical battens should be interrupted over floor zone by a gap of about 10 mm.

- Avoid carrying construction partly on the timber frame and partly on other construction which will not shrink similarly, eg masonry porch or store.

- Ensure that movement provisions are readily 'buildable' and are made very clear to site staff:
  — when considering buildability, note that erection sequences of timber frame are often different from those of traditional construction.

- Recognise that there will be relative movements between brickwork and either jambs or heads of windows, particularly at upper storeys:
  — sealants, if used, will shear in vertical joints and may need renewal after shrinkage of the timber frame, Figure 5;
  — certain horizontal joints, eg above window heads, may open and will require to be filled after shrinkage of the timber frame.

- Specify that timber and components delivered to site are to be kept as dry as possible at all stages.

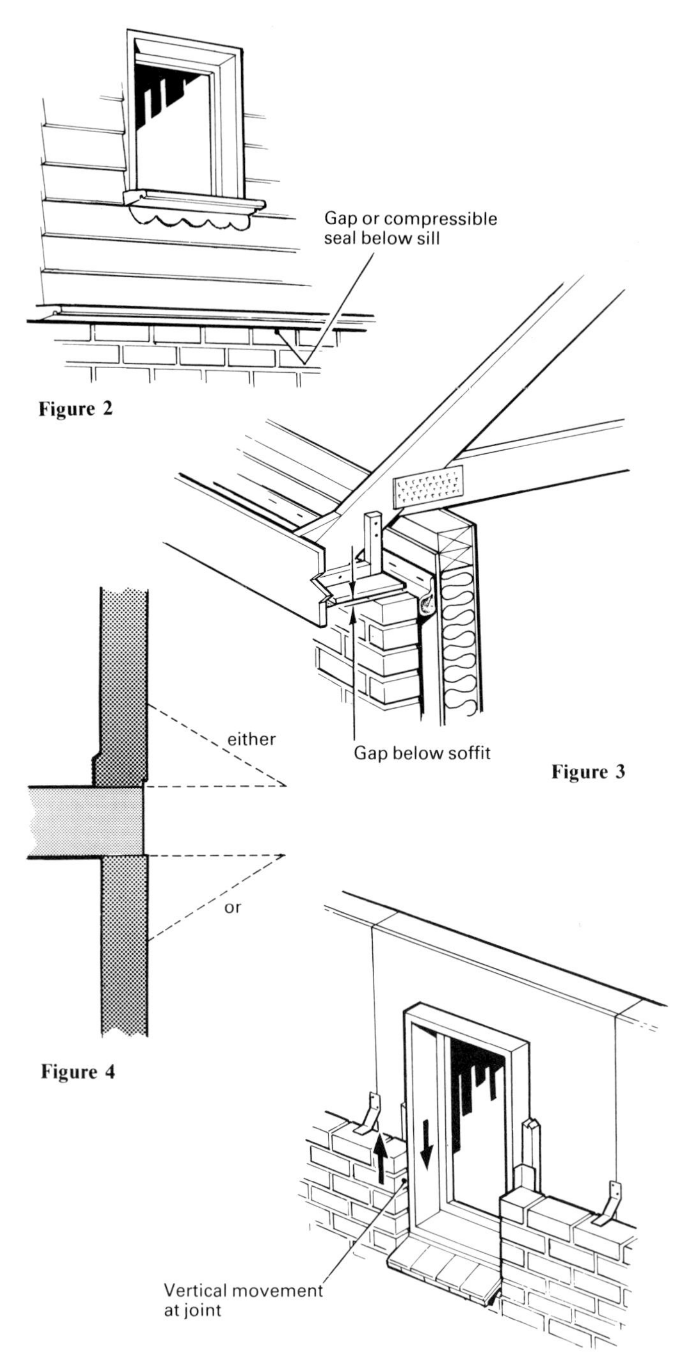

Figure 2

Figure 3

Figure 4

Figure 5

REFERENCES & FURTHER READING

1 BRE Digest 227/8/9 'Estimation of thermal and moisture movements and stresses'.
2 NHBC Practice Note 5

PRL Technical Note 38 'The movement of timbers'.

DAS 76
April 1986

Site
CI/SfB 8 (21)Z(F6)

External walls: brick cladding to timber frame — how to allow for movement

FAILURE: **Disruption at eaves, verges, sills or porches, following vertical shrinkage of frame relative to brickwork**

DEFECT: **Movement joints not built as needed**

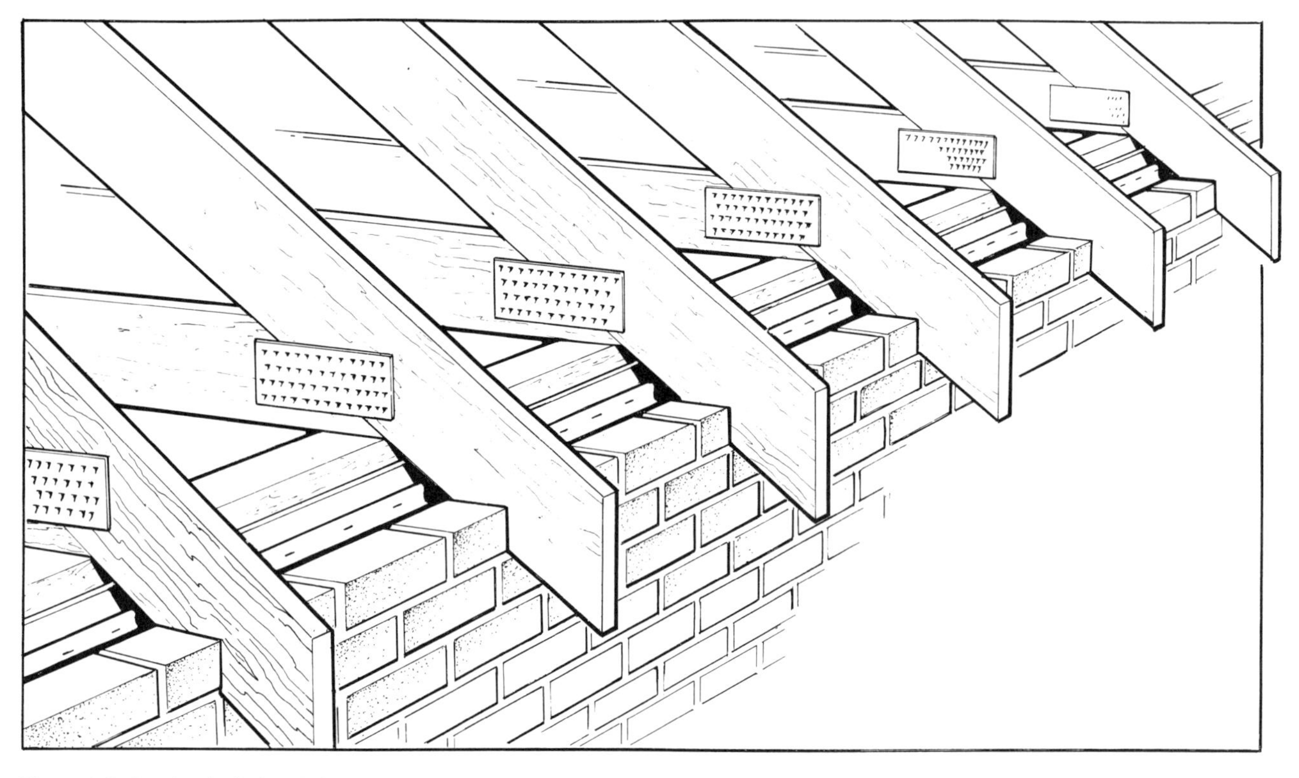

Figure 1 Rafter feet built into brickwork — no allowance for movement

The timber in the frame of a timber frame house shrinks as it dries down to the moisture content that it will have when the house is occupied. Wood shrinks most across the grain. Shrinkage in horizontal timber (sole plates, head binders, header joists etc) reduces the height of the shell relative to the brickwork.

If ***brickwork*** is built up tightly ***in contact with the underside of any part supported by the timber frame,*** as, for example, in Figure 1, distortion and ***damage*** may occur; sills may, for example, be tilted to a back-fall ***and rain penetration*** may follow.

PREVENTION

Principle — A timber frame shell must be free to shrink vertically without causing damage.

Practice

- Follow designer's requirements for gaps or 'soft joints' between brickwork and, for example, undersides of sills, gallows brackets, porch trusses, eaves soffits, verge undercloaks, flues, service entries:
  - — allowances are usually based on 6 mm shrinkage per storey and may therefore range from 3 mm below ground floor sills to over 20 mm at tops of 2 storey gables.
  - — Consult designer if provision has been overlooked.
- Ensure that nothing solid fills or is fixed across any horizontal joint or gap intended to permit movement, Figures 2, 3, or any zone where shrinkage will occur, Figure 4.
- Do not permit brickwork to be built up round rafter feet, Figure 1, or mortared up under eaves or verge soffits or undercloaks, Figure 5.
- Check whether the designer has made special provision for joints between brickwork and window jambs:
  - — these joints will be subject to 'shear' movements, particularly at upper storeys, which the usual corner mastic fillet will not tolerate.
- Ensure that timber and components delivered to site are kept as dry as possible at all stages.

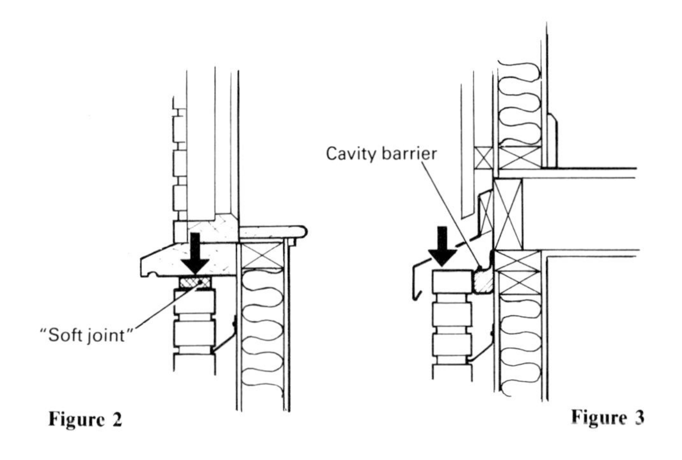

Figure 2

Figure 3

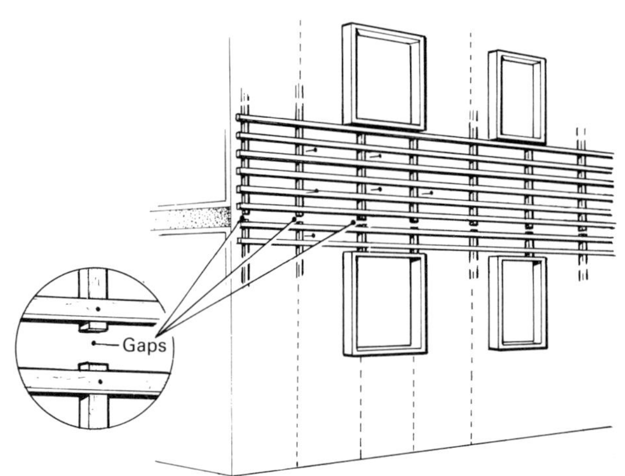

Figure 4

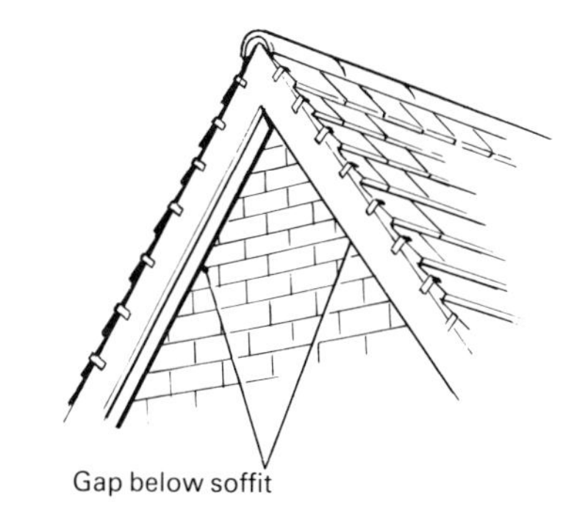

Figure 5

Reinforced-concrete framed flats: repair of disrupted brick cladding

FAILURES: Displacement of brick cladding; detachment of brick slips

DEFECTS: Contributory factors are inadequate provision for relative vertical movement, inaccurate floor edges and absent wall ties

In a recent survey by AMA disruption of brick cladding in high rise reinforced-concrete framed buildings was one of the most frequently mentioned problems (Figure 1), often accompanied by the detachment of brick slips. ***Vertical shrinkage of the concrete frame,*** sometimes accompanied by expansion of clay-brick infill cladding panels, are common causes of failures of this type (Figure 2) unless adequate provision has been made both to accommodate these movements and to support the cladding in a manner which caters for the unavoidable variability in alignment of in-situ concrete construction.

In some cases the design will have indicated a movement joint below the soffit of each floor slab. However, sometimes removal of the sealant will reveal that the ***back of the joint has been mortar filled,*** making it ineffective.

Figure 1 Disruption of brick cladding

In cases, where brick slips have been used on the frame, the movement joint may have been adequate to prevent the brick cladding panels from being compressed by the relative movements of the frame and the brickwork, but ***the movement joint may not have been taken to the face of the brick slips.*** In these circumstances, shrinkage of the concrete frame can have resulted in the self-weight of the brick panels being transferred from the frame to the brick slips. This will have given eccentric support to the brick cladding panels, leading to their displacement from the vertical plane, and may have caused ***spalling or displacement of the slips*** (Figure 3).

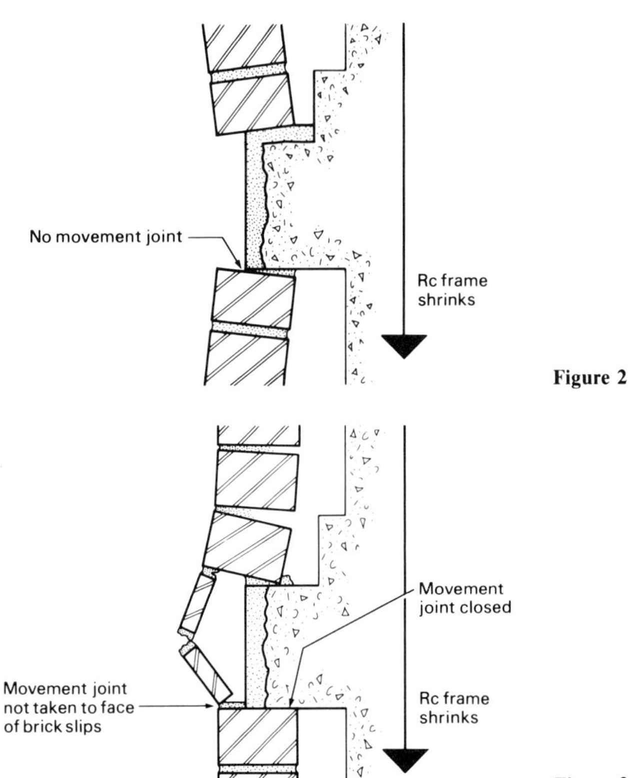

Figure 2

Figure 3

Reinforced-concrete framed flats: repair of disrupted brick cladding

The inevitable ***variability in alignment*** of in-situ concrete construction may have made the ***support for cladding panels unsatisfactory.*** British Standard BS 5606:1978 'Accuracy in building'[1] says that 'The inaccuracies of line at the edges, and of relative position on plan of successive floors .... are such that consistent satisfactory bearing for the over-sailing brickwork is unlikely to be achieved'. The resultant attempts during construction to improve alignment by hacking back or making up the edge of floor slabs can have introduced other defects, such as reduced cover to reinforcement, and made the support to the brick panels inadequate, as well as making ***remedial work more difficult.*** A further complicating factor can be the absence of or inadequacies in the wall ties.

REMEDIAL WORK

Principles Support and tie back brick cladding, making adequate provision for future movements.

Practice

- Temporary support during repair and subsequent renewed support for the cladding should be designed by a structural engineer and will depend upon the condition of the building and its constructional details.
- If shelf angles are introduced to provide new permanent support they and their fixings should be given full protection against corrosion or be non-corrodible (A, Figure 4.)
- Adequate movement joints should be provided. (B, Figure 4.) Initial moisture expansion of the original brickwork is likely to be complete, but relative movements between the brickwork panels and the concrete frame may still occur due to thermal movements of both, and bending and creep of the concrete frame
- Additional tying back of brick cladding panels may be required where new movement joints have been introduced (C, Figure 4.)
- If brick slips are to be retained in the remedial design they should be isolated by movement joints from the brick panels. Consider also using a high adhesion fixing system such as epoxy resin, polyester resin or SBR emulsion (D, Figure 4), or mechanical retention.
  Instruct site supervisors to ensure that, when new movement joints are installed, they are not made ineffective by, for example, mortar pointing.
  If, as part of the remedy, brick slips are to be replaced by rendering, movement joints must not be rendered over, or vertical compression of the rendering will cause it to become detached.
- If floor edges are hacked back to provide a fair line for new permanent support, consideration should be given to protection of the reinforcement.
- A cavity tray should be provided at the foot of each brick infill panel. It should be formed with an upstand at each end and provision made for drainage.

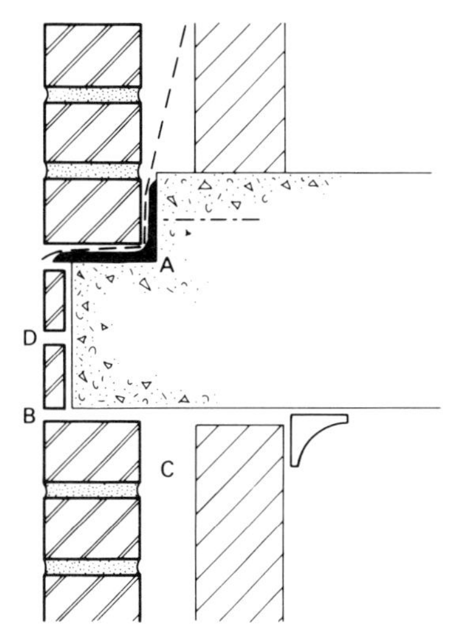

Figure 4

REFERENCES AND FURTHER READING

1 *British Standard* BS 5606:1978 'Accuracy in building.

BRE Digest 75 'Cracking in building'

BRE Digest 217 'Wall cladding defects and their diagnosis'

BRE Digest 223 'Wall cladding: designing to minimise defects due to inaccuracies and movements'.

GLC Development and Materials Bulletin No 101 (2nd series) Jan 1977 'Fixing of brick slips — feedback on a failure'.

GLC Development and Materials Bulletin No 105 (2nd series) May 1977 'The use of an SBR product for fixing of brick slips'.

Brick Development Association Technical Note 9 'Further observations on the design of brickwork cladding to multi-storey rc frame structures'.

Large concrete panel external walls: re-sealing butt joints

FAILURE: **Penetration of re-sealed joints by rain water, wind blown rain and snow**

DEFECTS: **Incorrect choice of sealant in relation to expected movement at joint; incorrect cross-section of applied sealant; incorrect or inadequate preparation of joint surfaces**

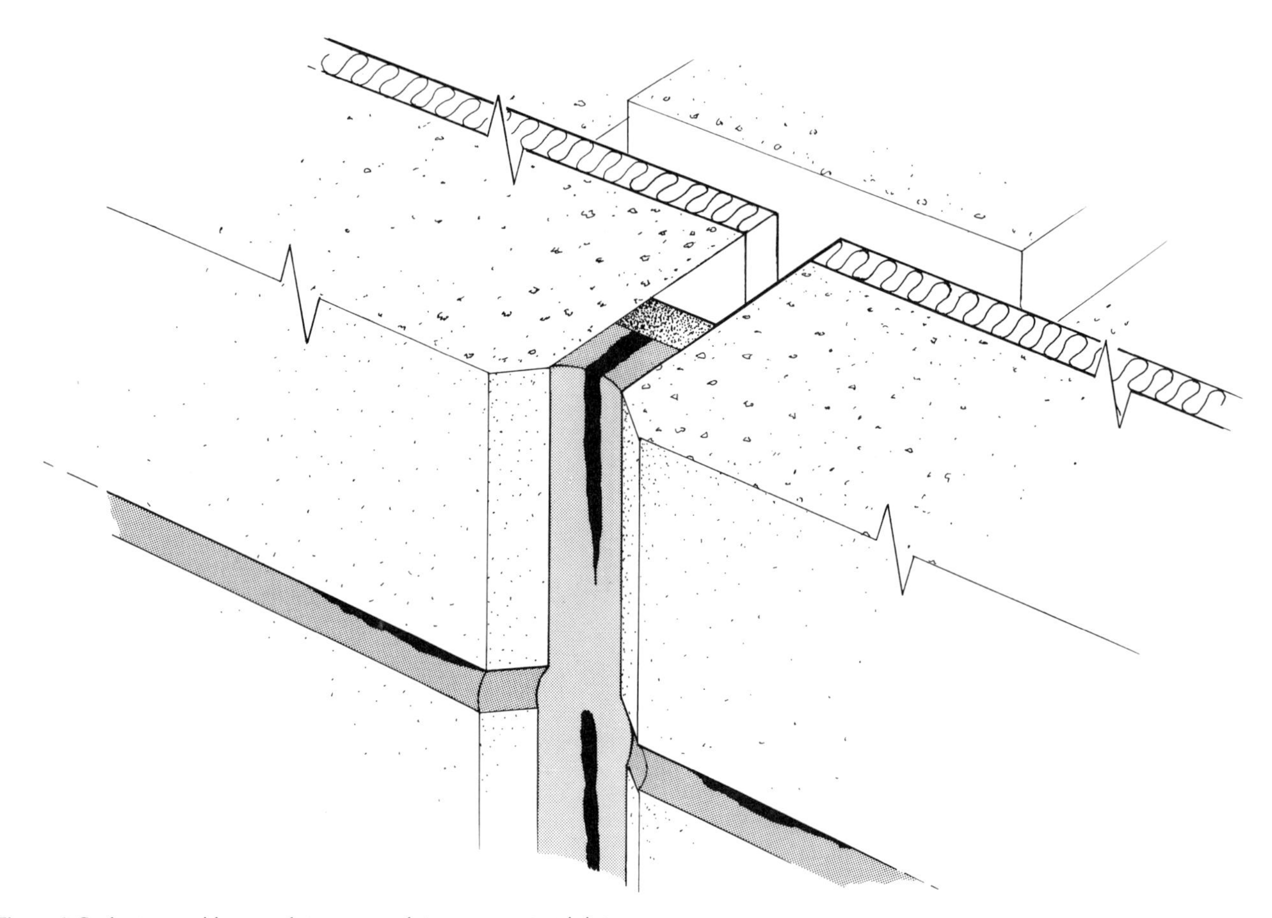

Figure 1 Sealant not wide enough to accomodate movement at joints

Re-sealing is one remedial option when sealed joints leak.

Failure of re-sealed joints occurs frequently. Replacement sealants may fail because their capability to accept movement was not considered when specifying. Often re-sealed joints fail because joint surfaces were inadequately prepared.

If a sealant is subjected to ***movement greater than the material can accept*** it will ***tear*** or ***lose adhesion***. If the ***old sealant*** is ***not fully removed*** the ***replacement sealant*** may ***not adhere***, and effective ***joint width*** may be ***reduced*** locally ***if hardened nodules*** of old sealant ***remain***.

REMEDIAL WORK

Principle — Replacement sealants must accommodate continual changes in joint width in service without loss of integrity.

Practice

- Identify, eg by examination of defective joints, the probable reasons for failure of original sealant, Figure 2.
  - — joints that are too narrow in relation to the movement experienced will show cohesion failure of sealant;
  - — movement of two adjoining panels may be concentrated at one joint (examine adjacent joints of similar width for signs of disruption; check panel fixings -Figure 3).
- Identify by inspection and measurement the widths of the narrowest and widest joints to be re-sealed;
  - — joints may taper, measure at their narrowest.
- Estimate the range of service conditions, especially temperature[1];
  - — dark coloured concrete panels can reach 65°C in summer and -20°C in winter.
- Establish the corresponding total movement occurring at each joint by calculation, or use the following approximation:
  - — for a 4 m wide concrete panel the range of movement will be about 4 mm (1 panel, Figure 3a) or 8 mm (2 panels, Figure 3b).
- Subtract total movement from narrowest measured joint width to give *potential minimum joint width in service* (because the joints are unlikely to have been measured when at their narrowest).
- Apply total movement to Figure 4 and select a sealant type having a minimum sealant width no less than the potential minimum joint width;
  - — if no sealant is suitable identify, from Figure 4, the minimum *sealant* width necessary, add to this the total movement and specify that joints are to be ground out at least to this width.
- Check that selected sealant can be applied to the widest joints to be re-sealed without risk of slumping (application to wide joints may need to be in stages). Check with manufacturer.
- Check life expectancy of selected sealants.
- Identify selected manufacturer's requirements for cross-section, Figure 5, back-up material, primers, bond-breakers etc, and specify accordingly;
  - — back-up material is needed to control depth of sealant
  - — bond-breakers should be specified wherever adhesion of the sealant is not wanted, eg as in Figure 6.
- Specify complete removal of original sealant by mechanical means. Do not permit solvents;
  - — old sealant should be cut out and joint surfaces should be cleaned by wire brushing or by similar abrasion;
  - — old sealant may not be chemically compatible with the new; it may also contaminate a porous joint surface preventing adhesion, and may interfere with the application of replacement sealant.
- Identify future maintenance needs of the re-sealed joint.

REFERENCES AND FURTHER READING

1 *BRE Digest* 228 'Estimation of thermal and moisture movements and stresses: Part 2', (Table 2).

□ BS 6093:1981: 'Code of practice for the design of joints and jointing in building construction'.

□ *BRE Information Paper* IP 25/81. 'The selection and performance of sealants'.

□ *British Standard* BS 6213:1982: 'Guide to selection of construction sealants'.

□ *BRE Information Paper* IP 10/86. 'Weatherproof joints in large panel systems: 3 Investigations and diagnosis of failures.

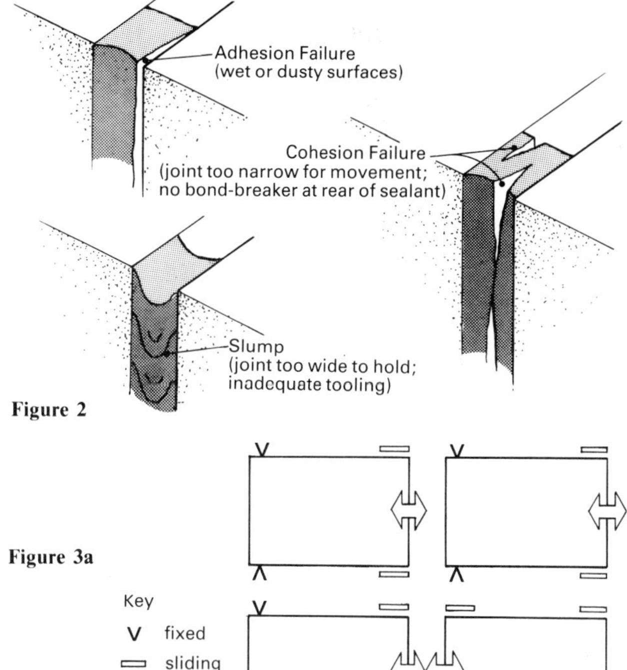

Figure 2

Figure 3a

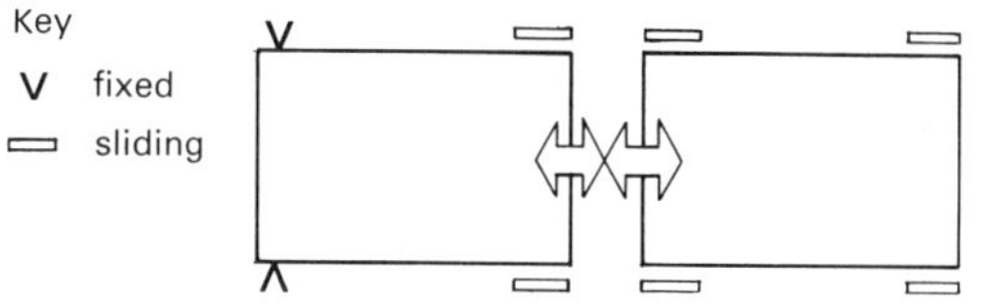

Figure 3b

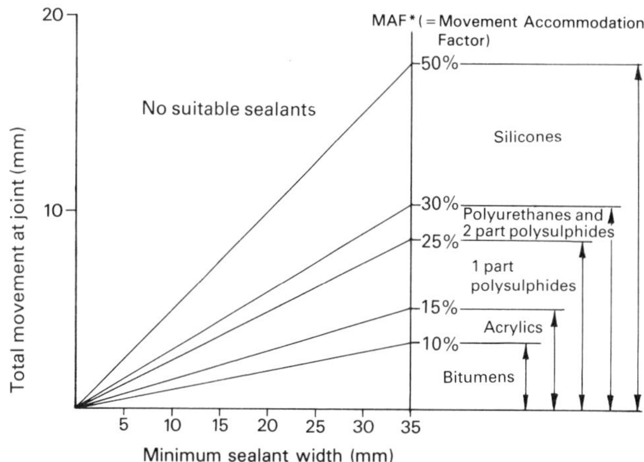

Figure 4

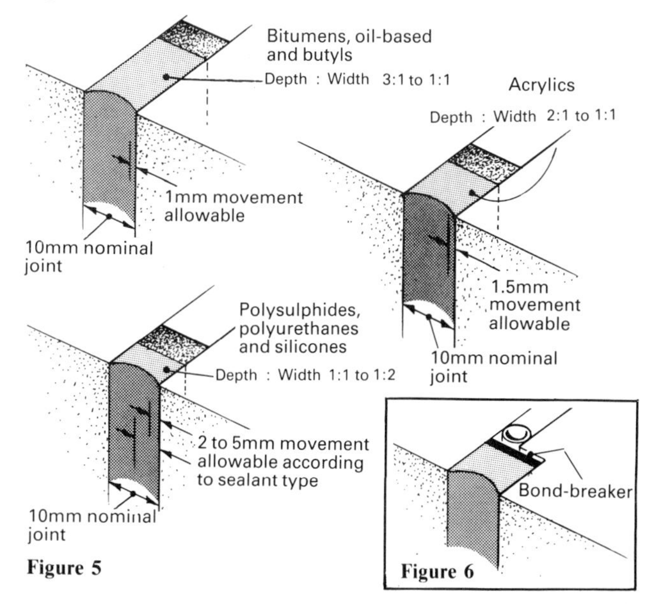

Figure 5

Figure 6

External masonry cavity walls: wall tie replacement

FAILURES: Instability due to corrosion or absence of wall ties; reduced weathertightness due to induced cracking

DEFECTS: Inadequately protected or missing ties

Figure 1 Corroded wall ties

The thickness of the zinc coating required by the British Standard for wall ties was increased in 1981 following earlier BRE research. This showed that galvanising to steel wall ties degraded at a faster rate than expected, and that some galvanised steel ties manufactured to the British Standard prior to that of 1981, could thus corrode prematurely. Risk of premature corrosion is greater for wire ties, since the zinc coating required was less than that for strip ties.

As corrosion products occupy about six times the volume of the original metal, the corrosion of strip ties produces a recognisable pattern of cracking at bed joints corresponding to wall tie levels. (Note: sulphate attack tends to produce cracking at all bed joints and can thus be distinguished from wall tie corrosion.) However, because there is less metal in them, corrosion of wire wall ties does not reveal itself in this way[1] and the need for inspection depends on circumstances such as those below.

Inspection should be carried out ***if brickwork is cracked at bed joints***, roughly every sixth course (corresponding to vertical spacing of wall ties). ***If black ash mortar has been used*** (do not confuse with black pigmented mortar), even if no bed joint cracks are apparent, ***wall ties*** of both strip and wire ***are at risk of corrosion.*** If outer leaf brickwork has become displaced relative to the inner leaf (eg has bulged locally) ties over a wider area may be unreliable as a result. Inspect also when tying action has failed on a closely similar building, if there is evidence that sub-standard ties may have been used, or if instability of the outer leaf on a building more than about 20 years old would present a high risk to life. Inspect ties if other maintenance work provides an opportunity and ensure that any maintenance that involves replacement of ties is reported to a competent person in the organisation so that its wider significance can be assessed.

REMEDIAL WORK

Principle — Restore full tying action between leaves, and weathertightness of outer leaf.

Practice

- Assess cracking for its structural significance — it may be due to some other cause — and decide whether wall tie replacement is appropriate and sufficient. (Note: where there is vertical expansion, the outer leaf may now carry unintended loads — eg from roof members — and the outer leaf will first need to be relieved of these).

If bed joints are cracked at tie spacing intervals vertically:

- Specify a trial inspection to confirm that strip ties are present and corroded.
- Specify use of a metal detector along disrupted bed joints to pin-point tie location.
- Specify bricks or blocks in the course below existing ties to be removed at tie locations to release remains of strip ties from outer leaf (Figure 2).
- Specify holes to be drilled into body of brick or block inner leaf so that new tie will bed one brick course lower than the original tie (it is better to put tie into body of brick or block because of the risk of variable mortar quality, and perpends and even bed joints may not be fully filled. Ties available include both strip and wire, Figure 3A. Specify that they are to be galvanised to BS 1243 or preferably to be of austenitic stainless steel).
- Specify keyed end of replacement tie to be cement- or resin-grouted into hole in inner leaf, splayed or cranked ends to be mortared into bed joints of replaced brick. Pointing to be made good (Figure 4).
- Beware hollow block or perforated brick inner leaves; if found, replace ties into same joint as original tie.

If black ash mortar used, or outer leaf displaced, or other reasons prompt examination (see Page 1):

- Specify removal of bricks on a sample basis to confirm that wire ties are corroded.
- Specify wire ties to be left in place: they are unlikely to produce sufficient further corrosion products to crack the joints. **Then:**
- Specify both leaves to be drilled at slight upward slope into body of brick or block using a drilling guide; specify that holes are not to be drilled into joints (unless hollow inner leaf units); specify patent replacement ties (Figure 3B or C) at frequency not less that that required by the Code of Practice[2], and that face is to be made good.
- In all cases specify care to avoid cavity-bridging with debris.

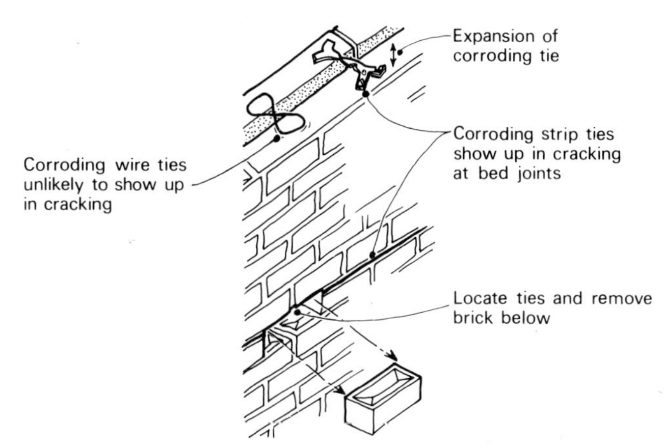

Figure 2

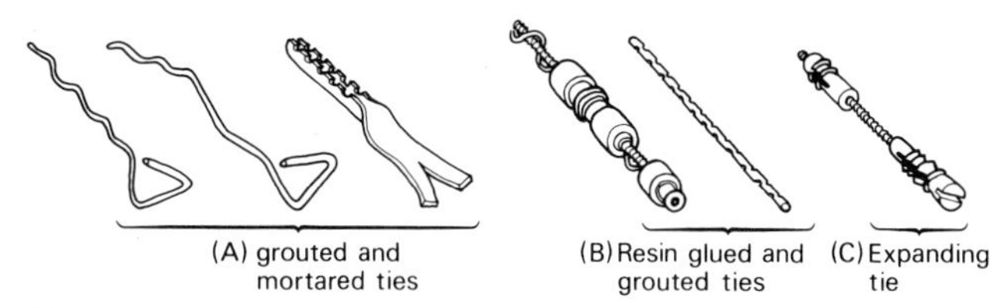

Figure 3

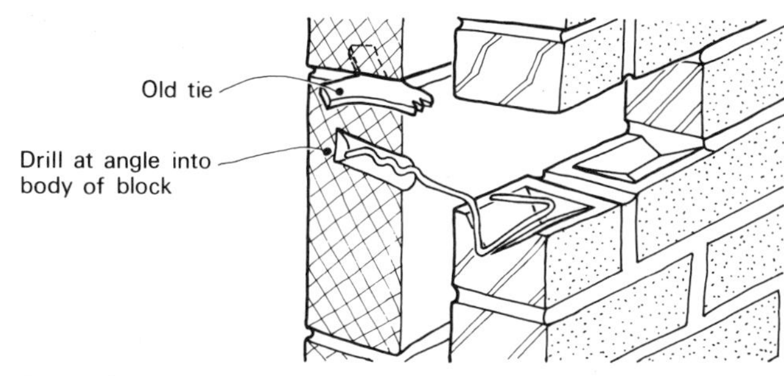

Figure 4

REFERENCES AND FURTHER READING

1 *BRE Information Paper* IP28/79 'Corrosion of steel wall ties: recognition, assessment and appropriate action'.

2 *British Standard Code of Practice* CP 121:Part 1:1973 'Brick and block masonry'.

BRE Information Paper IP29/79 'Replacement of cavity wall ties using resin-grouted stainless steel rods.

BRE Digest 237 'Installation of wall ties in existing construction'.

BRE Digest 160 'Mortars for bricklaying'.

BRE Digest 200 'Repairing brickwork'.

External masonry cavity walls: wall ties — selection and specification

FAILURE: Instability of external cavity walls

DEFECTS: Incorrect specification of type, size and frequency of wall ties

Figure 1 Wind damage to an inadequately tied cavity wall. Photo: *Kent Messenger*

Cavities 50 mm wide have been almost universal in low-rise post-war housing. However BRE surveys have shown that insufficient wall ties are often used — particularly at openings, where closer spacing is required than elsewhere.

Some design solutions to the need for increased thermal insulation require cavities wider than 50 mm. Designers will more often need to consider what size and type of tie is correct for the chosen cavity width. Requirements that may be unfamiliar to site workers and supervisors will need to be particularly clearly specified.

If the designer specifies **ties of incorrect type, length or frequency** in relation to cavity width and leaf thickness, **the wall may not be structurally sound.**

External masonry cavity walls: wall ties — selection and specification

DAS 115
June 1988

PREVENTION

Principle — The leaves of cavity walls must be adequately tied together for structural stability and load carrying.

Practice

- Specify ties of a type and size appropriate to cavity width and leaf thickness[1], Figure 2 gives examples of ties to BS 1243;
  - — select ties to BS 1243, or Agrément or other acceptable third party certificates;
  - — in conditions of severe or very severe exposure[1] consider specifying stainless steel or non-ferrous ties.

Note: Alternatively ties may now be specified by type according to 'end use'[2]; Table 1 provides guidance applicable to England and Wales only. Ties specified in this way should have been certified by a third party to be in accordance with DD140.

- Specify the required spacing and frequency of wall ties, Figure 2;
  - — ties should not be inserted within 450 mm of the internal corner of a masonry return, Figure 3.
  - — ties in the general run of walling should be staggered and evenly distributed.
- Specify that additional ties are to be used, at the sides of openings so as to provide, a vertical spacing not greater than 300 mm and a horizontal spacing not greater than 225 mm.
- Consider specifying that within 225 mm of verges a tie is to be inserted at least every 300 mm vertically, Figure 4. In practice this will mean at every block course.
- Do not specify ties at a frequency less than 2.5/m^2 whether the wall carries vertical loads or not.
- Specify that ties must be embedded at least 50 mm in both leaves[1].

Table 1 Classification of wall ties by 'end use', according to DD140

Classification	Type 1 Masonry Heavy Duty	Type 2 Masonry General purpose	Type 3 Masonry Basic	Type 4 Masonry Light Duty
Geographical location	Suitable for use anywhere in England and Wales		Suitable for use anywhere approximately south of Leicester and east of Bristol[2] and in towns elsewhere in England and Wales	
Height limitations	None	3 storeys and 15 m		2 storeys and 10 m
Suitability	Suitable for most cavity walls where flexibility is not required	Suitable for cavity walls comprising two leaves of brick or blockwork of similar thickness in the range 90 – 150 mm		As for 2 and 3 provided the masonry has both leaves returned at their ends to flank walls or buttresses[2]

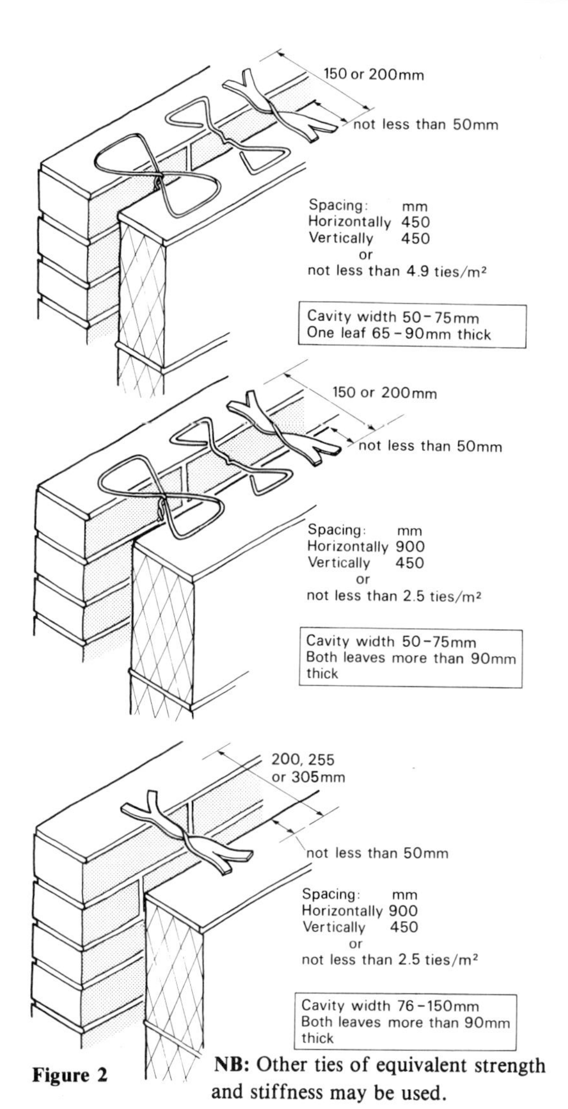

Figure 2 **NB:** Other ties of equivalent strength and stiffness may be used.

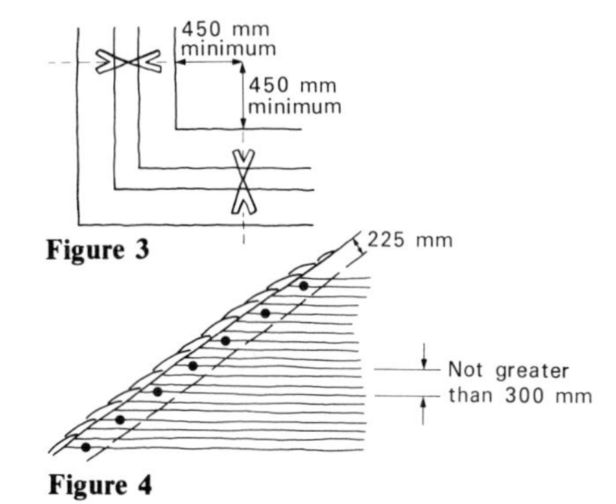

Figure 3

Figure 4

REFERENCES AND FURTHER READING

1 **BSI** *British Standard* BS 5628:Part 3:1985. 'Use of masonry' — Materials, components, design and workmanship.

2 **BSI** *British Standard* DD140:Part 2:1987. 'Recommendations for design of wall ties'.

☐ **BSI** *British Standard* BS 5628:Part 1:1978. 'Unreinforced masonry'.

☐ **BSI** *British Standard* BS 1243:1978. 'Specification for metal ties in cavity walls'.

External masonry cavity walls: wall ties — installation

FAILURE: **Instability and rain penetration of external cavity walls**

DEFECTS: **Insufficient ties to comply with Standards both generally and at openings; insufficient embedment; ties pushed into 'green' mortar; ties sloped the wrong way; drips not in the centre of the cavity; mortar on ties.**

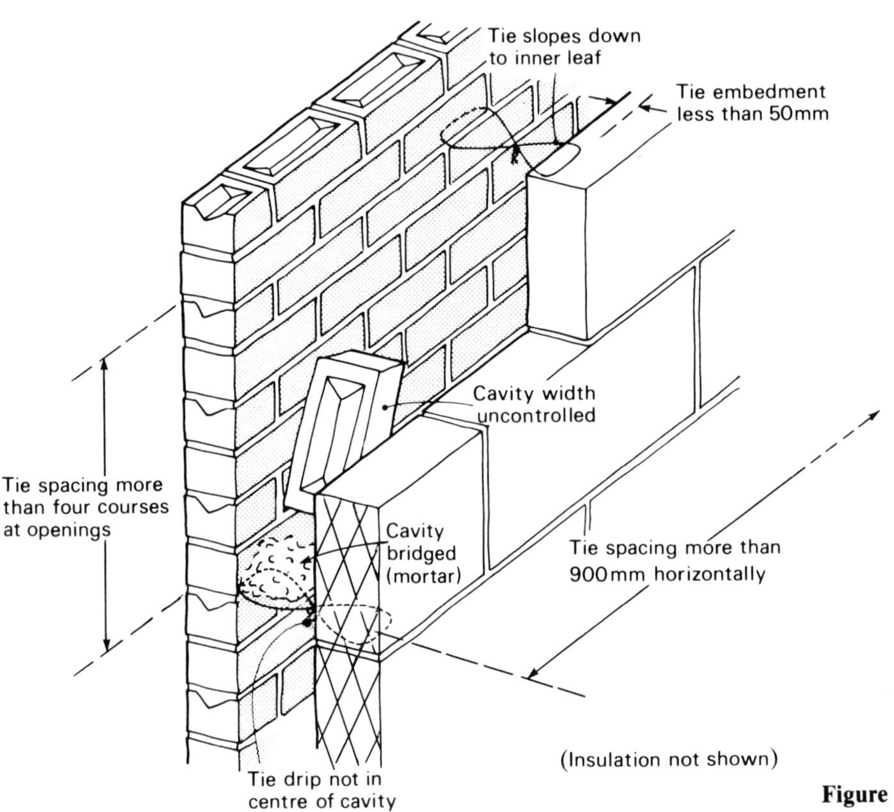

Figure 1 Typical wall tie faults

In a BRE survey of conventional low-rise housing with cavity masonry walls, nearly half the sites inspected had insufficient wall ties in straight runs of walling but particularly at openings. (The problem was made worse where those ties proved to be sub-standard with respect to gauge and coating thickness.)

If too few wall ties are used (especially at openings), if they are ***not long enough*** to lap both leaves by at least 50 mm, ***or if ties are pushed into mortar,*** the wall may ***not be structurally sound.***

Wall ties bridge the cavity for structural reasons, but they must not lead water across.

If ties ***slope down to the inner leaf,*** or have ***drips off-centre*** in the cavity, or are ***fouled by mortar, rain-water can cross to the inner leaf.***

PREVENTION

Principle — Ties must adequately bond the leaves together for stability and load carrying but must not lead water across the cavity.

Practice

- Find out what type, size and spacings of wall ties have been specified and ensure that the bricklayers understand what is needed.

- Check that the ties actually supplied to site conform to the specification. (The bundles should be labelled.)

Note: where the specification calls for ties classified as Type 1, Type 2 etc, check that the ties are certified to be in accordance with DD 140.

- Tell the designer if cavities are too wide for 50 mm embedment of ties and with his agreement use a longer tie.

- Check that there is a row of ties at every sixth course of (standard) bricks;
  — ties should be staggered and evenly distributed.

- Check that the horizontal spacing then gives the specified number of ties per square metre, Figure 2.

- Check that there is a tie at every block course within 225 mm of openings and verges, Figures 2 and 3.

- Ensure that ties are laid with the bed joint, not pushed in subsequently.

- Check that ties are horizontal or that they fall to the outer leaf, with drips in the centre of the cavity and no mortar bridging the cavity;
  — mortar falling on wall ties must be cleaned off.

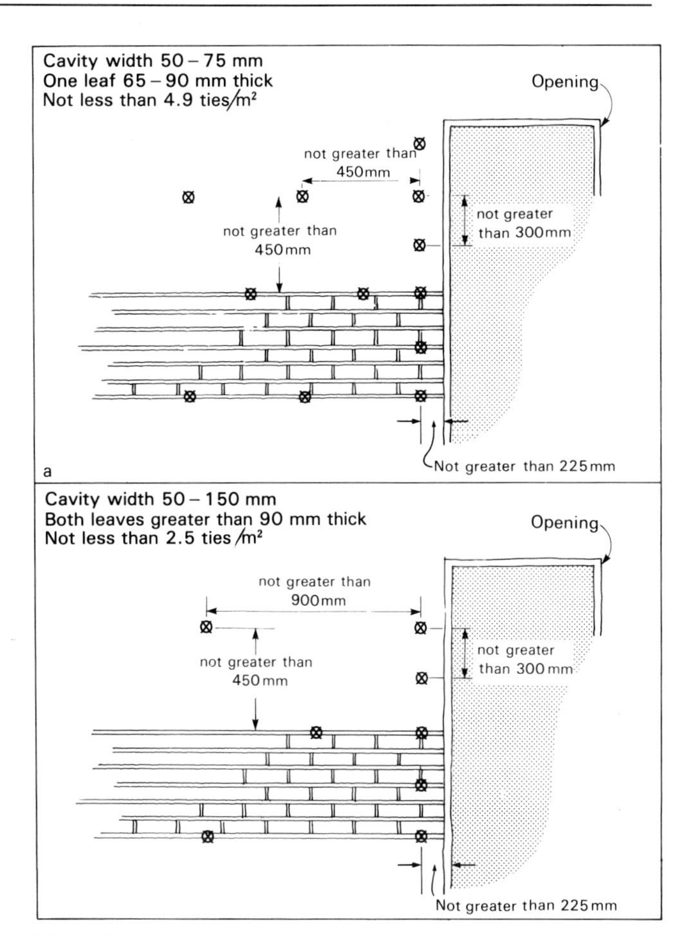

Figure 2

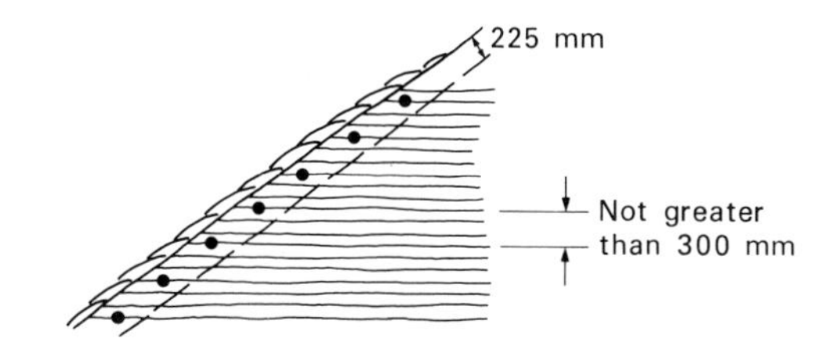

Figure 3

Freestanding masonry boundary walls: stability and movement

FAILURE: Part or complete collapse of the wall structure

DEFECTS: Lateral stability not considered in design, piers inadequate, poor foundation, poor bonding

Figure 1

Freestanding masonry walls up to 2.1 m (brick), 1.8 m (block)[1], dependent on wall thickness, may be constructed without the need for 'design input'; however, higher walls, walls subjected to forces in addition to wind and walls acting partly as earth-retaining structures need to be structurally designed for stability and durability[2].

Walls should be designed to withstand the effects of, for example, wind, accidental impact or public assembly, otherwise progressive or sudden collapse may follow, Figure 1.

Piers, reinforced or plain, at intervals along a wall are the most common and usually the most economical way of providing lateral stability. Piers on one side provide mainly additional stability in one direction only. Movement due to temperature, moisture and settlement needs to be accommodated by provision of vertical joints to prevent unsightly cracking and weakening of the structure.

If a ***freestanding masonry wall*** is ***not adequately designed*** for ***lateral stability*** and for ***movement*** the wall ***may collapse***, which can be ***dangerous***, ***expensive*** and may lead to ***legal action.***

PREVENTION

Principle — Freestanding walls must be capable of withstanding all forces to which they may be expected to be subjected. Unreinforced walls must be securely bedded onto their foundations. Reinforcement used in walls should be anchored to the foundations.

PRACTICE

- Check whether wall can be built without engineering design input as follows:
  - — a freestanding, half brick or 100 mm dense-concrete block wall laid to bond[1] and subjected to wind forces only, may be constructed up to 600 mm in height;
  - — a freestanding one brick or 200 mm dense-concrete block wall laid to bond may be constructed up to 1.3 m in height.

    Note

    Piers (215 × 225 mm) each side of the wall at maximum centres of 1.575 m (brick) and (440 × 210 mm) at maximum centres of 1.575 m (block) will enable the panel to be reduced to half brick or 100 mm block, Figure 2.

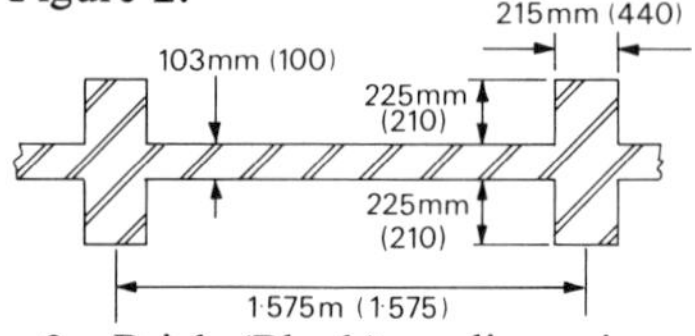

Figure 2 Brick (Block) — dimensions

  - — a freestanding one and a half brick or 300 mm dense concrete block wall laid to bond may be constructed up to 2.1 m (brick) or 1.8 m (block) in height.

    Note

    Piers (550 × 225 m) each side of the wall at maximum centres of 3.15 m (brick) and (440 × 210 mm) at maximum centres 2.7 m (block) will enable the panel to be reduced to one brick or 200 mm block, Figure 3.

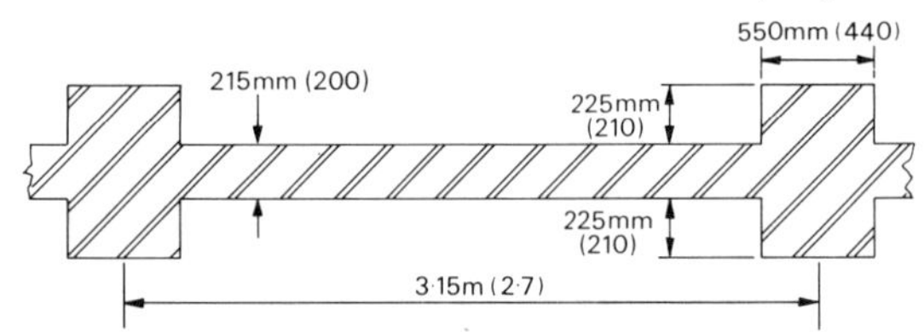

Figure 3 Brick (Block) — dimensions

- Check whether for higher walls and for walls subjected to other forces the design can be more economically achieved by employing different plan forms, Figure 4, or by the addition of a suitable pier arrangement, by cellular/diaphragm construction or by the addition of reinforcement. Such walls should be structurally designed.

- Select foundation to suit the wall height and type of ground. The width of the foundation should be not less than the wall width plus 150 mm and widened at piers:
  - walls up to approximately 3 m high on undisturbed soil will require a foundation thickness of at least 150 mm. The trench bottom should be located at 500 mm below ground level and should, for economy of construction, be filled with concrete to within 3 brick courses or 1 block course of ground level;
  - walls over 3 m high will need deeper foundations and should be structurally designed;
  - foundations for walls over 600 mm high on cohesive soils (clay) will need to be structurally designed;
  - foundations where the allowable bearing pressure is 50 kN/m^2 or less (eg made-up ground) will require reinforcement. Such foundations should be structurally designed.

  Note
  Walls near to trees or large shrubs will also need special consideration. Thick roots will need to be bridged.

- Specify that foundations are continuous beneath all vertical walls joints.

- Specify that the excavation is blinded or that concreting of the trench is carried out within 2 – 3 days.

- Specify piers at the wall ends and at the edge of large openings:
  - such piers to be built on the centre line of the wall;
  - consider reinforcing piers[2], particularly where gates are to be hung, eg Figure 5.

- Design to resist vandal damage. Top corners are the weakest points and most vulnerable. Where end-cramps are required specify galvanised or stainless steel, eg for brick-on-edge copings, Figure 6.

- Specify a vertical movement joint at intervals not exceeding 6 m (concrete masonry — simple butt-joints), or 7.5 to 9 m (calcium silicate bricks — 10 mm joints), or 10 to 12 m (clay bricks — 16 mm joints):
  - a vertical movement joint should be provided where the wall meets an existing building.

- Specify a horizontal dpc under coping. Lower dpc, if specified, must develop full bond strength of wall as designed and be rigid units (bricks or slates bedded in cement mortar) to BS 743[3]; other materials should not be used:
  - concrete masonry does not require a lower dpc.

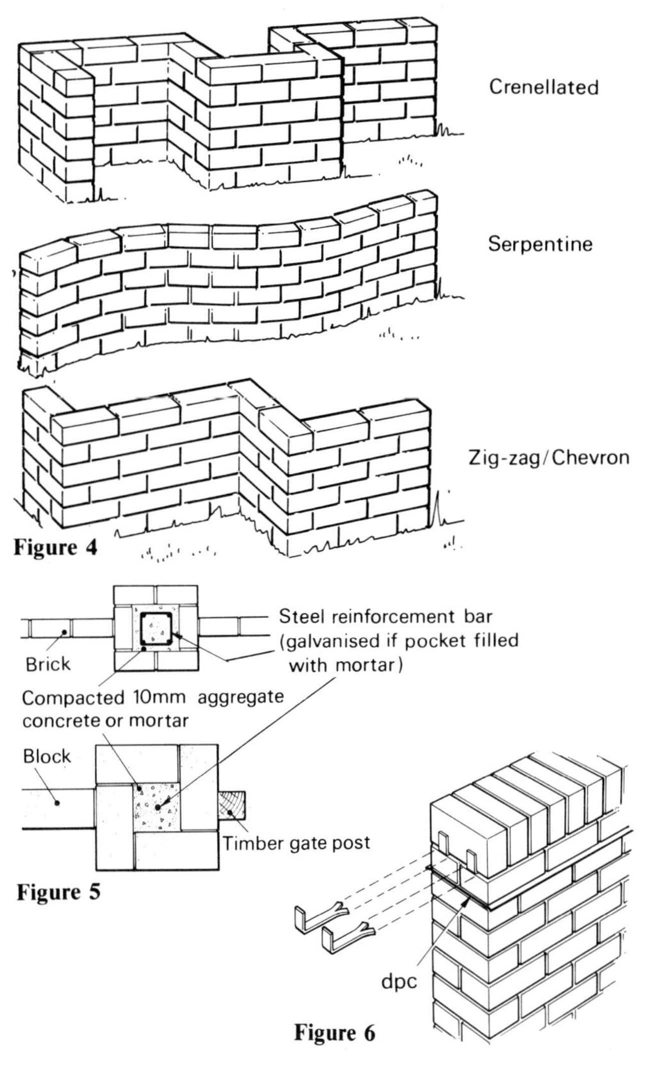

Figure 4

Figure 5

Figure 6

REFERENCES AND FURTHER READING

1 **British Standards Institution.** Code of practice for use of masonry. Part 3. Materials and components, design and workmanship. *British Standard* BS 5628:Part 3:1985. London, BSI, 1985.

2 **Building Research Establishment.** Freestanding boundary walls:materials and construction. *Defect Action Sheet* DAS 130. Garston, BRE, 1989.

3 **British Standards Institution.** Specification for materials for damp-proof courses. *British Standard* BS 743:1970. London, BSI, 1970.

British Standards Institution. Code of practice for use of masonry. Part 1. Structural use of unreinforced masonry. *British Standard* BS 5628:Part 1:1978. London, BSI, 1978.

British Standards Institution. Code of practice for use of masonry. Part 2. Structural use of reinforced and prestressed masonry. *British Standard* BS 5628:Part 2:1985. London, BSI, 1985.

British Standards Institution. Code of practice for foundations. *British Standard* BS 8004:1986. London, BSI, 1986.

British Standards Institution. Code of practice for trees in relation to construction. *British Standard* BS 5837:1980. London, BSI, 1980.

Korff J O A. *Design of free-standing walls.* The Brick Development Association, 1984.

Freestanding masonry boundary walls: materials and construction

FAILURE: Masonry deterioration, wall leaning-over or collapsed, cracking of wall

DEFECTS: Unsuitable bricks, mortar, dpc or coping

Figure 1

Freestanding boundary walls are subjected to the full effects of the weather on both sides. To cater for durability the materials must be carefully selected.

The construction and, especially, the bond of the wall should also be considered to ensure that the design in respect of stability and movement[1], is adequate.

If a ***wall*** is ***constructed*** without sufficient thought to the materials and method, ***unsuitable bricks*** or ***mortar, weak*** zones at ***dpc, poor water shedding*** at copings, ***poor foundations,*** the wall ***may fail prematurely.***

PREVENTION

Principle — Freestanding boundary walls must be durable. Materials should be selected to withstand the exposure and severity of weather expected, such as continual saturation, and freeze/thaw cycles. Construction methods must meet the strength and stability requirements.

PRACTICE

FOUNDATIONS

- Specify a concrete according to BS 5328[2]. Mix C1OP with high workability and 20 mm aggregate is suitable for most foundation work:
  - the concrete mix[3] requires a minimum of 330 kg/m^3 of OPC or a sulphate resisting cement where there are aggressive soils and groundwaters containing sulphates.
- Specify that the excavation is blinded or that concreting of the trench is carried out within 2 — 3 days.

BRICKS AND BLOCKS

- Specify units to suit the exposure and guard against possible sulphate attack[4]:
  - clay bricks to BS 3921[5] type MN and ML may suffice with good overhanging coping, otherwise specify frost-resistant bricks, type F. Coping bricks should always be type FL or FN. Bricks from foundation to 150 mm above ground level should also be type F;
  - severe exposure will require type FL bricks with high frost resistance and low salts content;
  - calcium silicate bricks class 3 or stronger to BS 187[6] are suitable for all degrees of exposure. Coping bricks should be class 4 or stronger;
  - concrete masonry units should be to BS 6073[7]. Dense concrete facing blocks, bricks, screen walling or reconstructed stone are all suitable.
- Specify recently built masonry be protected against wind and rain, Figure 2:
  - bricks laid in cold weather or if night frosts are expected will need protecting with an insulating watertight blanket. A cement-rich mortar to give high early-strength should be considered;
  - high-suction bricks may be docked or sprayed in hot dry weather or the mix should be wetted-up slightly.
- Specify that maximum height of masonry constructed per day be restricted to 1.5 m to avoid excessive pressure on fresh mortar.

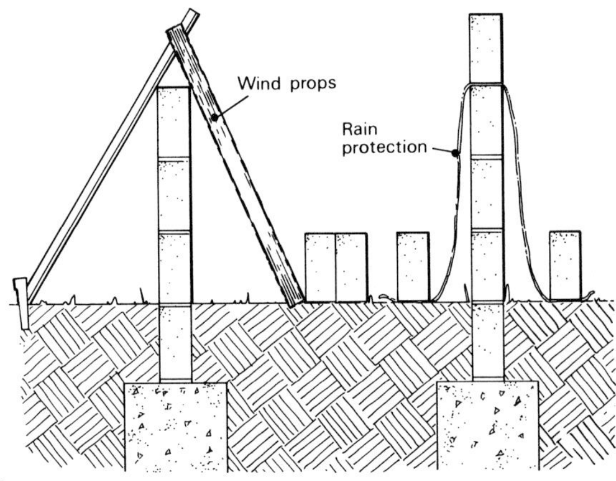

Figure 2

Freestanding masonry boundary walls: materials and construction

DAS 130
April 1989

MORTAR

- Specify mortar to Tables 13 and 15, BS 5628[8] to be cement:sand plus air-entraining plasticiser or cement:lime:sand unless site under close supervision, when mixes using masonry cement may be used:
  - — mix proportions not leaner than 1:5-6 plus plasticiser, or 1:1:5-6 (designation iii BS 5628)[8] for main wall with protective coping detail;
  - — mix proportions not leaner than 1:3-4 plus plasticiser, or 1:½:4-4½ (designation ii BS 5628)[8] for copings and area from foundation level to 150 mm above ground level, or for the complete wall if capped;
  - — discard all unretarded mortar over 2 hours old;
  - — SRPC (sulphate-resisting Portland cement) should be used from foundation level to 150 mm above ground level if sulphates are present in the soil and in all mortar used with type N bricks;
  - — all jointing to the masonry should be tooled, eg bucket handled. Weathered or struck joints are acceptable alternatives.

COPINGS

- Specify coping detail to throw the rain away from the wall, Figure 3:
  - — overhang should be at least 40 mm, provided with drip recess if possible;
  - — suitble coping bricks or slabs are shown in BS 4729[9] and BS 5642[10];
  - — concrete masonry requires coping to protect mortar joints against rain and frost damage.

DPCS

- Specify horizontal dpc at the lower level to be rigid units, DPC2 bricks to BS 3921 Table 4[5] or slates to BS 743[11]. At upper level below the coping, specify high bond material to BS 743 (this may be flexible) or two courses of tiles, Figure 4:
  - — concrete masonry requires no lower dpc but finish at coping can be similar to that for brickwork.

REINFORCEMENT

- Specify deformed carbon-steel reinforcing bars up to 16 mm diameter for pier design:
  - — starter bars should be 'hooked' into the foundation and project about 500 mm to give a lap onto continuation bars;
  - — cover to all reinforcement in concrete should be 50 mm, although this can be reduced to 40 mm in brickwork;
  - — infill of pockets around bars should use 10 mm aggregate concrete[8] and should be placed the day after laying the masonry. Infill should be compacted;
  - — if mortar is used as an infill material to the pocket the reinforcement must be galvanised.

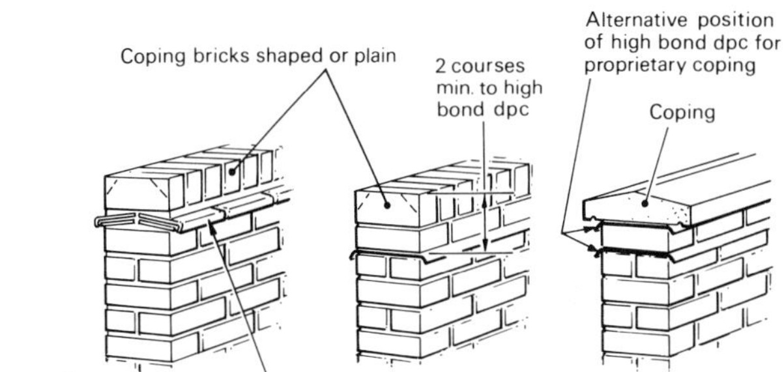

Figure 3

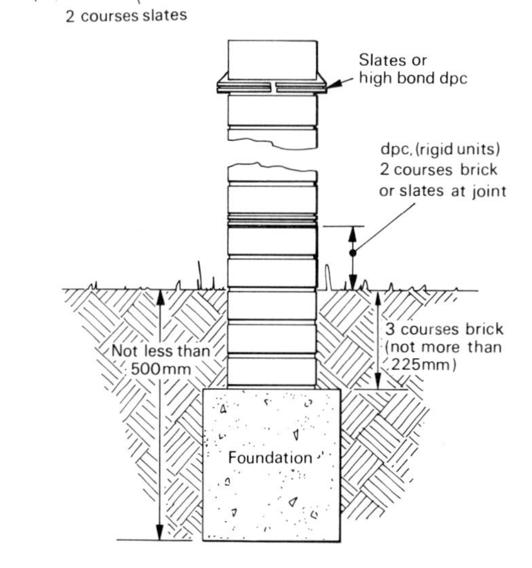

Figure 4

REFERENCES

1 **Building Research Establishment.** Freestanding masonry boundary walls: stability and movement. *Defect Action Sheet* DAS 129. Garston, BRE, 1989.

2 **British Standards Institution.** Methods for specifying concrete, including ready-mixed concrete. *British Standard* BS 5328:1981. London, BSI, 1981.

3 **Building Research Establishment.** Concrete in sulphate-bearing soils and groundwaters. *BRE Digest* 250. Garston, BRE, 1981.

4 **Building Research Establishment.** Sulphate attack on brickwork. *BRE Digest* 89. Garston, BRE, 1968.

5 **British Standards Institution.** Specification for clay bricks. *British Standard* 3291:1985. London, BSI, 1985.

6 **British Standards Institution.** Specification for calcium silicate (sandlime and flintlime) bricks. *British Standard* BS 187:1978. London, BSI, 1978.

7 **British Standards Institution.** Precast concrete masonry units. *British Standard* BS 6073:1981. London, BSI, 1981.

8 **British Standards Institution.** Code of practice for use of masonry. Part 3. Materials and components, design and workmanship. *British Standard* BS 5628:Part 3:1985. London, BSI, 1985.

9 **British Standards Institution.** Specification for shapes and dimensions of special bricks. *British Standard* BS 4729:1971.

10 **British Standards Institution.** Sills and copings. Part 2. Specification for copings of precast concrete, cast stone, clayware, slate and natural stone. *British Standard* BS 5642:Part 2:1983. London, BSI, 1983.

11 **British Standards Institution.** Specification for materials for damp-proof courses. *British Standard* BS 743:1970. London, BSI, 1970.

External and separating walls: lateral restraint at intermediate timber floors — specification

FAILURE: Instability of walls

DEFECTS: Lateral restraint connections not positioned or not fixed correctly; not durable

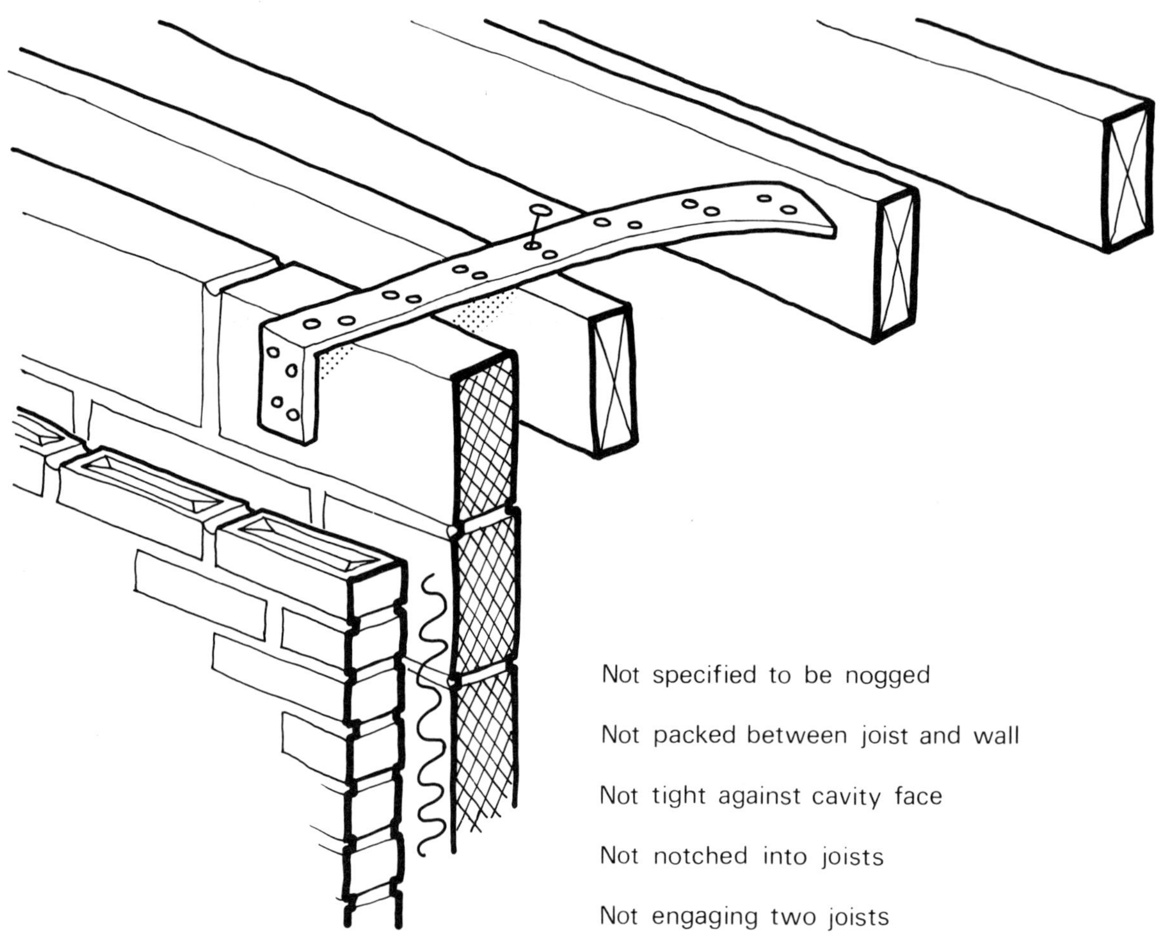

Not specified to be nogged
Not packed between joist and wall
Not tight against cavity face
Not notched into joists
Not engaging two joists

Figure 1 Strapping faults

This *Defect Action Sheet* applies only when lateral restraint is to be achieved by adopting the provisions of Schedule 7 of the *Building Regulations* 1976[1] (England and Wales)*. BRE surveys have revealed that lateral restraint connections are often badly made and much less effective than intended. ***If designers*** do not understand what is needed and ***do not provide clear instructions,*** strapping to Schedule 7 may be incorrectly positioned or fixed, adequate ***lateral restraint may not be achieved,*** and ***walls*** will then be ***unstable.***

*In Scotland the Building Standards (Scotland) Regulations 1981, Schedule 13.

External and separating walls: lateral restraint at intermediate timber floors — specification

DAS 25
April 1983

PREVENTION

Principle — Between floors and walls there must be adequate restraint connection, which operates in both tension and compression.

Practice

- Decide which walls need to be strapped to floors:
  - a wall in a house of not more than two storeys does not need to be strapped to the floor if joists, at not more than 1.2 m centres, have a full 90 mm bearing on it;
  - a wall does not need to be strapped where joists are carried on it by means of joist hangers, provided that 'restraint' hangers of the type shown in Figure 5 (and described in BS 5628[2]) are inporated at not more than 2 m centres.

Where straps are needed:

- Provide a layout drawing showing the position of every strap (Figure 2)
- Where straps are at right angles to joists, specify substantial nogging between joists at every strap position so that restraint will work in compression as well as tension. Floor boarding should not be relied on to provide sufficient resistance to compression.
- Specify packing between wall and adjacent parallel joist (Figure 3). This packing must be securely fixed so that it will not drop out if shrinkage occurs.
- Specify all straps to be of at least 30 mm × 5 mm cross-section mild steel, galvanised, with at least 260 gm/m^2 zinc coating. Straps at right angles to joists should be long enough to engage at least two joists and have a minimum 'turn down' length of 100 mm.
- Specify joists to be notched where straps are fixed to top or bottom edge of joist:
  - consider fixing straps to the underside of joists where joist deflection or shrinkage might otherwise cause the strap to disrupt flooring;
  - check that joist depths will be adequate after notching, particularly if notches will be in the underside or, if in the top, in the centre half of the span;
  - check that masonry coursing will provide a bed joint at strap level; otherwise locate straps on side of joists as in Figure 4.
- Specify that straps are not installed with their short leg turned upward in an external wall cavity: water might be led inward, particularly where full cavity insulation is installed.
- Specify all straps to be fixed with at least four 8 gauge × 50 mm counter sunk-head plated steel wood screws.
- Instruct site supervisors to ensure that 'turn downs' on straps are in tight contact with wall – tight contact is essential.

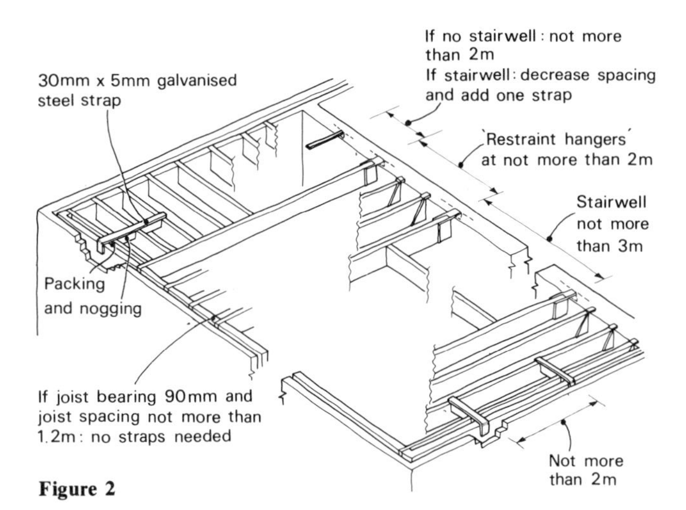

Figure 2

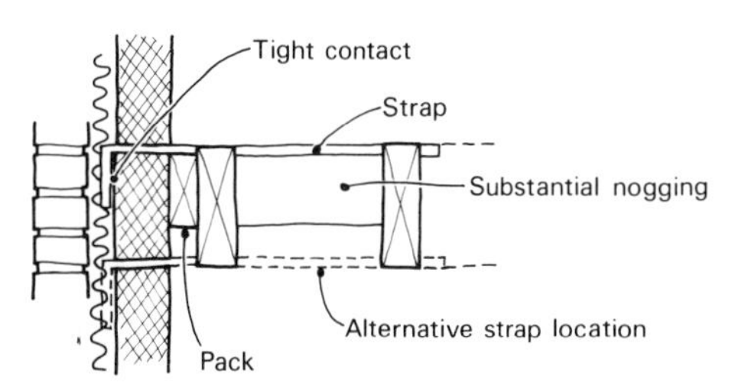

Figure 3

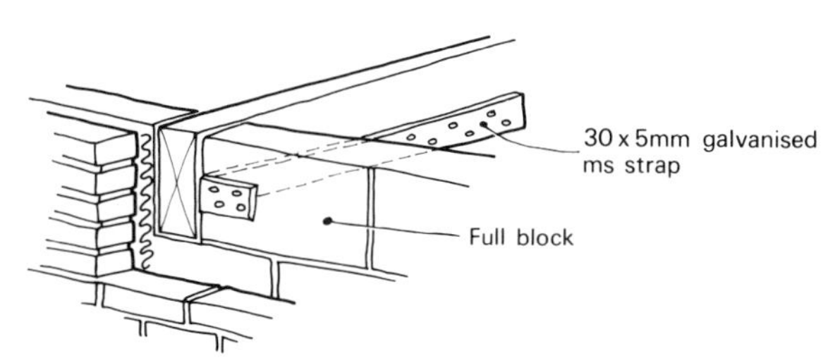

Figure 4

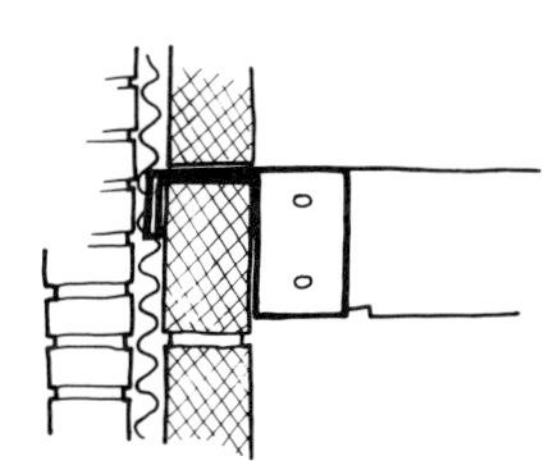

Figure 5

REFERENCES AND FURTHER READING

1 *Building Regulations* 1976 (England and Wales) Schedule 7.

2 *British Standard* BS 5628 'Code of practice for the structural use of masonry' Part 1:1978 'Unreinforced masonry'.

British Standard CP 111:1970 'Structural recommendations for loadbearing walls'.

Defect Action Sheet DAS 26 'External and separating walls: lateral restraint at intermediate timber floors – installation'.

External and separating walls: lateral restraint at intermediate timber floors — installation

FAILURE: **Instability of walls**

DEFECTS: **Straps missing, wrongly sized, mis-positioned, accidentally bent, inadequately fastened, not nogged; joists not packed against wall**

Figure 1 Strap bent, not nogged or packed

This *Defect Action Sheet* applies only when lateral restraint is to be achieved by adopting the provisions of Schedule 7 of the *Building Regulations* 1976[1] (England and Wales)*.

Floor to wall strapping is important for the structural stability of houses. The restraint has to work in compression as well as in tension. Where straps are fixed at right angles to joists, ***if the straps are not positioned on substantial nogging*** and with ***packing between wall and adjacent joist*** the connection ***will not work in compression. If the 'turn down' is not tight against the masonry*** and ***straps*** are ***not securely fixed,*** the connection ***will not work in tension.*** Straps which become bent accidentally must be removed altogether and replaced.

*In Scotland the Building Standards (Scotland) Regulations 1981, Schedule 13.

PREVENTION

Principle – Between walls and floors there must be adequate lateral restraint connection, which operates in both tension and compression.

Practice

- Check whether and where straps are required:
  - joists bearing at least 90 mm directly on walls, and at not more than 1.2 m spacing, do not need strapping.

Where straps are required

- Locate straps at not more than 2.0 m centres (Figure 2).
- Work out a practical sequence of installation:
  - ensure that bricklayers stop the brickwork at a bed joint at the strap level;
  - ensure that the straps are installed at that stage, not later as in Figure 4, or earlier as in Figure 5. (Note that this involves coordination between trades.)
- Ensure that straps are notched into joists (Figure 6). Design should have taken notching into account when sizing joists, but do not notch deeper than strap thickness.)
- Ensure that, where joists span parallel to wall to be strapped, packing is installed in the gap between joist and wall and securely fixed, and a nogging is installed between joists at strap locations (Figure 6).
- Ensure that all straps are fixed with a least four 8 gauge × 50 mm counter sunk-head plated steel wood screws.
- All straps should be of at least 30 mm × 5 mm cross-section galvanised mild steel. (Note: 'holding down' straps are often only 2.5 mm thick.) They should be long enough to engage at least two joists and have a minimum 'turn down' length of 100 mm.

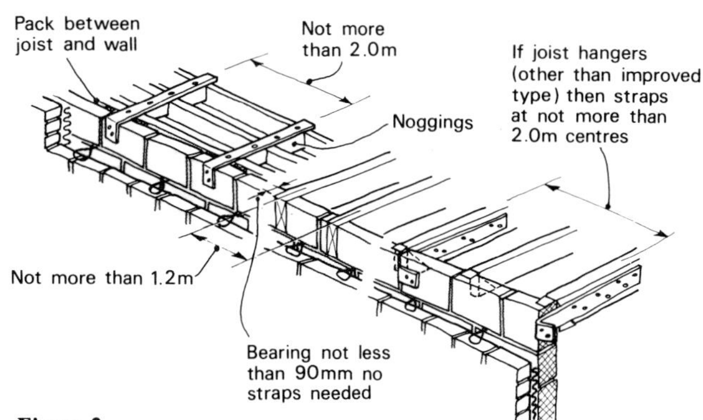

Figure 2

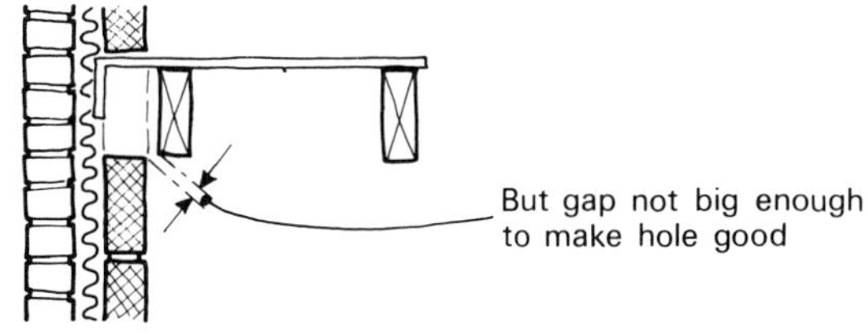

Figure 3

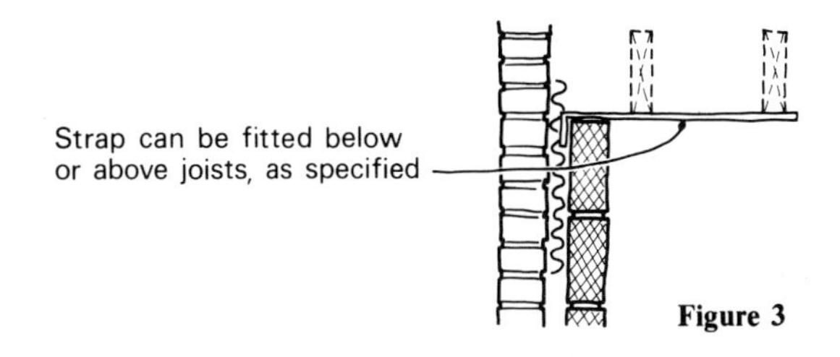

Figure 4

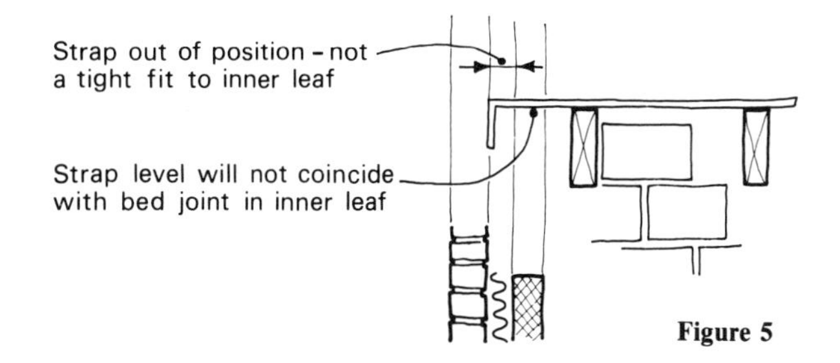

Figure 5

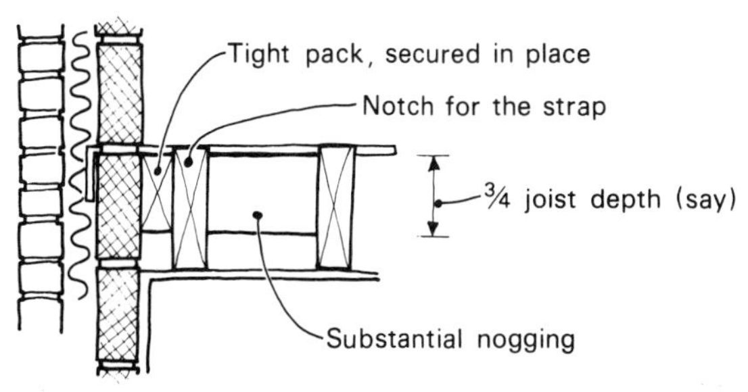

Figure 6

REFERENCES AND FURTHER READING

1 *The Building Regulations* 1976 (England and Wales), Schedule 7.

Defect Action Sheet DAS 25 'External and separating walls: lateral restraint at intermediate timber floors – specification'.

External and separating walls: lateral restraint at pitched roof level — specification

FAILURE: Instability of walls

DEFECT: Lateral restraint connections not positioned or not fixed correctly

Figure 1 Strapping not positioned correctly – requirements not clearly specified

This *Defect Action Sheet* applies only when lateral restraint is to be achieved by adopting the provisions of Schedule 7 of *Building Regulations* 1976[1] (England and Wales)*.

There is evidence that neither designers nor site staff are as yet familiar with the provisions of Schedule 7. Less than 10 houses out of more than 1000 inspected in BRE's site studies, complied in all respects with the provisions of Schedule 7. Straps at rafter level, and usually also at truss tie level are needed so that separating walls and gables, particularly above eaves level, receive adequate restraint.

If the requirements for lateral restraint *are not* fully considered and *clearly specified,* straps may be omitted, wrongly fixed or wrongly positioned, adequate *lateral restraint may not be achieved* and walls will then be unstable.

*In Scotland the Building Standards (Scotland) Regulations 1981, Schedule 13.

PREVENTION

Principle – There must be adequate lateral restraint connection, which operates in both tension and compression, between roof and walls.

Practice

- In all cases specify straps on both separating and gable walls at rafter level at not more than 2 m centres (measured on slope, **a**, Figure 2):
  - specify noggings and packings so that the restraint can operate in both compression and tension.

- Consider specifying that straps be fitted to underside of rafters to give a more substantial masonry connection. (Figure 3a and 3b.)

- Next, decide whether straps at truss tie level are required:
  - decide notional wall thickness t (Figure 2).
    For solid walls: t = wall actual thickness.
    For cavity walls: t = sum of thicknesses of two leaves plus 10mm;
    (In either case: t is to be not less than 190 mm.)
  - calculate X for both solid and cavity walls:
    X = height from bottom of truss tie to the height of the highest strap on roof slope (Figure 2).
  - calculate h:
    h = distance between underside of upper floor joists and lower edge of truss ties plus $X/2$
  - then, if h exceeds 16 t, straps are needed at truss tie level (at not more than 2 m centres).

- Consider using extra ceiling binders to take the straps, but make sure the bed joints will be at the right height.

- Specify twisted straps engaging perpends if coursing does not permit straps to engage bed joints (Figure 4).

- Instruct site supervisors to ensure that no trussed rafter members are notched to receive straps.

- Specify all straps to be of at least 30 mm × 5 mm cross-section mild steel, galvanised, with at least 260 gm/m² zinc coating. Straps should be long enough to engage at least two rafters and have a minimum 'turn down' length of 100 mm.

- Specify all straps to be fixed with at least four 8 gauge × 50 mm countersunk head plated steel wood screws.

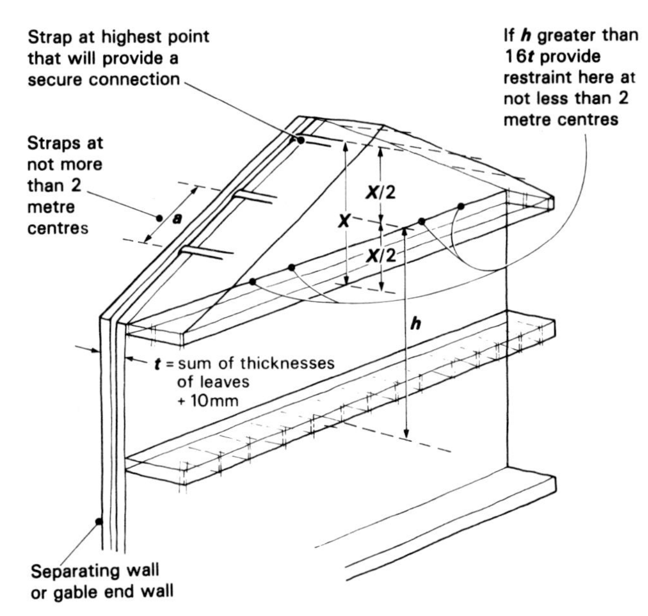

Figure 2

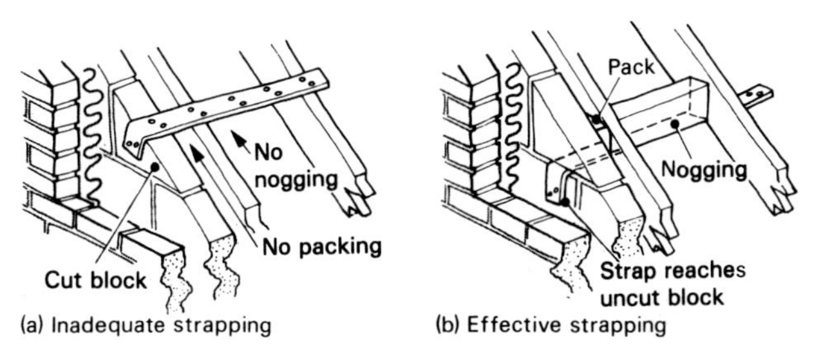

Figure 3

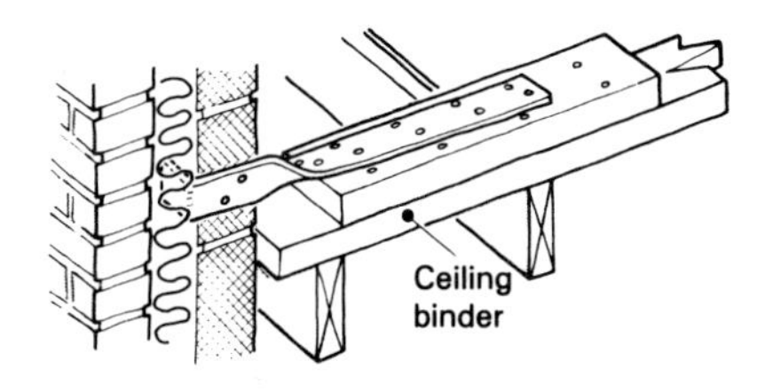

Figure 4

REFERENCES AND FURTHER READING

1 *The Building Regulations* 1976 (England and Wales) Schedule 7.

Defect Action Sheet DAS 28 'External and separating walls: lateral restraint at pitched roof level – installation'.

External and separating walls: lateral restraint at pitched roof level — installation

FAILURE: **Instability of walls**

DEFECT: **Lateral restraint connections not positioned or not fixed correctly**

Figure 1 Strapping installed without packing or nogging

This *Defect Action Sheet* applies only when lateral restraint is to be achieved by adopting the provisions of Schedule 7 of the *Building Regulations* 1976[1] (England and Wales)*. Roof to wall strapping is important for the structural stability of houses.

*In Scotland the Building Standards (Scotland) Regulations 1981 Schedule 13

The straps have to work in compression as well as in tension. If the ***straps*** are ***not*** positioned ***on substantial nogging between rafters or,*** in the case of straps at truss tie level, ***between ties, with packing between rafter (or tie) and wall,*** the connection ***will not work in compression. If the 'turn down' is not tight against the masonry*** and ***straps are not securely fixed,*** the connection ***will not work in tension.***

External and separating walls: lateral restraint at pitched roof level — installation

DAS 28
May 1983

PREVENTION

Principle – There must be adequate lateral restraint connection, which operates in both tension and compression, between roofs and walls.

Practice

- Always install straps at rafter level, at not more than 2 m centres measured on slope (Figure 2):
  - fix the straps on noggings, with not less than four 8 gauge × 50 mm countersunk-head plated steel wood screws, and pack between rafter and wall. (Where sarking is of rigid boards it may not be necessary to use noggings at strap locations.)
- Where specified, provide straps at truss tie level at not more than 2 m centres (Figure 3). Fix straps on noggings or binders, and pack between truss tie and wall.
- Watch that the coursing is appropriate to whichever strap position is specified (Figure 4) and check that bricklayers stop the brickwork at exactly the intended level for the straps.
- Do not bend straps to engage a bed joint at the wrong level (Figure 4): nog between truss ties and pack up to the required level.
- Check that straps at rafter level engage a substantial piece of blockwork.
- Do not notch trusses to take straps (trussed rafters should never be modified on site).
- All straps should be of at least 30 mm × 5 mm cross-section galvanised mild steel. They should be long enough to engage at least two rafters and have a minimum 'turn down' length of 100 mm.

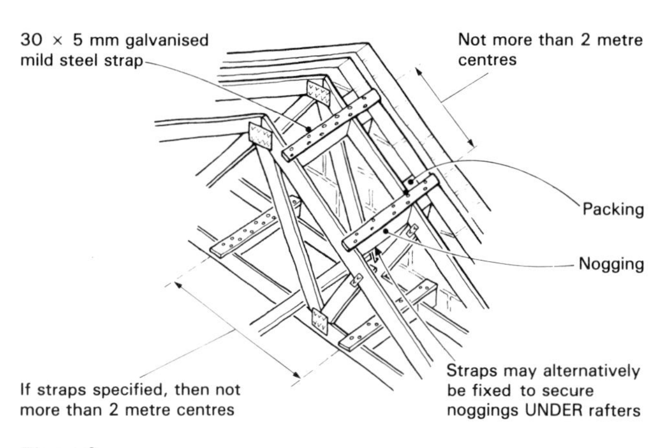

Figure 2

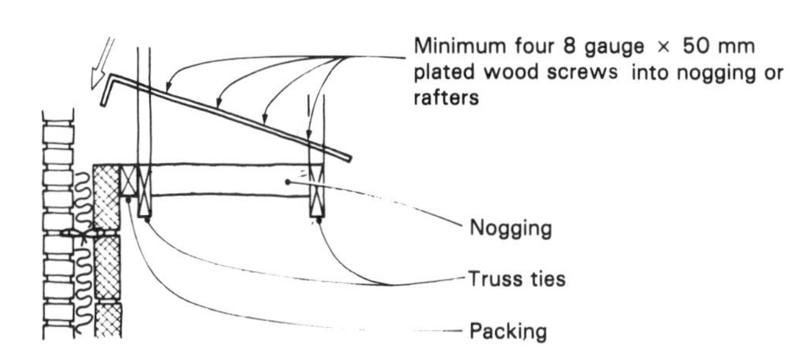

Figure 3

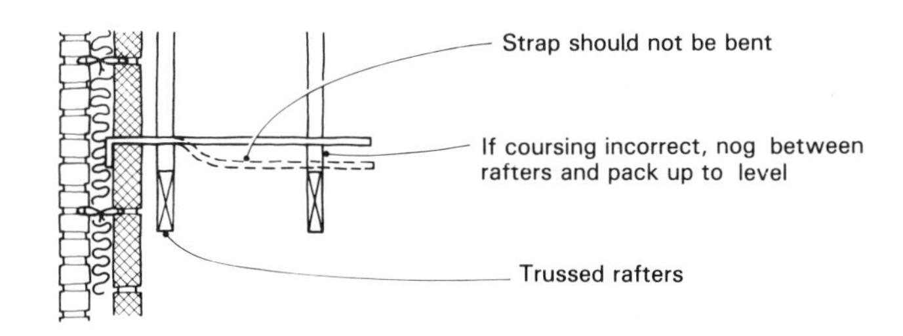

Figure 4

REFERENCES AND FURTHER READING

1 *The Building Regulations* 1976 (England and Wales). Schedule 7.
Defect Action Sheet DAS 27 'External and separating walls: lateral restraint at pitched roof level – specification'.

PART FIVE

Walls 2: Brickwork, Rendering and Decoration

DAS 64
August 1985

Site
CI/SfB 8 (21)(H121)

External walls — bricklaying and rendering when weather may be bad

FAILURES: **Frost damage; mortar washed out of joints by driving rain; weakened mortar; stained brickwork; detachment of rendering; walls blown down**

DEFECTS: **Bricklaying in frost; no protection of new brickwork against frost; new work not protected from driving rain; rendering applied to saturated walls; walls built when strong winds prevail; unstable new work not propped when strong wind forecast**

Figure 1 Bricklaying in bad weather — no precautions

BRE site inspections have often found evidence of ***mortar failure*** suspected to be ***due*** to ***frost action*** on ***green brickwork;*** damage often goes unnoticed until later lifts have been built, when remedial work becomes expensive. ***Loss of adhesion*** is common in ***renderings applied*** to ***saturated walls. Rain*** can ***weaken,*** or ***wash out,*** unset ***mortar. Mortar joints*** are readily ***disturbed*** when new walls are subjected to ***strong*** and gusty ***winds*** and occasionally walls may collapse.

PREVENTION

Principle — Checks and precautions become critically important if bad weather may occur immediately before, during, or immediately after bricklaying or rendering; in some conditions work should not proceed.

Practice

- Do not permit bricklaying or rendering when the air temperature is below 2°C (Reference 1), nor when it is expected to fall below 2°C within the next few hours.

- Check that aggregates are not frost-bound;
  — frost-bound aggregates must be thoroughly heated.

- Do not permit bricklaying or rendering in strong winds (Force 6):
  — in Force 6, force of wind makes it difficult to walk steadily;
  — note: walls that do not yet have adequate returns or other restraint against movement may need to be propped if such winds are forecast;
  — inspect for damage if strong winds occur soon after construction.

- Do not permit bricklaying or rendering when rainfall exceeds the lightest of showers:
  — rendering applied to saturated walls will almost certainly become detached;
  — rain can weaken mortar and stain face work;
  — (note need for adequate protection of materials prior to use).

- Ensure that green brickwork and render are protected from rain and possible overnight frost, Figure 2.

- Remember that pre-recorded forecasts are available — see 'Weather line' numbers in your dialling code booklet. Forecasts specific to contractors' needs can be obtained, for a fee, from your local Weather Centre (see directory).

- For your locality, the Meteorological Office can compute the proportion of working hours in which conditions will be unsuitable for the operations covered by this DAS – a charge is made – England and Wales, telephone 0344 420242, Extension 2278, Scotland, 031 334 9721, Extension 524, Northern Ireland, 0232 228457. A list of threshold conditions for many building operations (together with local data for Plymouth) is given in Reference 1.

REFERENCES AND FURTHER READING

1 *Building Research Establishment Report,* 'Climate and Construction Operations in the Plymouth area'; Keeble and Prior. (In preparation.)

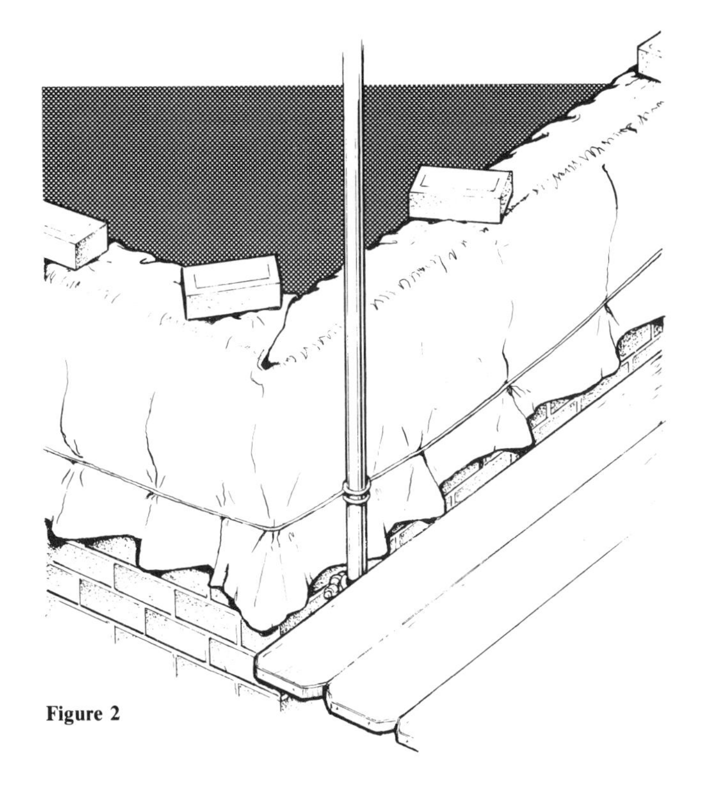

Figure 2

External walls: rendering — resisting rain penetration

FAILURES: Cracking, detachment; rain penetration into structure

DEFECTS: Specified mix too strong for background (or for preceding coat); requirements for keying to background not specified; rendering continuous over zones where relative movements occur in the background; associated detailing permits water penetration into background

Figure 1 Freestanding wall: detachment of rendering

Rendering failures are reported to BRE in increasing numbers. Poor specification or detailing, and poor workmanship, in about equal numbers, account for most failures (about 95% in Scotland).

The main function of a rendering is to help to exclude rainwater. ***If*** a ***rendering*** becomes ***cracked, rainwater will penetrate*** and ***cannot*** readily ***re-evaporate***. Also, ***if water enters, frost damage or expansion of brickwork mortar*** (sulphate attack) may produce extensive ***cracking or detachment***. Some constructions, such as ***parapets and freestanding walls***, are so exposed that ***alternatives to rendering should always be sought***, especially in regions of severe exposure.

PREVENTION

Principle — Rainwater must not saturate walls behind rendering; rendering must be isolated from relative movement in the wall.

Practice

- Check that weatherproofing details will keep water out of the background wall; provide sheltering features where possible (Figure 2).
- Check where relative movements in the background may occur, and specify action needed:
  - — Obvious locations are joints provided specifically for movement, but movements can occur at every change in background materials — eg brick in-filling between reinforced concrete columns. Renderings must either stop at joints or, if expected movement is small, be carried across the joint using, for example, metal lathing.
- Specify that rendering is to stop just above the ground level DPC (Figure 2).
- Specify how to achieve the bond to the background wall[1,2]:
  - — a good bond depends on both a good mechanical key (eg raked brickwork joints) and the right amount of absorbency. Too much absorbency can be dealt with by plasterers but appropriate action should be specified if low absorbency is expected (eg the use of bonding agents in spatter-dash coats, or wire brushing, grit blasting, bush hammering of concrete, etc). If metal lathing is used to provide a key, three rendering coats should be specified.
- Specify rendering mixes appropriate to strength of background wall or of earlier coats; no coat should be richer than the preceding coat or stronger than the background: Table 2 of Reference 2 (BRE Digest 196) recommends mixes for various backgrounds and exposures:
  - — Problems of inaccurate site-batching and misuse of additives can be reduced by specifying pre-mixed material.
- Specify sands complying with Table 1 of BS1199 (BS1199 also covers sands for internal plastering, and these are too fine for rendering mixes) or use proprietary pre-mixed materials.
- Specify required coat thickness:
  - — Undercoats should be between 8 and 16 mm thick — the upper limit should not be exceeded.
- Specify 3-coat work in all cases where exposure will be severe.
- Specify adequate curing (at least 3 days):
  - — In warm dry weather, spraying or protection by polythene sheets may be needed.
- Specify means of protection from rain and frost.
- Specify a textured finish in preference to a smooth dense finish; a rough-textured rendering gives more effective protection against rain penetration, and is less prone to cracking (Figure 3).

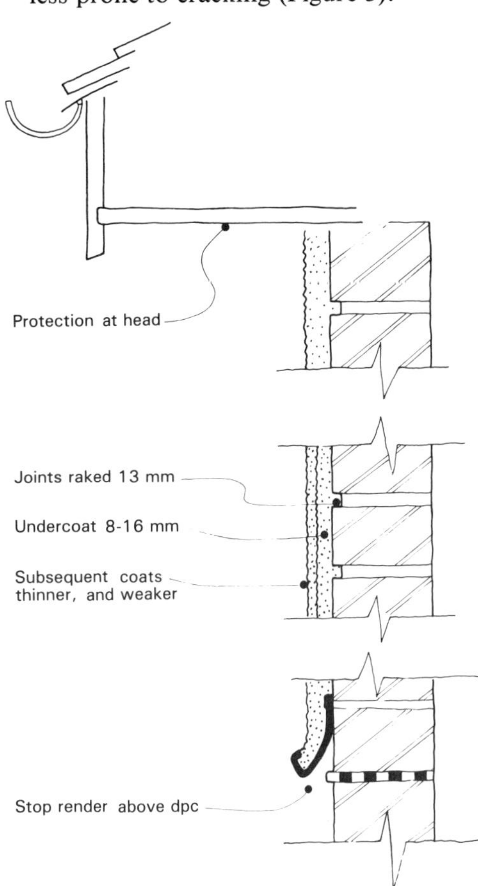

Figure 2

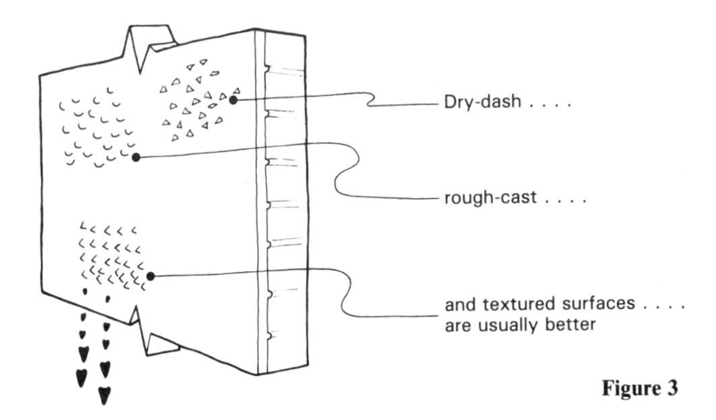

Figure 3

REFERENCES AND FURTHER READING

1 *British Standard* BS5262: 1976 'External rendered finishes'.
2 *BRE Digest* 196 'External rendered finishes'.
C & CA Publication 47.102: Appearance Matters No 2. 'External rendering'.

DAS 38
October 1983

Site
CI/SfB 8 (21)P(D4)

External walls: rendering — application

FAILURES: Cracking, detachment; rain penetration into structure

DEFECTS: Inadequate bond or key to wall; rendering stronger than background or preceding coats, too weak to exclude rain adequately, or too rich (or wet) to avoid cracking

Figure 1 Detachment of rendering from a dense wall with poor bond between coats and between rendering and background

Almost half of the rendering failures reported to BRE are attributed to bad site practices. Successful application of rendering requires skill and experience, beginning with assessment of the suction of the wall, and of the need for action to achieve a good bond. It may prove necessary to refer unforeseen problems back to the specifier.

If a ***good bond*** to the wall is ***not achieved, rendering*** may become ***detached***. If render coats are ***too rich or too wet***, or have too much fine sand, they ***crack***, water penetrates and ***failure*** may follow.

PREVENTION

Principle — Rendering must be well-bonded (both to the wall and between coats) and crack-free; it must be well supported at, but not bonded to, the wall at places where movement may occur between different parts of the wall.

Practice

- Check that the surface to be rendered is free from any contaminants and efflorescence, and is not too wet.
- Check the suction by splashing water onto surfaces:
  — If too much suction, spraying with water may be needed (or use a water-retaining additive in the render); if too little, a spatter-dash or stipple coat with bonding agent may be needed;
  — If background too wet, delay rendering until conditions improve.
- Check that a good mechanical key exists (always important but especially if suction is low, as on dense concrete, calcium silicate or engineering brick, etc):
  — Brickwork should have raked joints (at least 13 mm), Figure 2; concrete may need wire brushing whilst green or bush hammering or similar surface roughening treatment (Figure 2).
- Check that any metal lathing specified is well fixed; expanded metal should be fixed with the correct side towards the wall (see manufacturer's literature):
  — If metal lathing is used to bridge changes in background material a strip of, say, breather paper should be fixed behind the lathing so that render does not bond at the background joints; lathing should be spaced from the wall so that rendering can get behind it.
- Check that the sand does not contain excessive fine material:
  — If plasterers find they need to use a very wet mix, the sand probably has excessive fines — and renderings will shrink and crack (Figure 3).
- Check that mix proportions are as specified.
- Check that thicknesses of coats are as specified:
  — Check that admixtures are correctly used and that no unauthorised materials are added to the mix;
  — Undercoats between about 8 and 16 mm thick (dubbing out first if uneven background would otherwise need greater thickness; this ought not to be necessary if walls are built to normal accuracy). Succeeding coats should be progressively thinner, and never progressively stronger.
- Ensure that adequate curing is given (keep damp for at least 3 days and protect from sun and wind).
- Ensure that work in progress will not be damaged by rain or frost.

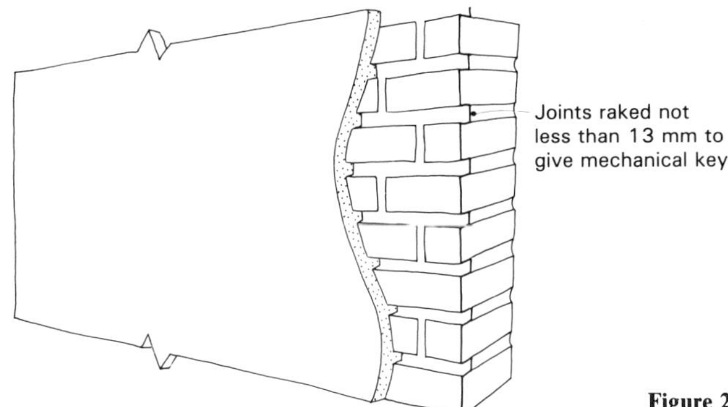

Figure 2

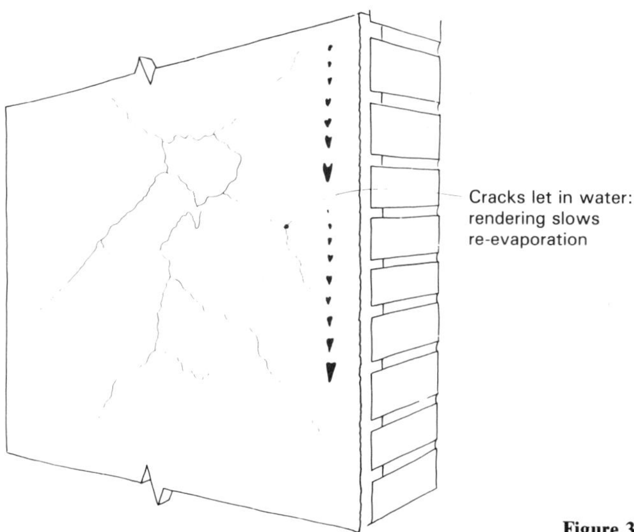

Figure 3

REFERENCES AND FURTHER READING

Defect Action Sheet DAS 37 'External walls: rendering — resisting rain penetration'.

External masonry walls: eroding mortars — repoint or rebuild?

FAILURE: Loss of brickwork strength following erosion of mortar

DEFECTS: Incorrect mortar mix; detachment of earlier pointing; long term ageing of mortar

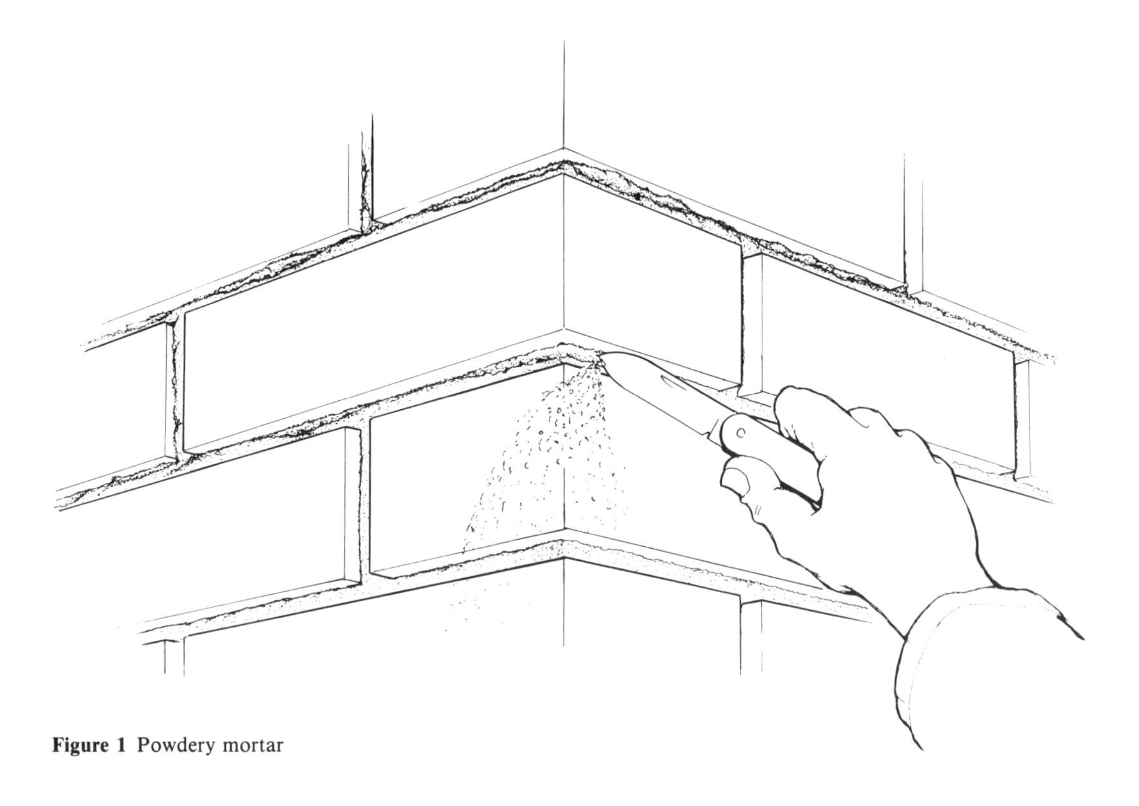

Figure 1 Powdery mortar

In rehabilitation work — and occasionally in relatively new properties — brickwork is sometimes found to be in poor condition, mortar powdery, easily rubbed away, or eroding so that joints become appreciably recessed. Owners must then decide whether repointing will adequately restore the strength of the wall and the durability of the joints.

Every case must be judged on its merits. Walls of dwellings up to two storeys are almost always amply strong. A structural assessment is always advisable for walls of three or more storeys, and may indicate the need for other measures in addition to repointing or even rebuilding.

Experimental evidence[1] shows that, in a typical case, halving the compressive strength of the mortar reduced brickwork strength by only about 15 per cent.

Brickwork strength is negligibly affected by unfilled perpends. However, ***deeply recessed bed joints*** can ***reduce*** compressive ***strength*** by about one third and lateral strength (eg against suction) ***by up to a half.***

External masonry walls: eroding mortars — repoint or rebuild?

DAS 70
February 1986

REMEDIAL WORK

Principle: If poor condition of brickwork is due only to erosion of mortar, repointing can restore adequate strength and rebuilding can be avoided.

Practice:

- Check that bricks are sound enough to justify repointing.

- Check that poor condition is not due wholly or partly to corrosion of wall ties, Figure 2 (see also DAS 21), settlement, or sulphate attack[2] (Figure 3); in either of these cases this DAS does not apply.

- Commission a formal structural assessment if the mortar lacks cohesion in the body of the joint, Figure 4, the brickwork is distorted, or if there is other reason to question structural stability, and for all walls of three or more storeys.

  — if analysis of mortar is needed in order to determine the mix[3] (and hence likely strength), take samples from inner half of outer leaf or middle third of solid walls, Figure 4;
  — estimate brickwork strength[4] in relation to points where stresses are greatest, Figure 5;
  — if distortion under load has occurred, identify what steps in addition to repointing, can be taken to avoid need for rebuilding — eg lateral restraint, additional ties.

- Ensure that repointing is done in good time;
  — if bricks are becoming frost-spalled, repoint urgently.

- Specify appropriate repointing mix (DAS 71).

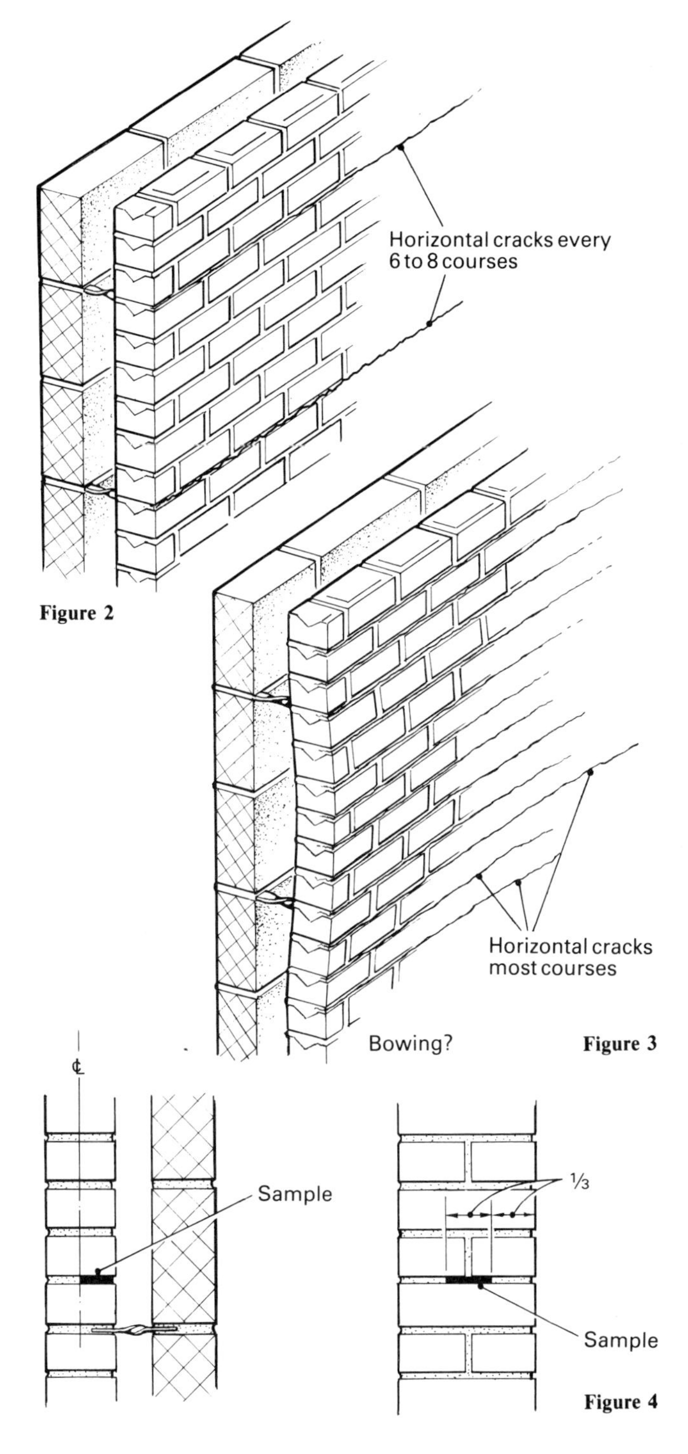

Figure 2

Figure 3

Figure 4

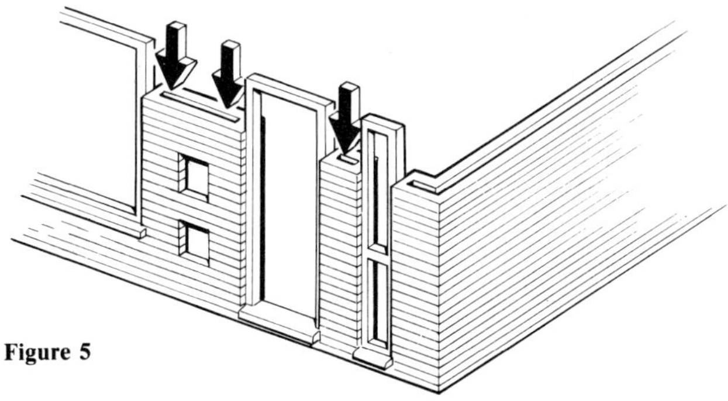

Figure 5

REFERENCES AND FURTHER READING

1 *BDA Technical Note* Vol 1, No 6, 1972 'Workmanship factors in brickwork strength'.
2 *BRE Digest* 89. 'Sulphate attack on brickwork'.
3 *British Standard* BS 4551:1980. ' Methods of testing mortars, screeds and plasters'.
4 *British Standard* BS 5628:Part 1:1973. 'Unreinforced masonry.' *BRE Digest* 246. 'Strength of brickwork and blockwork walls'.

DAS 71

February 1986

Design

CI/SfB 8 (21)F(D7)

External masonry walls: repointing — specification

FAILURES: **Brick edges spalled, rain penetration; replacement mortar eroded, cracked or frost-damaged**

DEFECTS: **Repointing mortar too strong; repointing mortar too weak or inadequately protected**

Figure 1 Repointing restores weathertightness to house on left

The specification of too strong a repointing mix, in the belief that it will be more durable and more weatherproof, is a common error. Mix specification should take account of existing brick and mortar strengths and the likelihood of frost occurring before repointing has hardened adequately.

Bricks may be ***spalled by repointing*** mortar that is appreciably ***stronger than*** the existing ***bed joint mortar,*** especially if the repointing is shallow. ***Cement-rich repointing*** mortar ***shrinks*** away from the bricks, producing fine cracks that encourage ***rain penetration***.

Repointing ***mortars containing lime*** have the advantage of being more flexible than ungauged cement mortars, but are ***more liable to frost damage*** during hardening.

PREVENTION

Principle: repointing mortars should contain enough cement to be durable but must not be stronger than the bricks.

Practice:

- Use Table 1 of BRE Digest 160, Mortars for bricklaying[1] to identify the mortar group appropriate to the brickwork.

- Select a mix, taking account of existing brick and mortar strengths (since the mortar is to be used for repointing rather than bricklaying), thus:
  — if bricks are old or weak or laid in very old lime mortar, specify the weakest practicable mix consistent with strength, eg:
  1:2:9 cement:lime:sand if no early frost likely (or protect for 7 days);
  1:6 masonry cement:sand or 1:8 cement:sand, plus plasticiser } if early frost likely.
  — if bricks are sand-lime, specify mixes as above;
  — if bricks are sound and medium or high strength, specify:
  1:1:6 cement:lime:sand if no early frost likely (or protect for 7 days);
  1:4 or 5 masonry cement:sand or 1:6 cement:sand plus plasticisers } if early frost likely

- Specify any limitations on areas to be raked out at one time;
  — eg locally highly stressed brickwork, or joints eroded (or raked out) more than 25 mm.

- Specify that joints are to be raked out squarely to a depth at least twice the width but not exceeding 35 mm.

- Specify ironed ('bucket handle') joints for best durability and weathertightness, Figure 2.

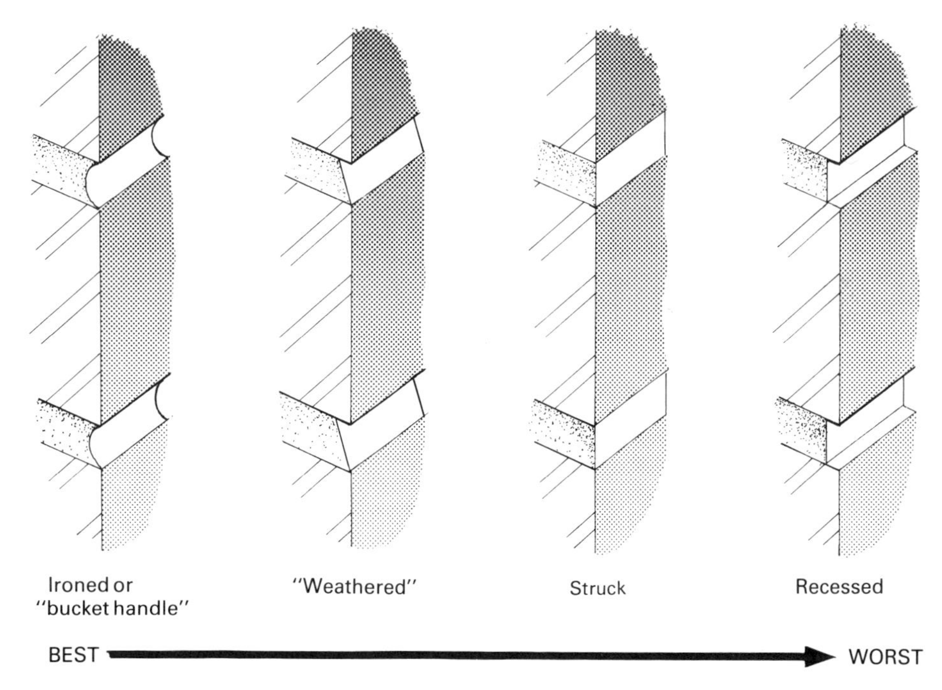

Figure 2

REFERENCE AND FURTHER READING

1 *BRE Digest* 160. 'Mortars for bricklaying.'
BRE Digest 200. 'Repairing brickwork.'

External masonry walls: repointing

FAILURES: **Brick edges spalled, rain penetration; replacement pointing mortar eroded, cracked or frost-damaged**

DEFECTS: **Inadequate joint preparation, repointing mortar too strong; repointing mortar too weak or inadequately protected; inconsistent repointing mix**

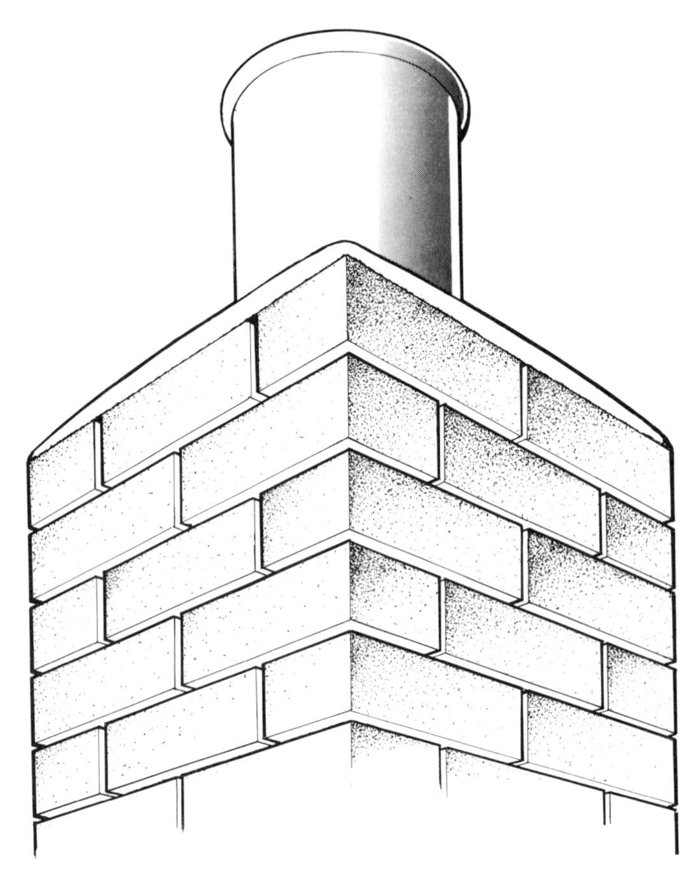

Figure 1 Repointing can restore strength

In repointing work it is important to keep closely to the mix proportions specified for the repointing mortar.

If ***repointing*** mortar is ***too strong,*** mortar may crack, ***bricks*** may ***spall*** — especially if repointing is shallow or the original mortar weak. If a ***lime-based mortar is used*** for repointing ***when frost may occur*** before it has fully hardened, ***repointing*** may be ***damaged.*** If ***joints*** are ***not raked*** to sufficient depth, if ***dust*** is ***not removed*** from raked joints, or if ***repointing*** is ***not firmly tooled,*** repointing ***mortar*** may become ***detached.*** If repointing ***mortar*** contains ***excess cement or fines,*** it will ***shrink*** and the ***wall may leak.***

PREVENTION

Principle — Repointing mortar must be strong enough to be durable but not stronger than the bricks; it must be firmly tooled and must not shrink.

Practice

- Ensure that cement:lime:sand mortars weaker than 1:1:5-6 are not used for repointing in winter months unless protected against frost for 7 days;
  — mortar containing air entraining agent providing good frost resistance.

- Check that joints are raked squarely to a depth of twice their width
  — depth should be about 15-25 mm, Figure 2, and never more than 35 mm;
  — observe any specified limits on area to be raked out at any one time; otherwise, if the depth is in the range 15-25 mm, raking out generally need not be restricted to limited areas; if deeper, raking out should be restricted to areas of about three courses high by three stretchers long, followed by repointing, Figure 3.

- Check that raked joints are brushed and washed, to remove dust and to ensure that repointing mortar bonds well.

- Check that repointing mortar is accurately batched, allowing for bulking of damp sand;
  — batched quantities must be small enough to be used within 2 hours.

- Ensure, if pigment is to be added, that the specified amount is not exceeded;
  — usually up to 10% by weight of cement but, if pigment is carbon black, proportion must not exceed 3%.

- Check that repointing mortar is firmly tooled to the specified profile ('bucket handle' is best if there is a choice);
  — repointing may need to be applied in two stages in deep joints to obtain good compaction.

- Ensure that newly-applied repointing is protected as necessary until hardened, (7 days).

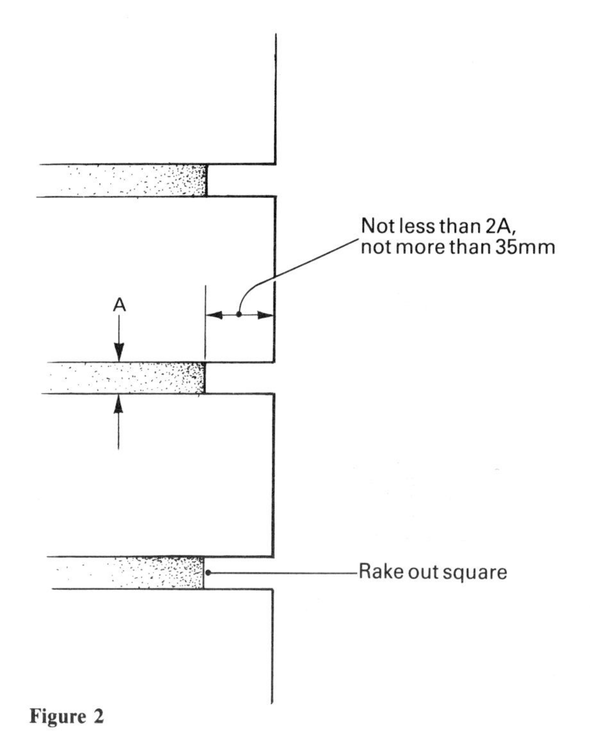

Figure 2

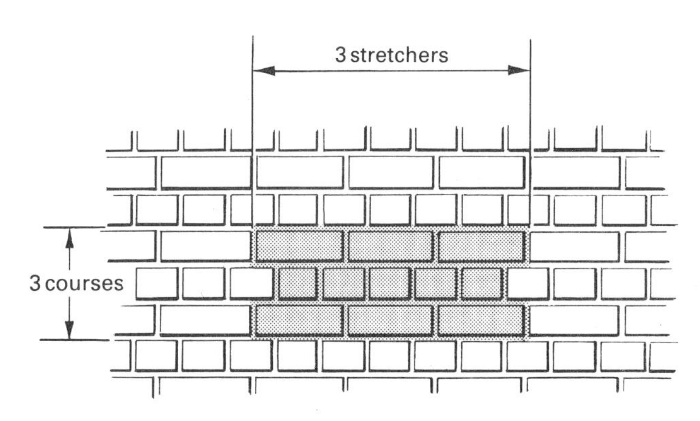

Figure 3

Brickwork: prevention of sulphate attack

FAILURE: **Expansion of jointing mortar, bowing of walls, oversailing of walls at dpc level**

DEFECTS: **Brickwork persistently wet, coupled with brickwork not isolated from presence of sulphates, and use of excessively weak mixes**

Figure 1

Sulphate attack in clay brickwork results from an expansive reaction between one of the constituents of Portland cement, and soluble sulphates.

Attack is **likely** to occur only when all the following conditions are found:

a) brickwork poorly detailed so that it can remain wet for long periods
b) neither strong Portland cement nor sulphate resisting cement mixes have been used
c) either bricks that are not low in sulphate content, or brickwork not isolated from sulphates from other sources (eg groundwater) are used.

Sulphate attack in clay brickwork can result in considerable damage sometimes occurring as early as two years after construction. Costly rebuilding is usually the only remedy and designers should therefore specify materials and details that will minimise the risk of sulphate attack.

If ***brickwork*** suffers from ***sulphate attack, expansion*** of the ***mortar jointing*** will ***disrupt*** the ***wall,*** Figure 1, and the accompanying ***expansion*** may damage ***adjoining construction.***

PREVENTION

Principle — Sulphate attack is **unlikely** to occur where any of the following conditions are found:
- brickwork remains as dry as can be achieved in service;
- bricks have a low sulphate content (designated L in BS 3921);
- adequate mixes of sulphate-resisting Portland cement or strong mixes of ordinary Portland cement are used.

Practice

- Check that detailing will guard against persistent wetting of brickwork, eg:
  - provide as generous a roof overhang as possible at eaves and verges;
  - provide flashings and damp-proof courses at sills;
  - provide parapet and other free-standing walls with copings having a generous overhang, throatings at least 25 mm clear of the wall, with a well supported and continuous dpc under coping[1];
    Note that brick subsills can be particularly at risk from sulphate attack, Figure 2;
  - provide earth-retaining walls with a dpm, protected against damage from fill.

- Specify[2], where there is a risk that brickwork may be persistently wet, bricks that are low in soluble salts.
  Note: BS 3921[3] limits the amount of soluble sulphates only in bricks designated L; bricks of normal quality (designated N) should not be used in these circumstances.

- Specify adequate mixes of sulphate-resisting Portland cement or strong mixes of ordinary Portland cement for walls below dpc level where sulphates are present in the groundwater.

- Specify, for brickwork earth-retaining walls, bricks wholly conforming to the 'frost resistant low-sulphate quality' (designated FL) of BS 3921 used in conjunction with sulphate-resisting mortar mixes (eg 1: ½: 4½ cement: lime: sand or stronger, preferably using a sulphate-resisting cement).
  - where the use of bricks containing an appreciable quantity of sulphate is unavoidable, eg to match existing brickwork, consider specifying an effective damp-proof membrane between the brickwork and the fill, Fig 3, or a retaining wall of in-situ concrete with facing brickwork separated by a ventilated cavity.

Figure 2

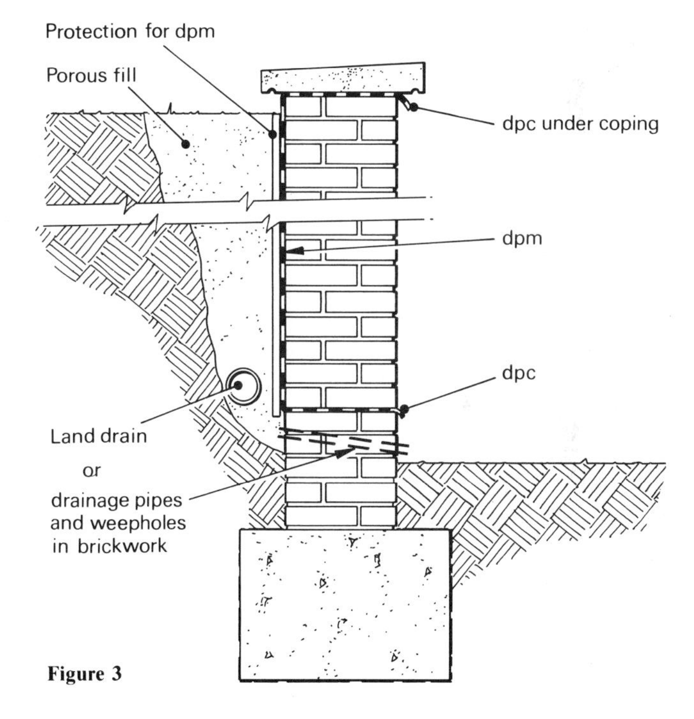

Figure 3

REFERENCES AND FURTHER READING

1 **Building Research Establishment.** Cavity parapets — avoiding rain penetration. *BRE Defect Action Sheet* DAS 106. Garston, BRE, 1987.

2 **British Standards Institution.** Code of practice for use of masonry. *British Standard* BS 5628:Part 3, Table 13: 1985. London, BSI, 1985.

3 **British Standards Institution.** Specification for clay bricks. *British Standard* BS 3921:1985. London, BSI, 1985.

Building Research Establishment. Sulphate attack on brickwork. *BRE Digest* 89. Garston, BRE, 1968 (minor revisions 1971).

Brick walls: injected dpcs

FAILURE: **Reappearance of rising damp following dpc injection**

DEFECTS: **Dpc injected at wrong level; insufficient fluid injected; fluid lost through cracks caused by poor drilling**

Figure 1 Damage from poor drilling permits fluid loss during injection

Properly installed, an injected dpc can provide an effective barrier to rising damp in walls with defective dpc's or built with no dpc's. BRE has seen many cases in which injected dpc's have not been correctly installed in walls that were suitable for such treatment.

If a ***dpc*** is injected ***at the wrong level*** in relation to ground or floor, ***or*** if ***fluid quantity*** is ***not*** correctly ***controlled,*** or if ***fluid*** can ***leak*** away ***via damaged brickwork*** during injection, the ***dpc*** may be ***incomplete*** and ***rising damp*** may ***reappear.***

PREVENTION

Principle — Injected dpc's must be installed so that moisture from the ground will not reach vulnerable materials or the inside of the building.

Practice

- Check that dampness is due to rising damp and not to other causes[1].

- Check that rising damp cannot be rectified by other means — eg removal of mortar, rendering, soil, paving etc, bridging an existing dpc, Figure 2.

- Specify that injected dpc is to be installed to BS 6576[2] using injection fluid with an acceptable third party certificate (eg British Board of Agrément Certificate).

- Check that injection fluid is compatible with any timber treatment applied or to be applied near the wall.

- Ensure that installers are aware of features that will determine the correct level at which to inject, eg other parts of construction should not bridge the dpc at any point, Figure 3, and, where possible, external ground or paving level should be at least 150 mm below line of injection.

- Check that installer's proposed drilling pattern matches fluid manufacturer's recommendations:
  — usually, for walls up to 120 mm thick, drilling should be from one side and to a depth of 65 to 75 mm, thicker walls are drilled from both sides; holes may be drilled horizontally in either bricks or bed joints; alternatively angled holes should slope downwards to terminate in bed joints.

- Do not permit percussion drilling in walls, or leaves, less than 110 mm thick[2] or in thicker walls where brickwork may be shattered.
  — injection fluid may be lost through damaged brickwork.

- Check installer's record of quantity of fluid used against manufacturer's recommendations for the length and thickness of wall treated.

- Check that external holes are suitably plugged to exclude rainwater.

- Check that the installer has reassured occupants that odours from volatile solvents should disperse within a short time.

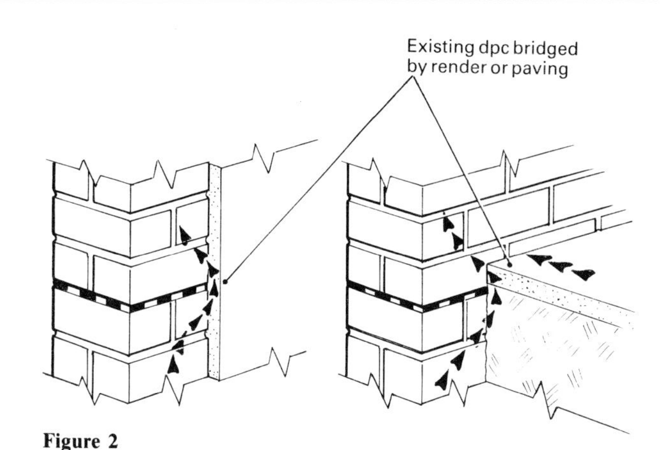

Figure 2

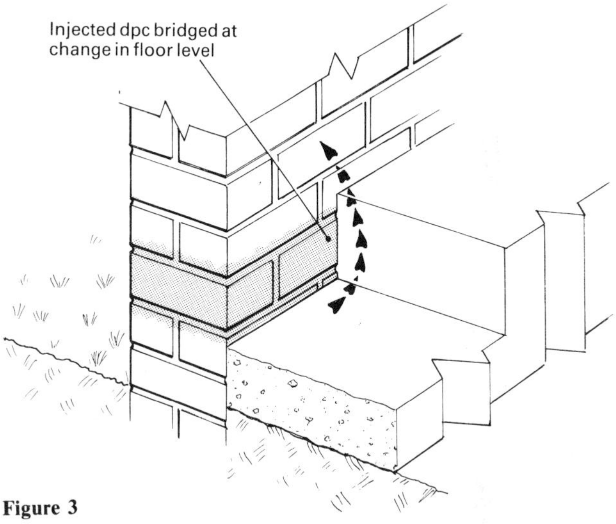

Injected dpc bridged at change in floor level

Figure 3

REFERENCE AND FURTHER READING

1 *BRE Digest* 245 Rising damp in walls: diagnosis and treatment.
2 *British Standard* BS 6576:1985. 'Code of practice for installation of chemical damp-proof courses'.
Defect Action Sheet DAS 86 'Brick walls: replastering following dpc injection'.

Brick walls: replastering following dpc injection

FAILURES: **Staining of new plaster and decorations by residual dampness in walls; staining of retained plaster; recurrence of rising damp**

DEFECTS: **Incorrect replastering system; plaster, contaminated by hygroscopic salts, not removed; injected dpc bridged by new plaster**

Figure 1 Recurrence of dampness

Damp walls will remain damp above dpc level for some time after dpc injection, and new plaster must resist the effects of this dampness and the passage of water and salts to the indoor surface. Building Research Advisory Service has seen many cases where, following dpc injection, new plaster and decorations have been spoiled by damp.

If ***salt contaminated plaster*** is ***not removed,*** or if ***new plaster cannot resist*** the effects of residual ***dampness*** in the wall, or if an injected ***dpc is bridged*** by new plaster, damp staining and ***deterioration may recur.***

PREVENTION

Principle — Residual water and salts in the wall must not damage replacement plaster.

Practice

- Specify that plaster is to be removed to at least 300 mm above the highest level at which dampness is detectable and in no case less than 1 m above the level of the dpc.
- Specify that brickwork joints are to be raked out 10 mm deep to provide a key.
- Specify a 1:3 cement:sand undercoat and a Class B gypsum plaster finish;
  - — or specify a remedial plaster system formulated for this purpose and having an acceptable third party certificate (eg BBA certificate).
  - — ordinary gypsum-based undercoats must not be used — they can act as a 'poultice' and accumulate hygroscopic salts from the wall.
- Specify cement:sand undercoat to be nowhere less than 10 mm thick;
  - — where very uneven walls dictate a much greater thickness it should be applied in successive coats.
- Check proposed remedial work to ensure that injected dpc will not be bridged, eg by external render or, as in Figure 2, by new screed or plaster.
- Check that new plaster will not be in contact with any new screed: (for example, design as in Figure 3) — otherwise water from the screed mix can wick up into plaster.
- Inform finishing trades or occupants that, before applying an impervious wall covering, the wall must be allowed to dry;
  - — at least six months, and sometimes twelve, will be necessary.

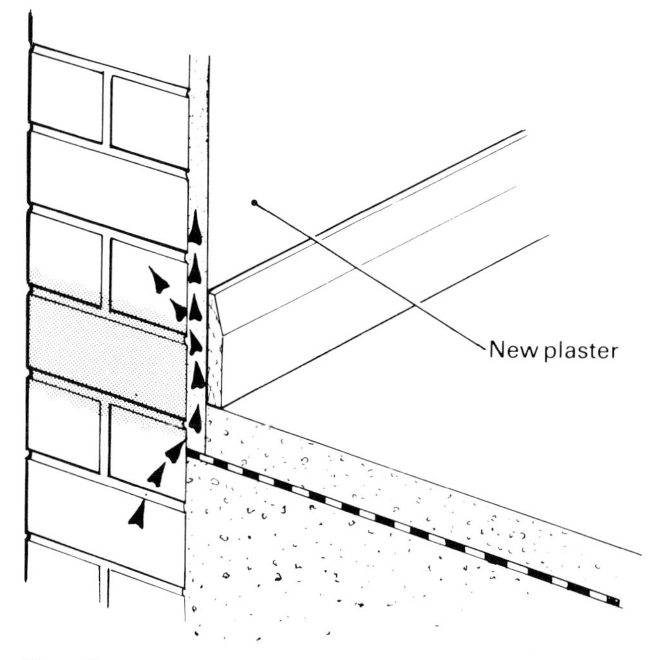

Figure 2

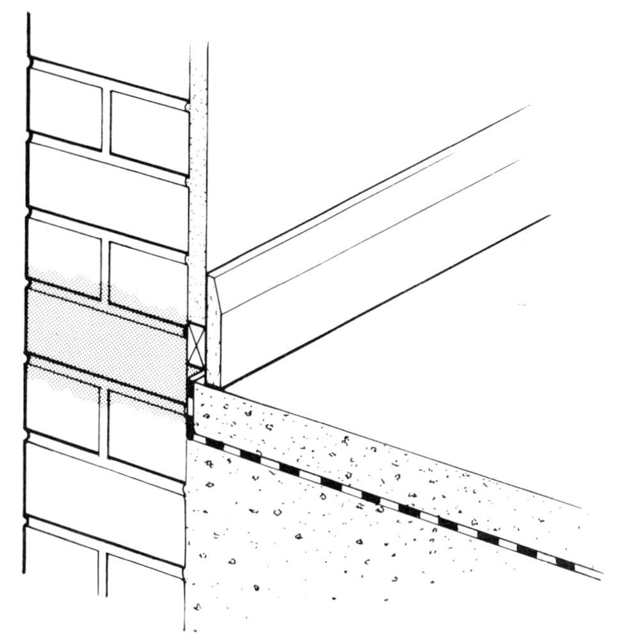

Figure 3

REFERENCES AND FURTHER READING

British Standard BS 6576:1985. 'Code of practice for installation of chemical damp-proof courses'.
BRE Digest 245. 'Rising damp in walls: diagnosis and treatment'.
Defect Action Sheet 'DAS 85 Brick walls: injected dpc'.

External masonry painting

FAILURE: Breakdown of paint film, mould growth, discoloration

DEFECTS: Unsuitable substrate; friable, cracks, uneven, dirty, dampness, containing soluble salts, alkalis, or already painted with incompatible coatings; unsuitable choice of paint; unsuitable conditions for application.

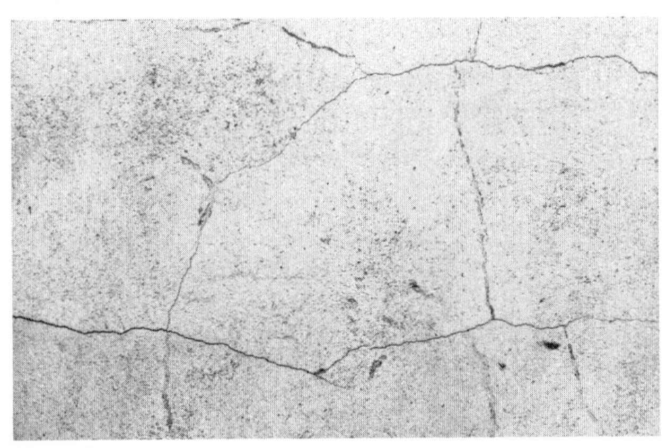

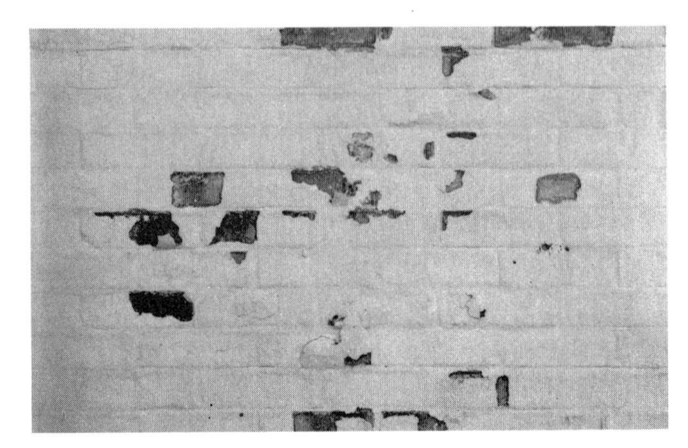

Figure 1

Many surfaces of concrete, block and brick construction, especially when rendered, are painted. Brickwork and stonework are not usually designed to be painted, but are often coated to improve appearance, to hide surface blemishes, to cover graffiti, to reduce rain penetration*, or because they have been painted previously. Painting is sometimes specified to cover a disintegrating surface. Although ***paint*** may hide hairline cracks or other blemishes it will ***not*** be ***durable unless*** the ***surface*** below is ***sound.*** Wide cracks will need special treatment. Mortar joints frequently show the first signs of a breakdown.

Painting to reduce rain penetration is not a certain cure. ***Moisture*** in the substrate ***may increase*** after painting. If rain enters through joints, faulty details, cracks or pinholes, the ***paint may severely restrict*** the ***drying-out*** of masonry. This increases the ***risk of frost damage*** leading to ***spalling,*** particularly with some bricks[1]. Such damage may require to be dealt with first. Ideally a coating should resist rain but allow the masonry to 'breathe' and dry out. It needs high water resistance combined with high vapour permeance. There are some paints, eg cement paints, emulsions and some pigmented solvent-borne types formulated in this way.

*This *Defect Action Sheet* does not include water-repellent or non decorative waterproofing treatments.

Once painted most masonry is very difficult to strip because of its porosity and surface roughness. Solvent removers are usually slow, expensive and inefficient. Mechanical abrasion (grinding or blasting) is rather more effective, especially on larger areas, but may damage the surfaces. It is difficult to remove coatings completely and restore the brickwork or stonework to its original appearance. A decision to paint previously unpainted masonry will be a commitment to regular repainting.

PREVENTION

Principle — Substrate (ie wall, including previous paints or coatings) must be suitably sound for painting and properly prepared; the appropriate paint system must be specified; it must be applied following manufacturer's recommendations and under suitable conditions[2].

Practice

- Specify masonry to be thoroughly cleaned removing all loose material. If there is paint with poor adhesion it should be removed down to a sound surface:
  - — any mould, lichen or algal growth should also be removed and the affected areas treated with an approved masonry biocide[3,4];
  - — graffiti or other contamination which may affect subsequent paint, must be removed or effectively sealed. Specialist advice may be required in some cases.

- Specify that the masonry be examined for cracks or other defects[5,6,7], efflorescence[8], dampness, etc:
  — hollow rendering, large cracks, failed pointing, etc, should be made good using a mortar mix similar to the original mortar[9];
  — efflorescence should be brushed off and the source of dampness, which is bringing the salts to the surface, identified.

- Specify paint and primer (if needed) with regard to the masonry material, alkalinity, porosity, or type of paint used previously[10]. If unsure seek expert advice. See Table 1.

- Specify coating system, sequence and number of coats, coverage, etc, according to the manufacturer's instructions:
  — broad cracks will need to be bridged with reinforcing tape or scrim;
  — due regard should be taken of the masonry surface[10], weather and site conditions before and during painting, as recommended in the Code of Practice[2].

REFERENCES AND FURTHER READING

1 *BRE Information Paper* IP22/79 'Difficulties in painting Fletton bricks'.

2 *British Standard* BS 6150: 1982 'Code of Practice for Painting of Buildings'.

3 **Ministry of Agriculture Fisheries and Food and also Health and Safety Executive** *Pesticides 1988: Reference Book 500,* 'Pesticides approved under the Control of Pesticides Regulations' 1986, HMSO.

4 *BRE Digest* 139 'Control of lichens, moulds and similar growths'.

5 *Defect Action Sheet* DAS 18 'External masonry walls: vertical joints for thermal and moisture movements.

6 *Defect Action Sheets* 122 and 123 'Windows and doors: reconstituted stone non-structural components; 'plastic' repair using Portland cement mortar'.

7 *BRE Digests* 263, 264 and 265: The durability of steel in concrete:- Parts 1, 2 and 3.

8 *British Standard* BS 6270 'Code of Practice for Cleaning and Repair of Buildings'.

9 *Defect Action Sheet* DAS 70, 71 and 72 'External masonry walls: eroding mortars — repoint or rebuild?; repointing specification; repointing'.

10 *BRE Digest* 197 and 198 'Painting walls: Part 1 — Choice of paint; and Part 2 — Failures and remedies'.

Table 1 Selection of paint types

APPEARANCE	EXTERIOR FINISH TYPES	ALKALI RESISTANCE	WATER RESISTANCE	WATER VAPOUR PERMEANCE	DURABILITY	SUITABILITY FOR: BRICKWORK	RENDER	CONCRETE	BLOCKWORK	NORMAL APPLICATION	REPAINT TYPES (#)	COMMENTS
Glossy, Semi-gloss, or Smooth	1 Alkyds, Oil paints	L	M	M	L-M	(a)M	(a)G	(a)M	(a)M	B	No. 1,2,4 6,7,8 (x)	**Compatible filler and/or sealer may be needed for finish types 1 – 4**
	2 Vinyl, or Acrylic Copolymer solutions	G	H	M	G	M	H	G	G	B/S	No. 1,2,4, 6,7,8,11 (x)	
	3 Polyurethanes	H	H	L	H	M	H	H	G	B	No.1,2,3,4 5,6,7,8 (x)	1- and 2-pack products
	4 Acrylic or Copolymer Emulsions	G	G	G	G-H(d)	G	G	G	G	B/R	No. 4,8,12	Protect from frost in can and during painting
Irregular gloss, Matt, or Fine-textured ('Masonry or Stone-paint')	5 Polyurethanes or Epoxy-polyurethanes	H	H	L	H	M	H	H	H	B	No. 1,2,3,4 5,6,7,8,11	1— and 2-pack products
	6 Acrylic or Co-polymer solutions	G	H	M	G-H	M	H	H	G	B/S	No. 1,2,4 6,7,8,11	
	7 Chlorinated rubber solutions	G	H	M	G-H	M	H	H	G	S/B	No. 2,4,6 7,8	
	8 Acrylic or Co-polymer Emulsions	G	M	H	G-H(d)	G	H	G	H	B/R	No. 4,8,12	Protect from frost in can and during painting
	9 Cement paints	G	G	H	L-M(d)	G	H	G	L-M	B	No. 4,8,9	Tyrolean textured coatings available
Medium or Heavy-textured	10 Alkyds or Oil paints (filled)	L	G	G	H(d)	(a)M	(a)H	(a)H	(a)H	S	No. 6,8 10,12	**Heavy coatings for finish types 10 – 12 need stable substrates**
	11 Acrylic or Co-polymer solutions	G	G	G	G(d)	M	G	G	H	B/S	No. 4,6,8 11,12	
	12 Acrylic or Co-polymer Emulsions	G	M	H	G-H(d)	G	G	G	H	R/S	No. 8,12	Protect from frost in can and during painting

KEY

Resistance/Permeance/Suitability/Durability

H	High	H over 10 years
G	Good	G 5 - 10 years
M	Moderate	M 4 - 7 years
L	Low	L less than 5 years

Application

B Brush
S Spay
R Roller

Notes
(a) Alkali-resistant primer or pretreatment needed
(d) May become dirty
(x) Hard glossy surfaces may need abrasion for adhesion of new paint
(#) Trial patch advisable to check satisfactory application and drying of new treatment

DAS 137
November 1989

Site
CI/SfB 8 (22)S g2 (D6)

Internal walls: ceramic wall tiles — loss of adhesion

FAILURE **Loss of adhesion: tiles to adhesive, adhesive to paint/plaster/render, paint/plaster/render to background. Plaster or plasterboard deteriorates when wetted, tiling subsequently bulges or becomes detached.**

DEFECTS: **Unsuitable adhesive, adhesive incorrectly spread, tiling insufficiently tamped, fixing delay, surface incorrectly prepared or primed, substrate bond poor, lack of movement joints in large tiled areas.**

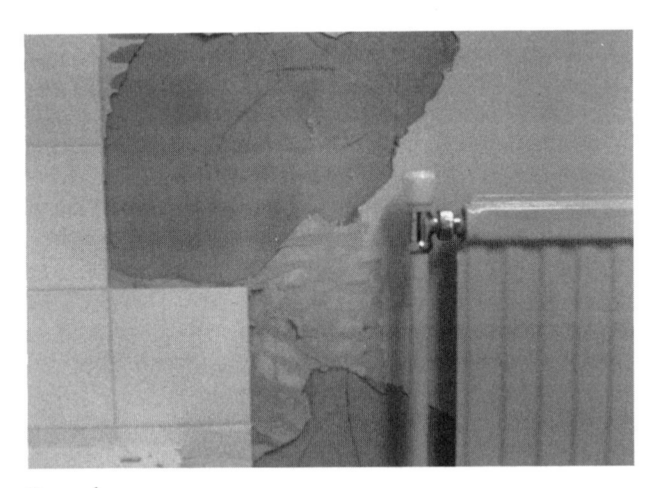

Figure 1

Problems continue to arise due to both short term and long term failure of the bond of tiling to its background or adhesive. Almost every failure of adhesion is the result of movement breaking the bond: either (i) the bond is unduly weak and so is broken by normal small thermal and other movements and/or (ii) excessive movement takes place, sufficient to disrupt what would otherwise have been a satisfactory bond.

The differential thermal and moisture movements associated with the background masonry, render and/or plaster substrate and tiles can disrupt or break an inadequate bond within the tiling system. Ceramic tiling expands rapidly immediately after manufacture but can continue to expand very slowly for many years. This long term expansion may cause failure of the bond after many years of satisfactory service.

If the ***relative movements*** of the ***tiling*** and ***background*** are not taken into account, particularly in large tiled areas, or if the ***adhesive*** is ***not suitable*** for the ***background*** both as a material and in thickness, or the adhesion between each layer of construction is poor the ***bond may fail*** and the ***tiles*** may ***bulge.***

PREVENTION

Principle — Effective adhesion of tiling requires: good adhesion of tiles to substrate; use of suitable adhesive for substrate and conditions; backing and substrate to be sound and dry; provision of movement joints in large areas of tiling.

Practice

Note: Today nearly all tiling in domestic situations is fixed by proprietary thin bed or thick bed adhesives.

- Check that the background texture is suitable for the adhesive. An intermediate substrate of cement/sand render or plaster may be required to provide a true surface with adequate suction and key.
  - cement/sand render (not richer than 1:3 cement/sand by volume) using a sand to BS 1199 provides *the preferred* substrate for fixing tiles. The render should not, however, be stronger than the background.
  - plasterwork should be carried out in accordance with BS 5492:1977 and excessive trowelling which might give a dusty surface should be avoided. The plaster may require a special primer before applying the adhesive.
- Check the trueness of the surface to be tiled using a 2m straight edge (Figure 2):
  - any gaps under the straight edge should not exceed 3mm for the 'thin-bed' adhesives;
  - any gaps under the straight edge should not exceed 6mm for the 'thick-bed' adhesives;
  - local thickening is possible with some adhesives to build up specific hollows (refer to the manufacturers specification) otherwise the surface will need to be rendered.
- Ensure that the background surface finish and its condition are compatible with the adhesive:
  - organic based adhesives, either water or solvent based, are suitable for use on most backgrounds but can be affected by damp conditions. Solvent based adhesives may react with the background (eg paint);

— cement based adhesives are suitable for use on cement rendered surfaces, concrete and brickwork. Specially formulated 'thin-bed' cement based adhesives are also available for use on suitably primed plaster surfaces. The background and tiles should be dry.

Note: Tables 2 and 3 BS 5385:1976:Part 1[1] give more detailed information on choice of adhesive and preparation of backgrounds. Adhesives are continually being developed; for example, epoxy resins. The manufacturers' instructions should be followed.

- Check and stiffen up the background, if necessary, to give sufficient rigidity and strength to support the tiles:
  — stud walls may be stiffened by using two layers of plasterboard or where damp conditions are likely (eg shower areas) the use of a water resistant plywood or rendering on expanded metal lathing should be considered.
- Check that the background to which the tiles are to be bonded has dried out sufficiently to keep shrinkage to a minimum.
  At least:
  — 14 days should be allowed for a rendered finish;
  — 28 days should be allowed for a newly constructed wall;
  — 28 days should be allowed for a newly plastered finish.
- Ensure adhesive is applied correctly and in accordance with manufacturer's recommendations:
  — enough adhesive needs to be adequately spread;
  — tiles should be pressed firmly into place;
  — the chosen adhesive will have a limited useful life and should be discarded if it has stiffened due to poor storage or prolonged use;
  — treat very absorbent plaster surfaces with an appropriate primer;
  — the area of adhesive should be limited to the working time available for the prepared adhesive, which may be only 15 – 20 minutes.
- Ensure that substrate is integral and suitable for the conditions to which it will be subjected:
  — avoid use of gypsum plasters where repeated or persistent wetting may occur;
  — avoid use of gypsum plasters where repeated or persistent heating occurs (flues, heaters etc);
  — check that there is adequate adhesion between the gypsum plaster skim coat and cement/sand undercoat.
- Ensure provision is made to accommodate differential thermal and moisture movements in the background, substrate and tiling:
  — movement joints should be built into the tiling at centres not exceeding 4.5m both vertically and horizontally and also at all vertical corners in large tiled areas. The movement joint (Figure 3) should be a minimum of 6mm wide, and should extend through the tile and bedding material;
  — check that tiles without spacer lugs or without universal edges are never fixed with a butt-joint as an adequate width of joint is necessary for the relief of any local stress. When universal or spacer lug tiles are used, correct spacing is automatically provided between tiles;
  — where substantial movement is expected the joint should additionally extend through the plaster or render (Figure 4);
  — joints may be needed where different substrate materials abut or where tiles abut a different material;
  — ensure that no movement/structural joint has been bridged, (Figure 4) and that the movement joint sealant only adheres to the tile and not to the background, as the ability to absorb movement may be seriously impaired (Figures 3 and 4);
  — any grouting materials used should be compressible.
- Check that the movement joint sealant is non-rigid:
  — silicone has a very good water resistance and is suitable for internal tiling;
  — sealants as above are available with additives to inhibit mould growth, and should be used in bathrooms and shower areas.
- Ensure that the adhesive has adequate time to set before grouting. This should be at least 24 hours and preferably 72 hours. Longer periods may result in the joints being filled with dust and deleterious material:
  — grouting materials resistant to cleaning agents, mould growths, or heat, may be needed in some locations.

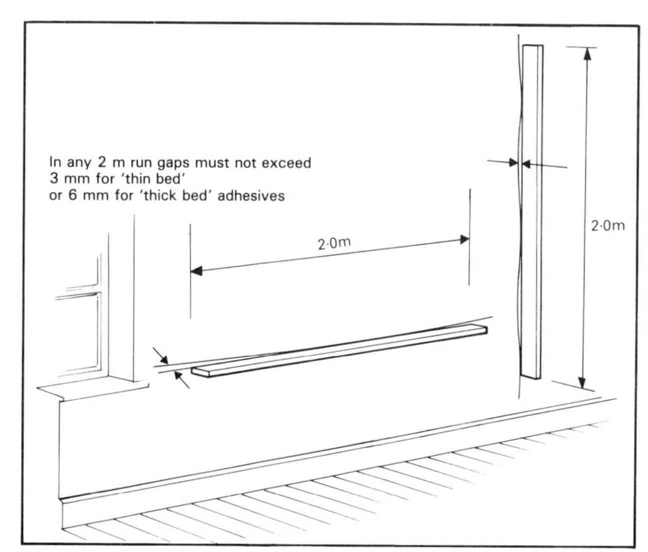

Figure 2

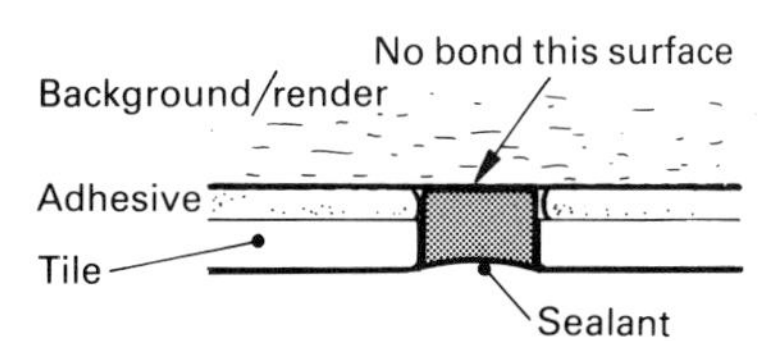

Figure 3 Typical movement joint

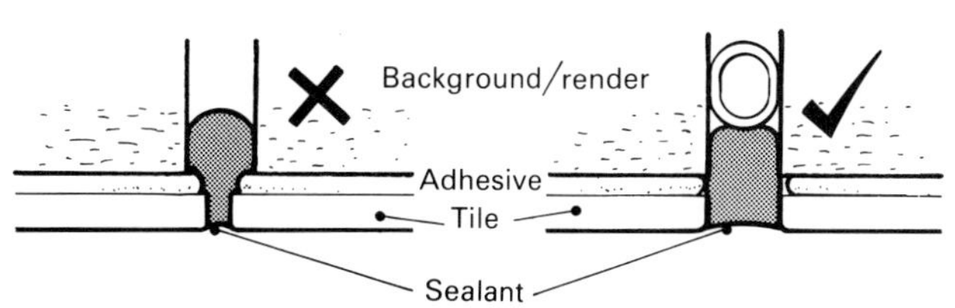

Figure 4 Deep movement/structural joint

REFERENCES AND FURTHER READING

1 **British Standards** BS 5385:Part1:1976: Code of practice for internal ceramic wall tiling and mosaics in normal conditions.

PART SIX

Walls 3: Thermal Insulation, Condensation and Moisture

External walls: reducing the risk from interstitial condensation

FAILURES: **Rot in timber; corrosion of steel framing, reinforcement, fixings; wetting of insulation**

DEFECTS: **House poorly ventilated; vapour checks omitted, poorly sealed, wrongly sited; walls not ventilated or drained**

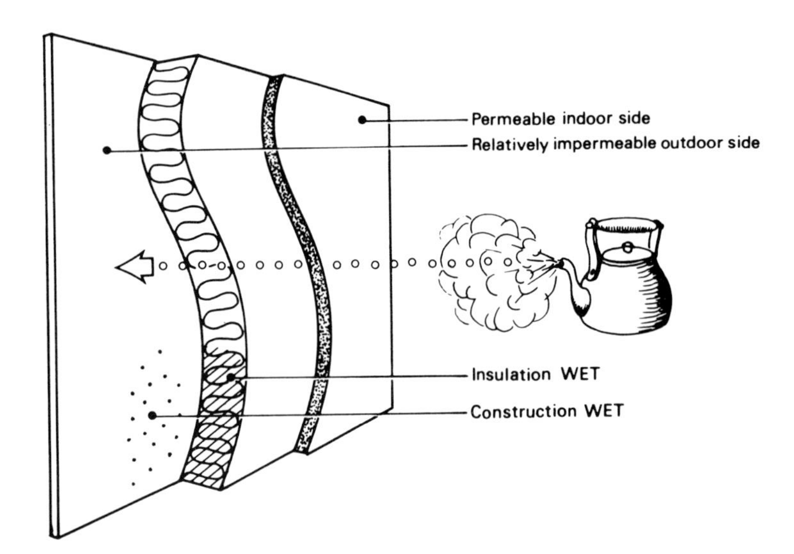

Figure 1
Interstitial condensation

Condensation is likely to occur within the thickness of most types of external wall under some conditions (interstitial condensation Figure 1). In traditional masonry construction it may not matter, but some other types of wall construction may be adversely affected. Appropriate design can minimise the risk of damage, but BRE surveys have shown that proper precautions are not always taken.

Activities in houses, particularly in kitchens and bathrooms, add large amounts of ***water vapour*** to that already present in incoming air. Much of this water vapour can, and should, be removed near its source by ventilation or extraction, but some is likely to find its way ***into the structure of the external walls;*** it can permeate through porous materials such as plasterboard, or be carried by air movements through cracks and holes. Even if a vapour check has been provided the practical difficulties of installation make it likely that some vapour will leak into the wall.

If vapour, from indoors, ***reaches a relatively cold and impermeable layer*** of, say, cladding or sheathing, or a wrongly sited vapour barrier, it is likely to ***condense*** unless there is good ventilation at that point. Water which condenses may drain harmlessly away or, if it soaks into porous materials, be subsequently dried out by ventilation without causing any damage. ***However, if no such drainage or ventilation has been provided and the water accumulates in or around timber or metal components, rot or corrosion may occur.*** Insulation also may become wet, reducing its effectiveness even to the point where condensation and mould growth appear on the indoor surface of the wall.

PREVENTION

Principles — Restrict the entry of water vapour from the house to the external walls to levels which can be adequately dispersed from the walls by diffusion, ventilation and drainage.

Practice

- Provide for good, controllable ventilation to the house and particularly to kitchens and bathrooms (Figure 2).
- Whenever possible, design external walls having materials of increasing permeability towards the outside (Figure 3).
- Provide a vapour check on the warm side of the insulation for all external walls containing materials liable to be adversely affected by damp (Figure 4). No vapour check is required for traditional masonry cavity walls.
- Do not rely on a vapour check to prevent interstitial condensation if there is another relatively impermeable layer in the wall, (for example, a fairly impermeable cladding material). Provide in addition ventilation and drainage from its inside face to the outside (Figures 5 and 6). If the construction materials can absorb water, ventilation is more important than drainage.
- If a water-shedding layer is needed behind cladding such as vertical tiling, use a breather paper to BS 4016:1972[1]. Do not use polythene or building paper.
- The greater the risk of interstitial condensation, and the greater the difficulty in providing ventilation or drainage where condensation may occur, the greater the efforts that should be made to ensure that the vapour check is as effective as possible. Pay particular attention to maintaining the continuity of the vapour check at joints and at difficult details such as the junctions between elements of the building, and where services penetrate. Try to locate services so that they do not penetrate the vapour check.

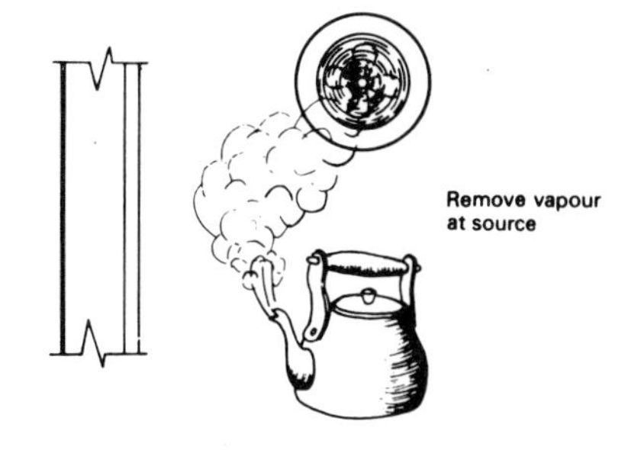

Figure 2

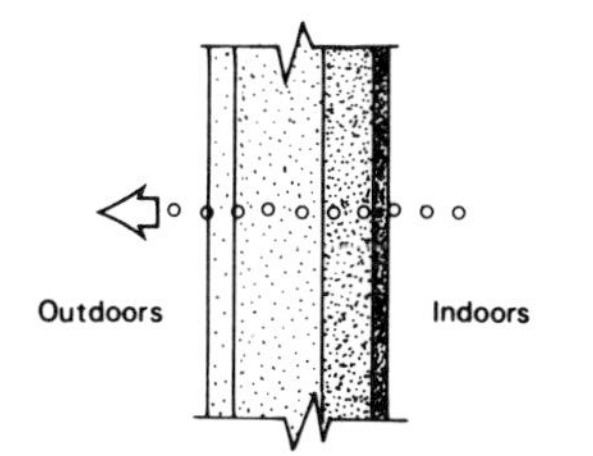

Figure 3

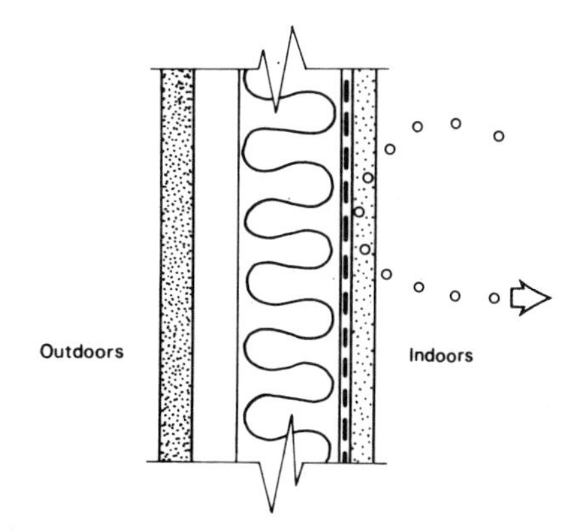

Figure 4

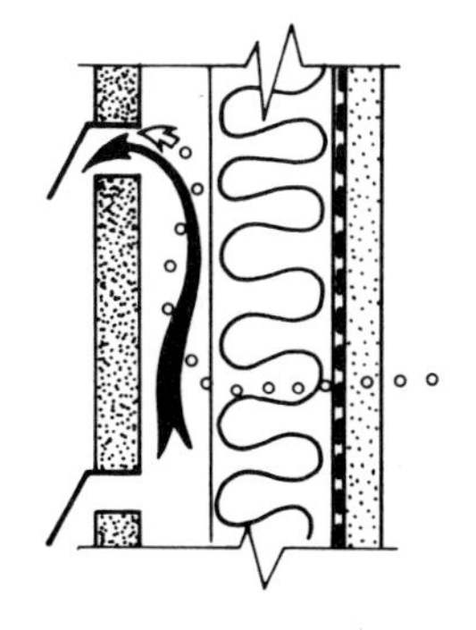

Figure 5

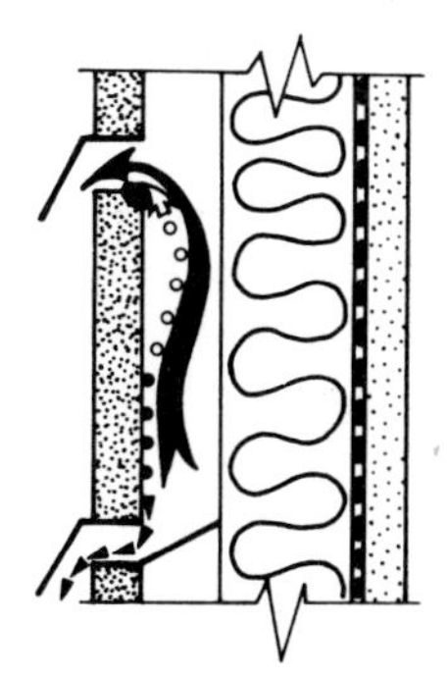

Figure 6

REFERENCES AND FURTHER READING

1 *British Standard* 4016:1972 'Building papers (breather type)'.

British Standard 5250:1975 'Code of basic data for the design of buildings: the control of condensation in dwellings'.

BRE Information Paper IP 1/81 'Vapour diffusion through timber framed walls'. Obtainable from Distribution Unit, Building Research Establishment, Garston, Watford WD2 7JR.

Cavity external walls: cold bridges around windows and doors

FAILURE: Condensation, staining and mould growth on walls around windows and doors

DEFECTS: Insufficient insulation; discontinuities in insulation ('cold bridges')

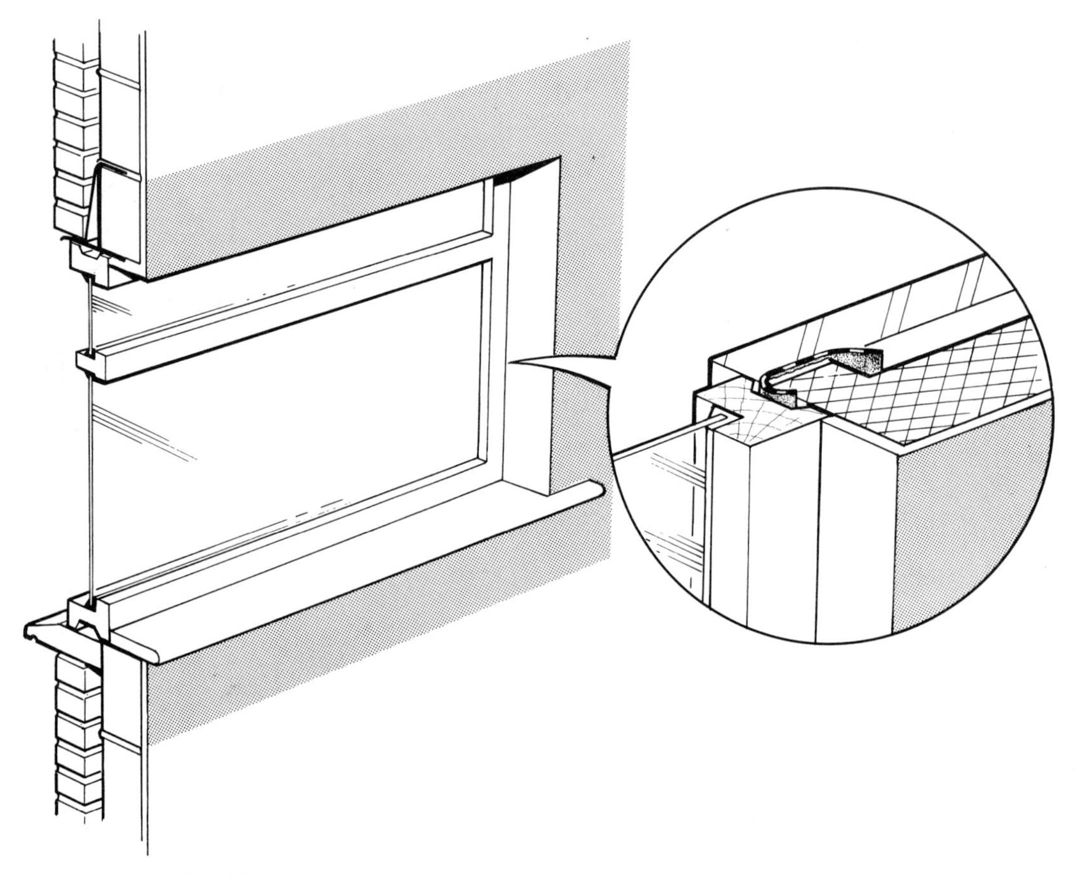

Figure 1 The vulnerable areas

Condensation, with consequential staining and possible mould growth, often occurs on window and door reveals and heads, and under window sills. 'Cold bridges' can occur at openings even when the U-value of the surrounding wall is better than the 0.6 W/m²K permitted by the Building Regulations 1985 (England and Wales). Lintels, jambs and sills may be regarded as part of the opening when calculating the average U-value of the wall and thus the reveals of the opening may have high U-values locally*. The thermal resistance at these locations can be improved and risk of localised condensation reduced by adding even a small amount of insulation at the indoor surface.

If the ***wall around an opening permits*** substantially ***greater heat flow*** than elsewhere, its indoor ***surface temperature*** may locally ***fall below the dew point,*** leading to ***condensation, staining and mould growth.***

*For England and Wales, Approved Document L 2/3 of the Building Regulations 1985 states that 'in some circumstances it may .... be desirable to limit the U-value .... to 1.2 W/m²K. Building Regulations for Scotland and Northern Ireland specifically limit the U-value for these locations to 1.2 W/m²K.

PREVENTION

Principle — Walls at lintels, jambs and sills should be thermally insulated sufficiently to minimise the risk of local surface condensation.

Practice

- Design walls to provide an overall U-value of 0.6 or better, treating lintels, jambs and sills as though part of each opening in a wall:
  - provide 10 mm of insulation near the indoor surface, when detailing reveals at lintels, jambs and sills; when heating is intermittent, surface condensation risk is reduced if insulation is near the indoor face.

OR

- Check whether reveals, head and window sill will provide a total resistance not less than 0.65 m^2K/W (which value, together with surface resistances, gives a U-value of 1.2):
  - using Table 1, sum the resistances of the construction lying in the heat loss path, Figures 2 and 3 (ignoring surface resistances). If the sum of resistances is less than 0.65 (ie there is a 'resistance deficit') additional insulation will be needed to improve the U-value to 1.2. Table 2 gives appropriate thicknesses.

EXAMPLE 1 — heat loss at the reveal

Use Table 1 to sum the resistances of the materials in the heat loss path shown in Figure 2.

	Resistance
12 mm plasterboard	0.08
Inner leaf block (density 1100 kg/m^3), for 50 mm path length .... 50/100 × 0.29 =	0.15
Outer leaf brick	0.14
Sum of resistances	0.37
Total resistance needed for U-value of 1.2	0.65
'Resistance deficit' .... 0.65 - 0.37 =	0.28

Approximately 11 mm of EPS or 8 mm of polyurethane board (refer to Table 2) would be needed to raise the total resistance of the construction of the reveal to 0.65 and give a U-value of 1.2

EXAMPLE 2 — heat loss at window head

Sum the resistances as in Example 1: Figure 3 gives the heat loss path.

	Resistance
12 mm plasterboard	0.08
Inner leaf block (density 1100 kg/m^3), for 25 mm path length .... 25/100 × 0.29 =	0.07
Air space	0.18
Outer leaf brick	0.14
Sum of resistances	0.47
Total resistance needed for U-value of 1.2	0.65
'Resistance deficit' .... 0.65 - 0.47 =	0.18

Approximately 7 mm of EPS or 5 mm of polyurethane board (refer to Table 2) would be needed to raise the total resistance of the construction of the reveal to 0.65 and give a U-value of 1.2

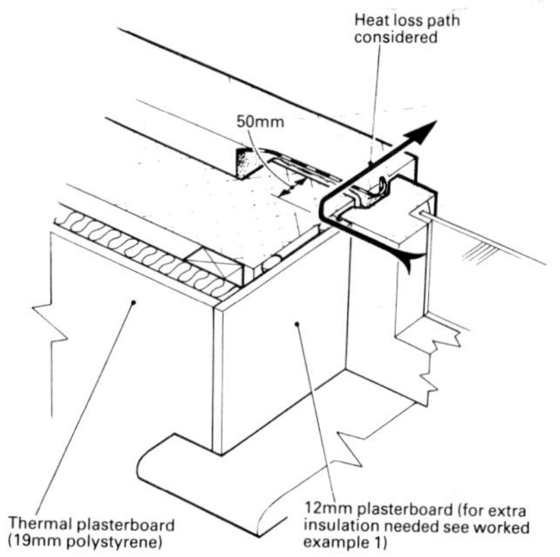

Figure 2

Heat loss path considered
25mm
For extra insulation needed see worked example 2

Figure 3

Table 1

Materials/construction	kg/m³	W/m K	Resistance m^2 K/W (for thickness given in col 1)
Outer leaf brick (105 mm thick, 1600 kg/m^3)			0.14
Inner leaf block (per 100 mm thickness)			
High density	1700	0.76	0.13
	1400	0.51	0.20
	1100	0.34	0.29
Lightweight	800	0.22	0.46
	600	0.19	0.53
Ultra lightweight	400	0.15	0.67
Dense concrete (2200 kg/m^3) per 100 mm			0.06
Cavity or airspace (not less than 25 mm)			0.18
Wood (per 25 mm)			0.18
Plaster, lightweight (19 mm thick)			0.04
Plasterboard: 12 mm thick			0.08
Plasterboard: 19 mm thick			0.12
Mineral fibre batts 25 mm thick			0.69
U/F foam 50 mm thick			1.39

Table 2

Resistance deficit ≤	Minimum additional insulation thickness (mm): EPS or mineral fibre slab	Minimum additional insulation thickness (mm): Polyurethane board
0.1	3	3
0.2	7	5
0.3	11	8
0.4	14	10
0.5	18	13
0.6	21	16
0.65	23	17

External walls — dry lining: avoiding cold bridges

FAILURE: **Localised condensation and mould growth**

DEFECT: **Avoidable discontinuities in applied internal insulation**

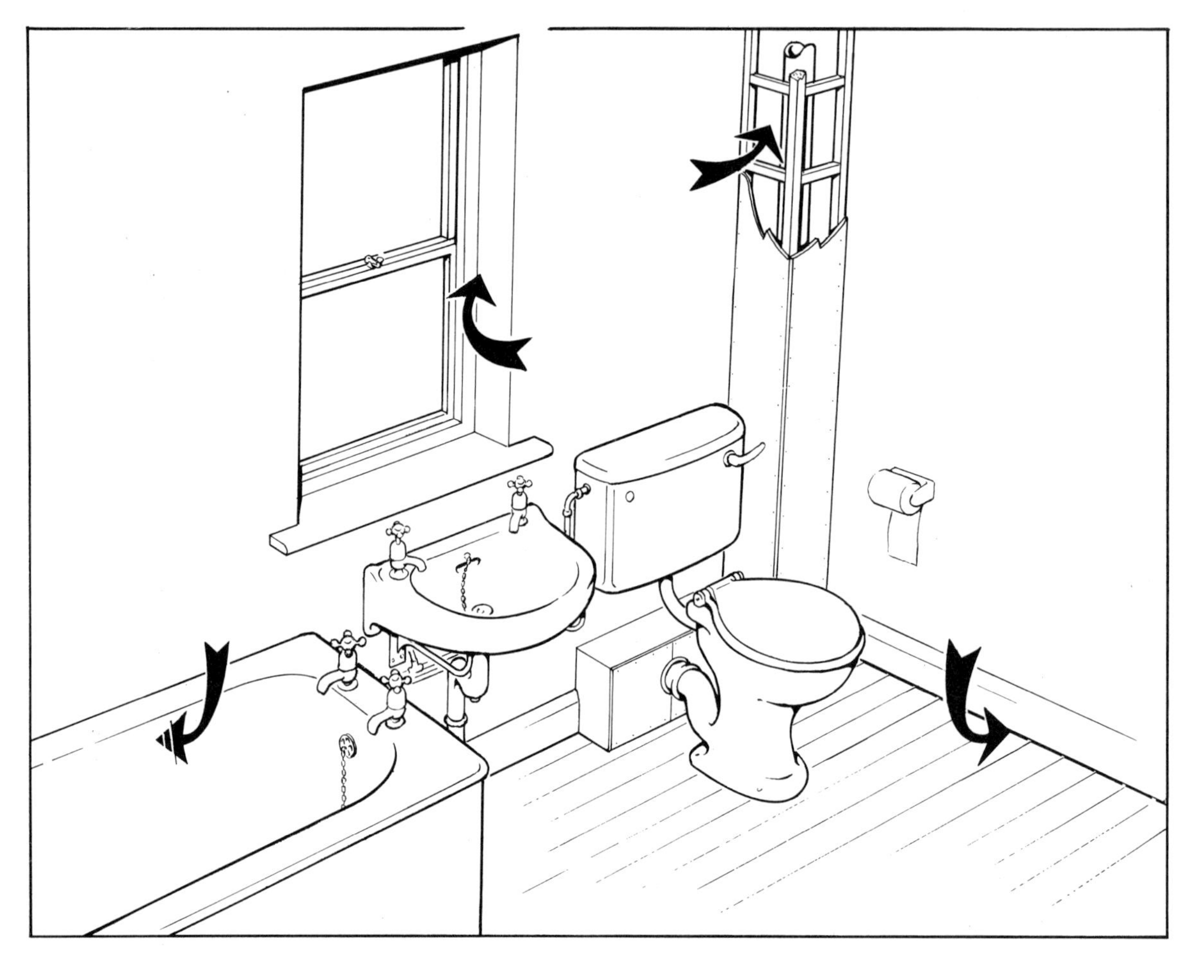

Figure 1 Places where cold bridges may remain after dry lining

In rehabilitation work, insulating dry lining is sometimes used to improve thermal resistance, particularly of solid external walls including walls in large concrete panel systems. Infra-red thermography has shown that cold bridges can remain if the insulation is not continuous at edges, corners and salient features such as beams and columns. The effect of such discontinuities then becomes especially marked, because surface temperatures at such points will be significantly lower than elsewhere; pattern staining, condensation and mould may result.

If applied ***insulation*** does ***not extend far enough*** or is ***not continuous*** at edges, corners and over salient features, ***local paths of high heat loss*** may remain with risk of pattern staining, ***condensation*** and ***mould growth.***

PREVENTION:
Principle — All heat loss paths should have sufficient thermal resistance to minimise the risk of local surface condensation.

Practice

- Identify features which are possible heat loss paths, eg:
  - — junction of external walls with internal walls, floors, columns, beams and balconies;
  - — windows and external door perimeters;
  - — external wall areas 'concealed' within cupboards, below stair flights etc;
  - — walls between accommodation and common access or other unheated areas such as lift shafts or refuse shutes.
- Check by calculation (see DAS 77) that no 'cold bridges' will remain after insulating those areas. Extend insulation as far as is necessary to achieve sufficient thermal resistance in all heat loss paths; even a long heat loss path, Figure 2, through uninsulated construction may have a relatively low resistance.
- Design all edge and corner details to eliminate discontinuities in insulation at fixings, Figure 3, eg use adhesives at such points:
  - — it may be necessary to specify that plaster be hacked from reveals to accommodate insulation without oversailing window frames.
- Use insulation laminates incorporating a vapour check or with good intrinsic resistance to vapour passage. Remember that, where a separate vapour check is provided, it must be on the warm side of insulation. In the absence of an effective vapour barrier condensation may occur within the insulation or on, for example, dense or gloss-painted substrates.
- Locate socket outlets elsewhere than on external walls or specify surface mounted boxes.
- Remember that cables running behind applied insulation may need to be derated[1].

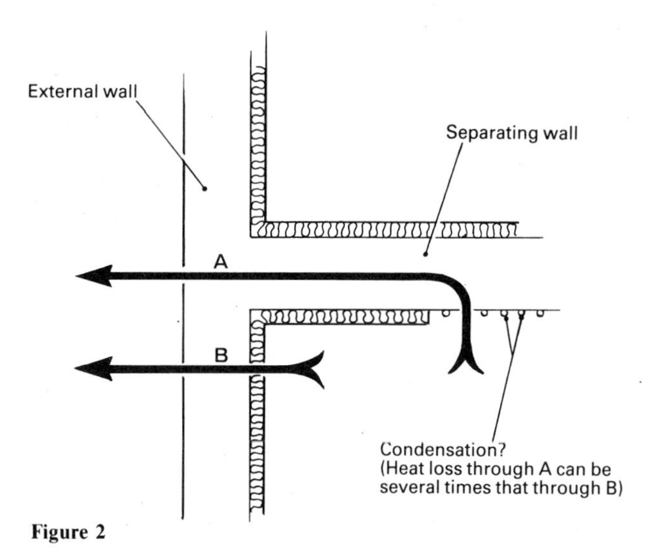

Figure 2

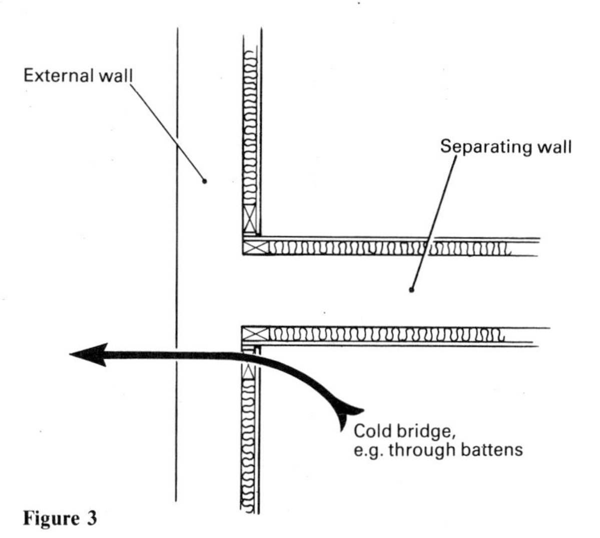

Figure 3

REFERENCES AND FURTHER READING
1 *Defect Action Sheet* DAS 62.

Solid external walls: internal dry-lining – preventing summer condensation

FAILURE: **Condensation on back of vapour control layer, staining of internal finishes, rot in studs or joists, wetting of insulation**

DEFECTS: **No external protection; no ventilation to outside from behind dry-lining**

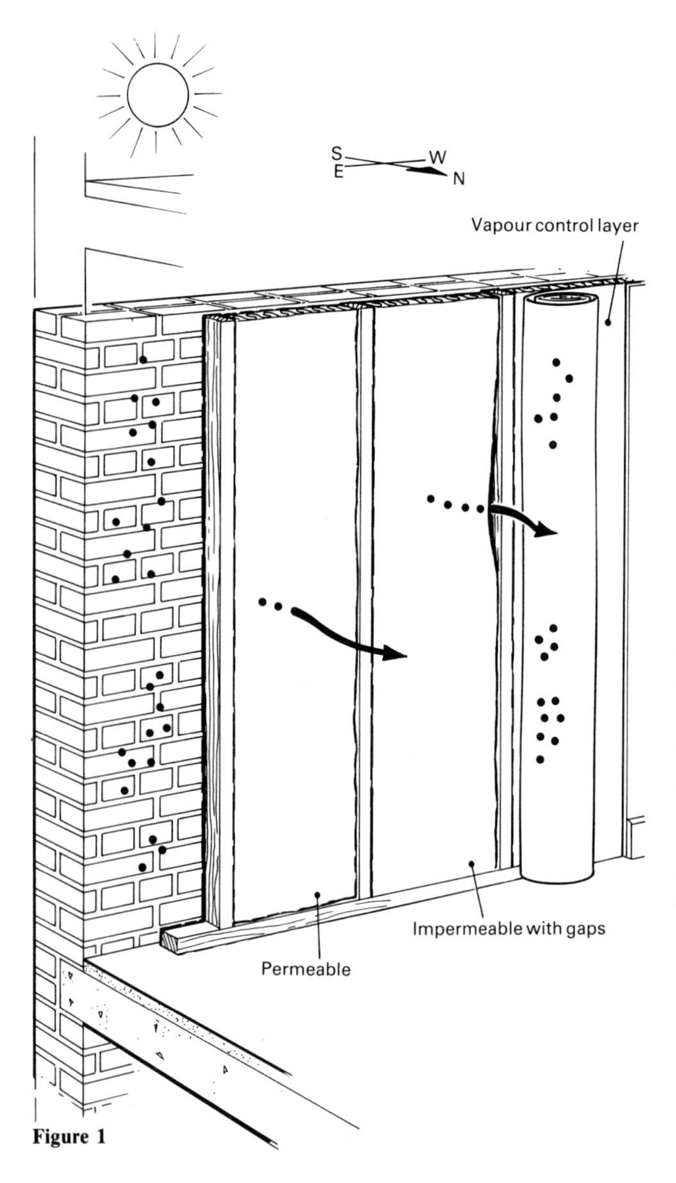

Figure 1

During cold weather interstitial condensation occurring within internally insulated solid walls is normally prevented by a vapour control layer on the warm (internal) side of the insulation. However, in summer, strong sun on unprotected walls can drive moisture towards the inside of the building and through permeable insulation or through gaps between less permeable insulation, to condense on the outside face of the vapour control layer. Moisture contents of the external wall need not be high for this failure to occur. Once condensation has occurred it stays for many days, as transfer out is much slower than transfer in. The condensation can be prevented by internal heating, but this is not an option for the summer. Since the condensation occurs behind a waterproof vapour control layer it is often not noticed.

Strong ***sun on*** damp solid external ***walls*** can cause ***condensation*** to occur, which can trickle down the vapour control layer, and ***wet*** timber.

Solid external walls: internal dry-lining — preventing summer condensation

DAS 133
June 1989

PREVENTION

Principle — The moisture must be reduced to a level at which condensation will not occur, by external protection to the wall or by ventilation.

Practice

- Assess risk before deciding to insulate thermally solid external walls on the inside;
  — external insulation avoids this type of failure.

- Decide which walls are at risk, Figure 2; ie, walls facing East South East to West South West.
  EITHER
- Specify shading for walls exposed to direct sun or specify covering with tile hanging or other claddings;
  OR
- Specify a cavity ventilated to the outside, Figure 3;
  — specify for solid masonry walls, vertical joints to be raked out through the full thickness of the wall at top and bottom of each storey height, or specify slot ventilators angled to shed water outwards;
  — horizontal spacing not to exceed 1.5 metre, Figure 4;
  — specify for existing walls of other materials, ventilation holes to give an equivalent open area of 500 mm² per metre run.

Note: Specifying the omission of a vapour control layer is not an acceptable solution. Summer condensation could be deposited behind low permeability internal finishes.

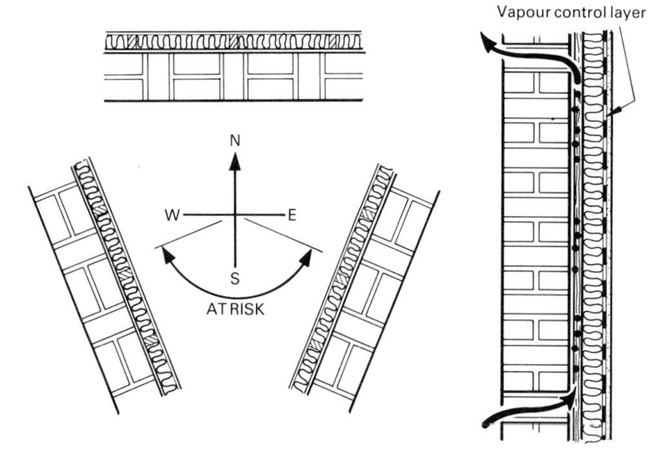

Figure 2

Figure 3

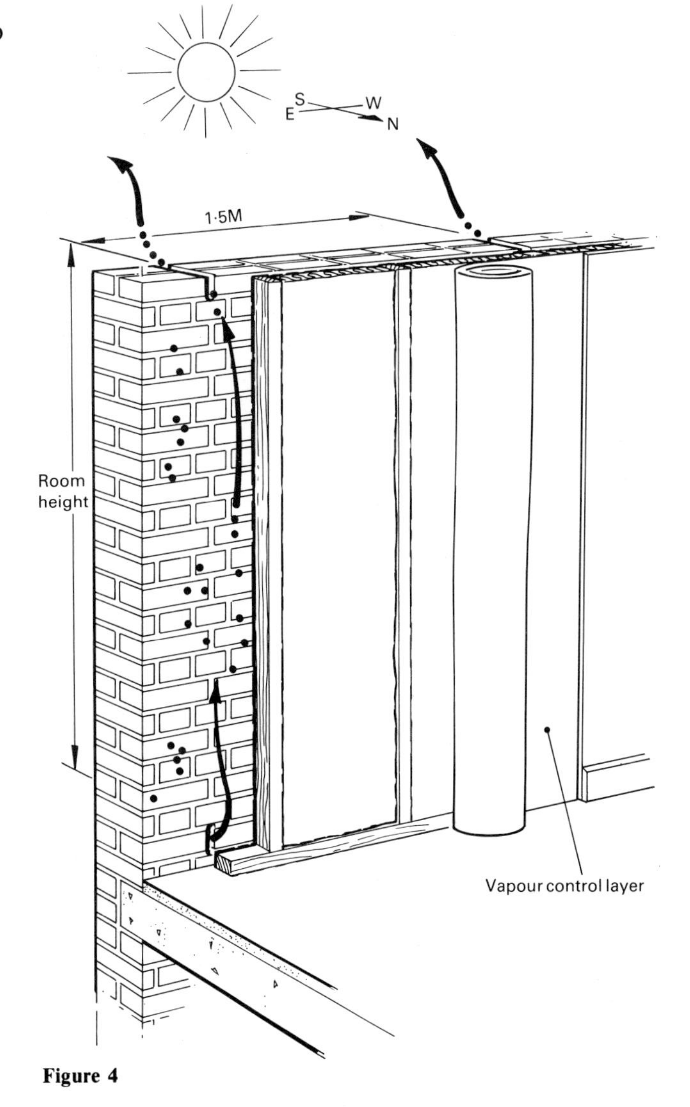

Figure 4

REFERENCES AND FURTHER READING

☐ **Southern J R.** Condensation on the vapour check of battened internally insulated solid walls. *BRE Information Paper* IP 12/88, 1988.

☐ **Building Research Establishment** External walls : reducing the risk from interstitial condensation. BRE *Defect Action Sheet* DAS 6. 1982.

External masonry walls insulated with mineral fibre cavity-width batts: resisting rain penetration

FAILURE: **Rain penetration of walls thermally insulated with cavity-width mineral fibre batts**

DEFECTS: **Mortar in horizontal joints between batts; 'make-up' pieces installed with laminations in the wrong direction**

Figure 1 Mortar trapped between batts

A maximum U-value of 0.6 has been required in external walls of dwellings[1] since April 1982*. One means of achieving this in masonry walls is to build in mineral fibre insulation batts** that completely fill the cavity. With this technique there is no longer a clear cavity to act as a defence against rain penetration. The batts made for this purpose are treated to make them water repellent but, ***if there is mortar*** (either droppings or extrusions from bed joints) ***in the horizontal joints between batts rainwater can penetrate to the inner leaf.*** The technique therefore demands particular care to keep both cavity faces free from mortar extrusions and to build in the batts without trapping mortar in the joints between them.

*In Scotland, from March 1983.

**In inner London the means of achieving thermal insulation is subject to the approval of the District Surveyor.

External masonry walls insulated with mineral fibre cavity-width batts: resisting rain penetration

DAS 17
February 1983

PREVENTION

Principle — Cavities in walls insulated with cavity-width batts should contain nothing that might lead water inward.

Practice:

- Ensure that batts are never under any circumstances pushed down into cavities (Figure 2).
- Build the leaves as follows:

  — first lift: lead with the outer leaf* (Figure 3).

  — subsequent lifts (if it is not possible to continue to lead with the outer leaf, eg because external scaffolding is used): lead with the inner leaf (Figure 4); if leading with the inner leaf consider using a 1-brick trough (Figure 4) which allows the mortar bed joint at the critical level of the batt joint to be struck clean on its cavity face.
- Ensure that excess mortar, on the cavity face of the leading leaf, is cleaned off totally: use only just sufficient mortar in the bed joints of the trailing leaf, to minimise mortar extrusions into the cavity.
- Ensure that the top edge of installed batts is covered by a cavity board to protect it from mortar droppings while the next lift of the leading leaf is built.
- Ensure that ties are kept free of mortar, do not slope down to the inner leaf and have their 'drips' at the centre of the cavity.
- Ensure that insulation is dry when installed; cover partly completed walls during rain and overnight.
- Ensure that there are no gaps in the insulation.
- Ensure that any 'make up' pieces are installed with their laminations parallel to the cavity faces.
- Provide weep holes over cavity trays and at foot of wall.

*Scottish practice is to lead with the outer leaf throughout.

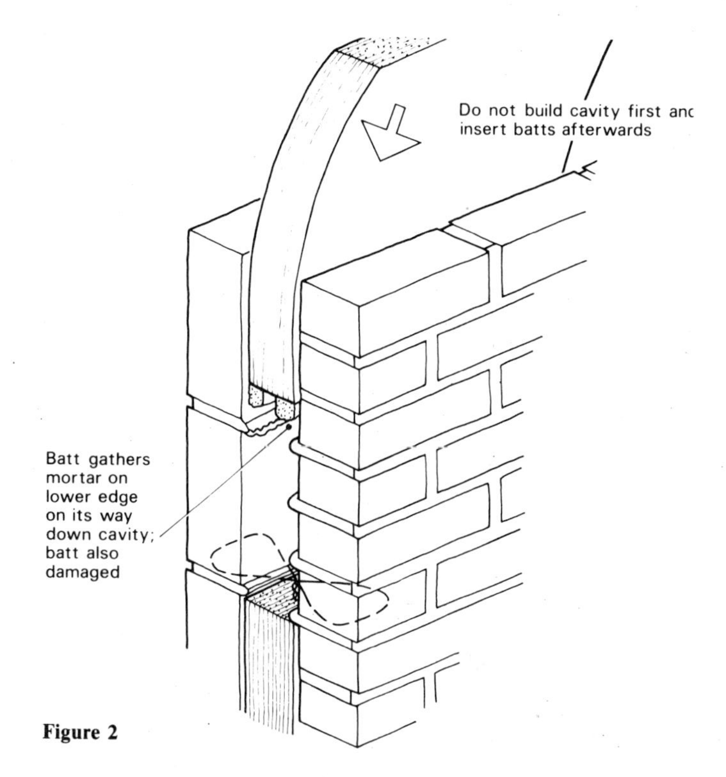

Figure 2

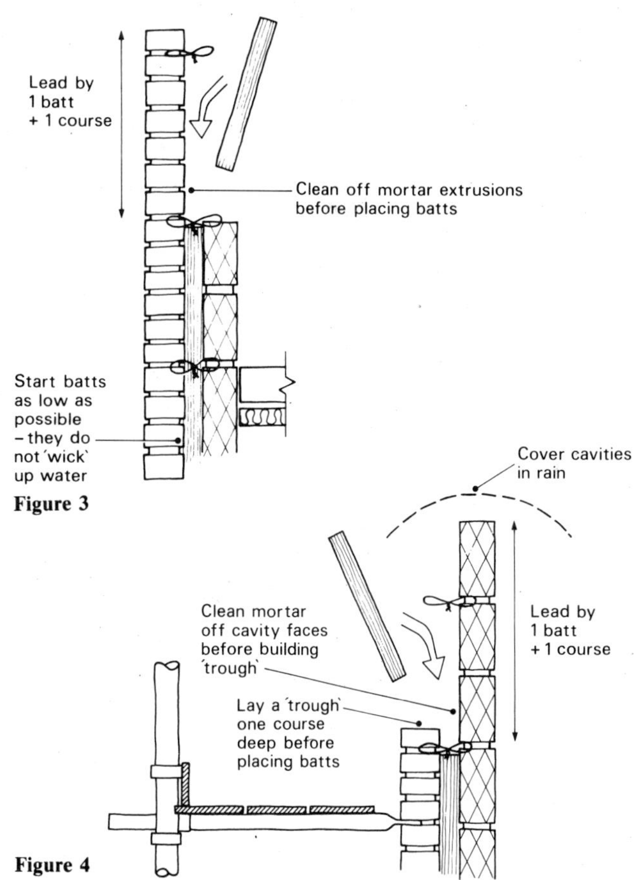

Figure 3

Figure 4

REFERENCE

1 The Building (Second Amendment) Regulations 1981.

DAS 79
June 1986

Design
CI/SfB 8 (21)R(D6)

External masonry walls: partial cavity fill insulation — resisting rain penetration

FAILURE: Rain penetration of walls

DEFECTS: Loose boards in cavity; deformed boards bridging cavity; mortar debris in cavity; wrongly installed wall ties

Figure 1 Board pushed off cavity face at corner: mortar snots and squeeze

One means of improving the thermal insulation of masonry walls is to employ partial cavity fill. In this technique rigid thermal insulation boards are secured against one of the masonry faces. Properly built, this technique maintains a clear residual cavity as a defence against rain penetration. It demands care to keep this cavity unobstructed by debris or loose boards. This does not preclude the need to render walls to reduce rain penetration in situations of 'severe' exposure where such rendering would be normal practice. Residual cavities must never be less than 25 mm and should preferably be 50 mm, particularly if insulation is against outer leaf. Many rigid insulation boards have Agrément Certificates and are approved for use in any exposure zone when the residual cavity is at least 50 mm. When this cavity is between 25 mm and 50 mm, boards are approved for use subject to certain restrictions — see the relevant Agrément Certificates.

If debris (eg mortar droppings or extrusions from bed joints, or brickbats), ***or loose*** insulation ***boards,*** are ***within*** the ***cavity,*** there is a ***risk of rainwater penetration*** to the inner leaf.

External masonry walls: partial cavity fill insulation — resisting rain penetration

DAS 79
June 1986

PREVENTION

Principle — The residual cavities in walls with partial cavity fill insulation should contain nothing that can lead water inwards.

Practice

- Ensure that boards are stored flat without bearers (Figure 2) or they may warp and be difficult to fix (Figure 3):
  - — protect expanded polystyrene boards from prolonged exposure to sunlight.
- Check, before they are used, that boards are not buckled.
- Ensure that leaves are built in this sequence:
  - — always lead with the leaf against which the insulation is to be secured; check which leaf is specified;
  - — fix the insulation flush against the cavity face of the leading leaf. (Combination wall ties: minimum 4 per board, Figure 4; nail fixings: minimum 6 per board);
  - — build up second leaf.
- Ensure that, unless otherwise specified, insulation begins below dpc level, overlapping any ground floor slab insulation — preferably by at least 150 mm:
  - — Agrément Certificates permit boards to be installed below dpc.
- Ensure that boards are **never under any circumstances** pushed down into cavities.
- Ensure that excess mortar, on the cavity face of the leading leaf, is cleaned off totally otherwise boards will not rest flush against it. Avoid excess mortar in the bed joints of the trailing leaf in order to minimise mortar extrusions into the cavity.
- Ensure that ties are kept free of mortar, have 'drips' in the centre of the residual cavity and do not slope down to the inner leaf.
- Ensure that no boards bridge the cavity.
- Ensure that there are no gaps in the insulation.
- Protect the top edge of insulation with a cavity tray (where the residual cavity is less than 50 mm) if it does not extend to the full height of the wall (Figure 5):
  - — a cavity tray is not needed to protect the top of the insulation if residual cavity exceeds 50 mm.
- Provide weep holes at all cavity trays and at foot of wall.

Figure 2

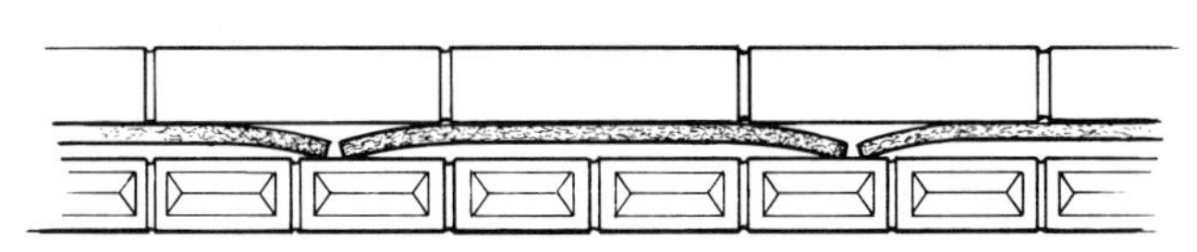

Figure 3

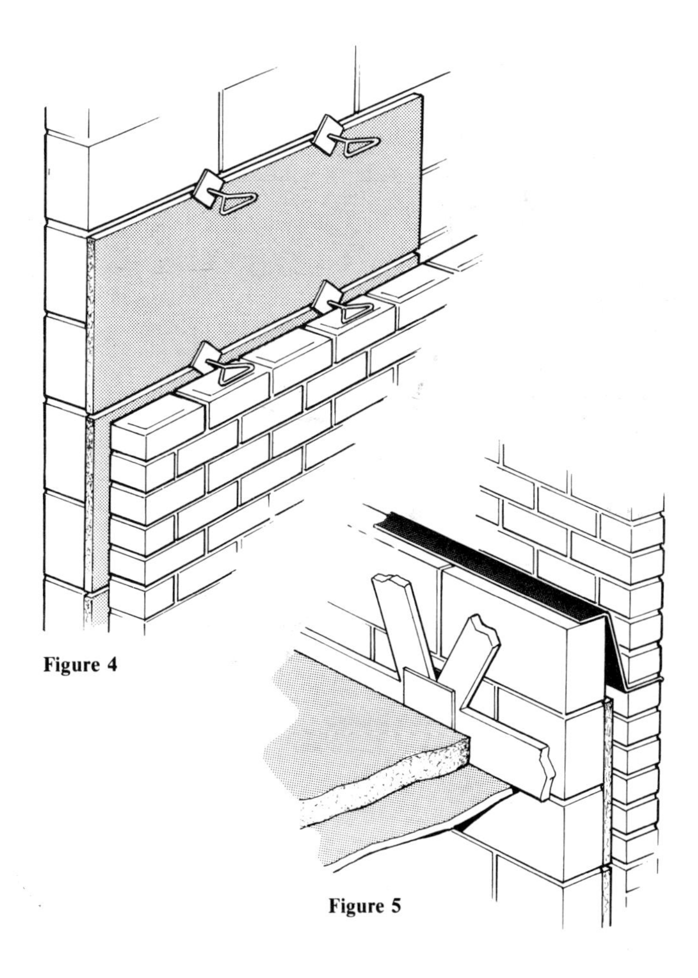

Figure 4

Figure 5

REFERENCES AND FURTHER READING

BRE *Digest* 227 'Built-in cavity wall insulation for housing'.

Cavity trays in external cavity walls: preventing water penetration

FAILURES: Damp penetrating to inner leaf of external cavity walls; damage to susceptible materials in the cavity

DEFECTS: Cavity trays omitted or poorly detailed

Figure 1 Cavity tray unsupported, sagging, lap not bonded

Water penetration through external walls was one of the most frequent problems reported in an AMA survey. Cavity trays (also called 'cavity gutters') play a vital part in preventing rain penetration in cavity walls, but a BRE survey found that they were frequently inadequate or even missing (Figure 1). This defect can be prohibitively expensive to correct later.

Building Regulation C8 requires external walls to be constructed to resist the penetration to the inside of the building of moisture from rain or snow, and to prevent its being transmitted to any part of the building where it might cause damage.

Rainwater will penetrate the brick outer leaf of a cavity wall and run down its cavity face. Any construction which obstructs its path has the potential to lead water to the inner leaf and could itself, unless protected, be subject to persistent damp. ***Susceptible materials in the cavity may be damaged and dampness appear inside the building.***

Items of construction, within the cavity, that could be damaged by water or lead water inwards, need a cavity tray above them to shed water from the cavity to the outside. Where the outer leaf of an external wall becomes part of an internal wall at a lower level, for example a separating wall in a stepped or staggered terrace, ***cavity trays*** are needed ***to shed water outward above the point where the wall becomes internal.*** Where the change from external to internal wall occurs at the abutment of a pitched roof (and where the overhang of a roof at a higher level does not provide adequate protection to the wall), stepped cavity trays are needed; ***a flashing alone is inadequate*.***

*Cavity trays are also needed in parapets, but the detailing is not dealt with in this DAS.

Cavity trays in external cavity walls: preventing water penetration

DAS 12
December 1982

PREVENTION

Principle — Water on the cavity face of an external leaf must be led outwards wherever it might otherwise reach the inner leaf, or susceptible materials in the cavity.

Practice

- **Provide cavity trays at:**
  - lintels. A tray should cover, and extend beyond, the cavity closer: Some lintels are said to need no tray: However, it is important that the head of each jamb dpc is adequately cloaked (Figure 2) and this may well demand a longer lintel than that required to give the minimum structural bearing, (Figure 3): Alternatively a superimposed tray of sufficient length must be provided;
  - air bricks, ducts, meter boxes;
  - horizontal cavity barriers;
  - top of cavity insulation batts if there is an exposed external leaf above them, such as at gables;
  - places where floor, balcony or other construction crosses the cavity;
  - places where an external leaf becomes internal wall below;
- **Cavity trays should:**
  - be higher (usually not less than 150 mm) at the inner leaf to protect it and drain water outwards;
  - have sealed upstand ends where they abut constructions such as columns to divert rainwater (Figure 4)
  - have at least one upstand end if they are stepped trays at a roof/wall abutment: Preferably such trays would have two upstand ends, particularly if there is insulation within the cavity. The lowest tray should always have two upstand ends;
  - dress over, not under, any associated flashings or vertical membranes so as to drain water outwards;
  - be bonded at any lap joints in the tray;
  - be provided with weep holes, such as open perpends at every fourth joint in brickwork; weep holes should not be drilled out later: the tray will be damaged (short trays, such as those over air bricks, do not need weep holes);
  - be supported, wherever practicable, to avoid sagging and to reduce risk of damage when mortar droppings are cleaned out;
- **Remember that, where 'steps' or 'staggers' in terraces are small and space for cavity trays is therefore limited:**
  - the need for trays can be avoided by cladding the exposed wall above (and this may be a more economic solution);
  - 'last minute' changes in stagger on plan may further reduce the step and thus the space available for trays (Figure 5);
- **Remember that pre-formed trays remove some of the risks associated with site work.**

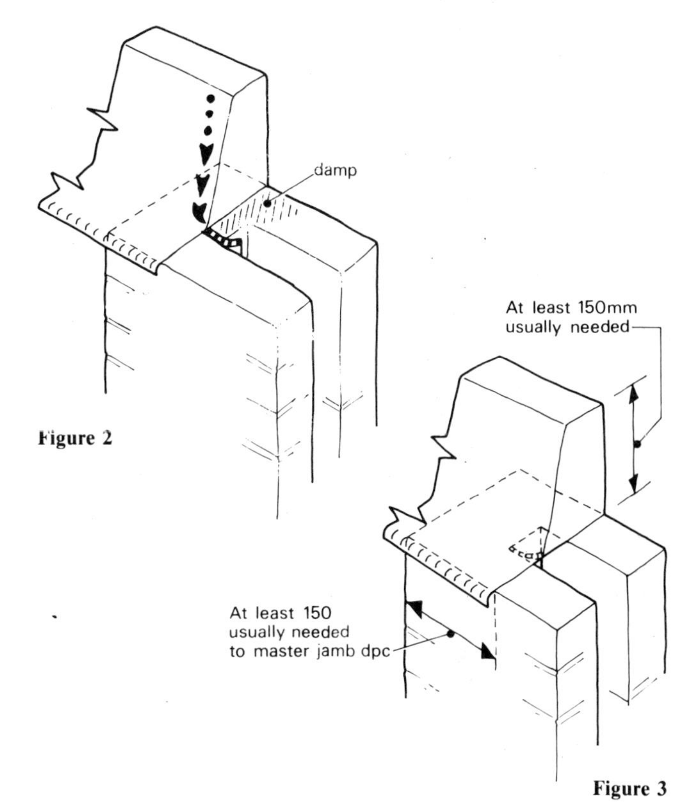

Figure 2

Figure 3

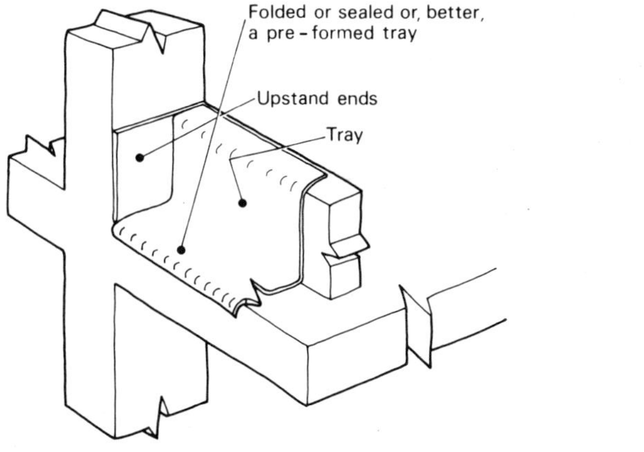

Figure 4

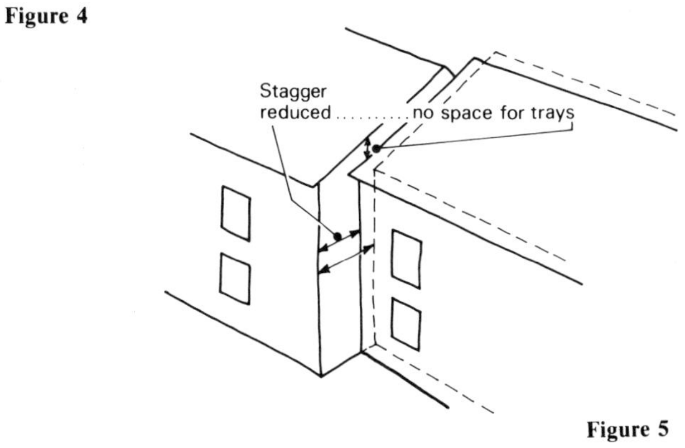

Figure 5

FURTHER READING

BRE Digest 77 'Damp-proof courses'.

Cavity parapets — avoiding rain penetration

FAILURE: Rain penetration into roofs and internal walls; sulphate attack of brickwork

DEFECTS: Inadequate detailing at coping, cavity tray and roof to wall junction

Figure 1 Dpc unsupported, not wide enough and cover flashing loose

Parapets are severely exposed to the elements in all but the most sheltered locations. Many instances have been seen where flush copings and inadequate detailing of damp proof courses and trays has resulted in dampness appearing on the wall below.

If the ***coping*** has an ***inadequate overhang*** and is ***not throated,*** or if the ***damp proof course*** below a ***coping*** is ***not supported, lapped and sealed*** or if a ***cavity tray*** at ***roof level*** is ***not correctly installed,*** then ***rainwater*** may ***penetrate*** into the structure and ***wet the inner leaf.*** If parapet ***brickwork*** is ***not protected,*** by overhangs in the coping, ***from constant dampness, sulphate attack*** may follow.

Cavity parapets — avoiding rain penetration

DAS 106
August 1987

PREVENTION

Principle — Weather proofing arrangements at parapet and roof level must prevent water passing into the roof or into the inner leaf.

Practice

- Consider whether use of a parapet can be avoided;
  — the risk of failure is high; a well-designed roof overhang is much less likely to lead to rain penetration problems and offers better weather protection to the wall below.
- Specify a coping of sufficient width to ensure that throatings are positioned a minimum of 40 mm away from the faces of the wall, Figure 2;
  — see Table 1 for British Standards for copings.
- Specify a continuous dpc bedded in designation (i) mortar[1] where parapet is built of fired-clay units or designation (ii) mortar[1] for calcium silicate or concrete units;
  — if the parapet ends at an abutment a flashing should be provided, Figure 3a.
- Specify a rigid support for the dpc across the top of the cavity;
  — for example, either slate or a proprietary rigid plastics item.
- Specify that dpcs and trays are to project a minimum of 5 mm beyond the face of the wall.
- Specify that a roof cover flashing is to be set *under* the dpc — see DAS 34.
- Specify that all joins in dpc materials shall be lapped at least 150 mm and sealed.
- Specify a cavity tray at the base of a parapet wall stepping downwards at least 150 mm towards the inner or outer part of the wall;
  — designers should consider which way to slope the dpc tray: externally water will be discharged down the face of the brickwork and may cause run-off staining; internally the vulnerable roof perimeter junction is put at additional risk from water discharging over it;
  — if parapet ends at an abutment, a stop end should be provided, Figure 3b.
- Specify dpcs to BS 6398 or choose a material with an acceptable third party certificate;
  — specify that the tray is to be dressed down the inner leaf and supported where it crosses the cavity horizontally, Figure 2;
  — trays need purpose made units at corners.
- Specify weep holes in the brickwork immediately above the cavity tray to be formed by leaving open perpend joints at not greater than 1 m intervals;
  — in very exposed locations proprietary weep tubes or weep slots are preferable to open perpends.

REFERENCES AND FURTHER READING

1 *British Standard* — BS 5628 Code of practice for use of masonry: Part 3: 1985 'Materials and components, design and workmanship'.

□ *Defect Action Sheet* — 34 'Flat roofs: built-up bitumen felt — remedying rain penetration at abutments and upstands'.

□ *British Standard* — BS 6398 1983 'Specification for bitumen damp-proof courses for masonry'.

Table 1 Copings should comply with the relevant British Standard

Material	Standard to be complied with	Recommended thickness
Aluminium*	BS 1470	0.9 mm
Cast stone	BS 5642 : Part 2	
Clay tile	BS 402	
Concrete (cast)	BS 5642 Part 2	
Concrete tile	BS 473 & 550 or BS 1197	
Copper	BS 2870, grades C 104 or C 106 in the O condition	
Lead	BS 1178	1.8 mm (code no. 4)
Natural stone	BS 5642 Part 2	
Zinc**	BS 849	0.8 mm

* Commercial or super purity quality aluminium should be used. Copings should preferably be preformed.

**Copings should preferably be preformed. In heavily polluted atmospheres it is advisable to use a heavier sheet, eg 1.0 mm thick or use another material.

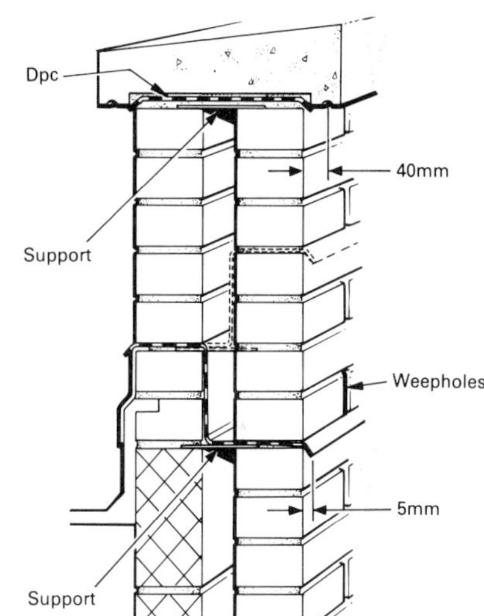

Figure 2

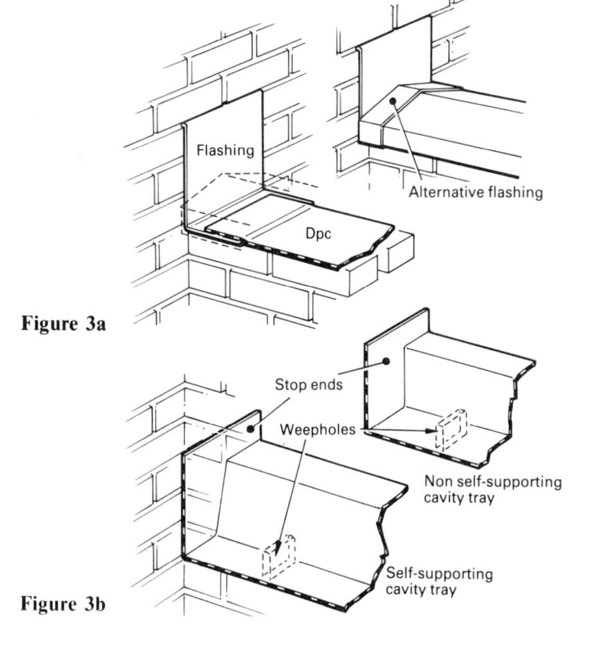

Figure 3a

Figure 3b

DAS 107
August 1987

Site
CI/SfB 8(21.8)(L34)(D6)

Cavity parapets — installation of copings, dpcs, trays and flashings

FAILURE: Dampness on internal walls and in roofs

DEFECTS: Dpcs and cavity trays not supported, not lapped enough, not sealed and flashings installed over dpcs instead of under them

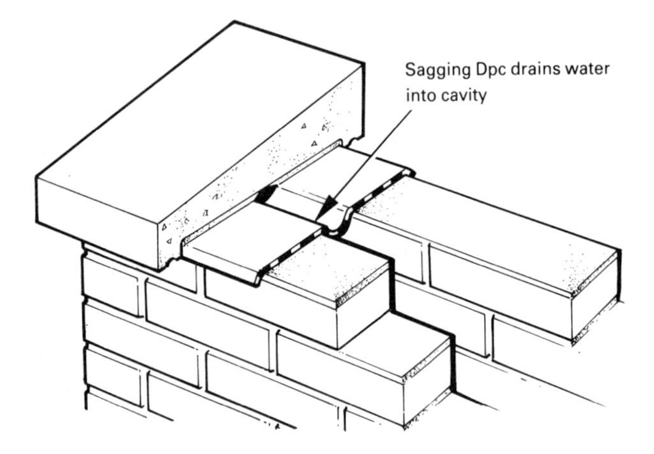

Figure 1

Parapets get very wet on both faces during periods of driving rain. Water can stream down both faces of the cavity. A sagging open joint anywhere in a cavity tray can feed large amounts of water into the roof and inner leaf of external walls.

If dpcs below copings are ***not lapped, sealed and supported*** across the top of the cavity, ***rain*** penetrating coping joints ***will enter the cavity; if cavity trays*** at roof level are ***not lapped, sealed,*** dressed over (NOT UNDER) a flashing to the roof covering, ***and arranged to drain water out of the cavity*** via weep holes, the ***roof*** construction and ***internal walls*** will become ***wet.***

PREVENTION

Principle — Damp proof courses and cavity trays must be installed so that no rainwater can wet the inside of the building.

Practice

- Check that material to be used for the dpc beneath the coping and for the cavity tray is that specified (BS 6398 or third party certificate.)
- Ensure that the dpc at the top of the cavity is supported beneath the coping so that it cannot sag, Figure 2.
- Ensure that the dpc beneath the coping projects beyong the face of the brickwork, Figure 3.
- Ensure that laps in the dpc are at least 150 mm in length and sealed.
- Ensure that laps in the cavity tray are at least 150 mm in length and sealed, that the cavity tray is watertight and fully supported and that any purpose-made corner units are sealed to the tray lengths.
- Ensure that all dpcs and trays are laid on a bed of fresh mortar.
- Check that weep holes are formed by leaving open perpend joints at not more than 1 m intervals;
  - — care must be taken to avoid blocking holes with mortar droppings; where necessary they should be cleared out taking great care to avoid damag ing the cavity tray;
  - — alternatively purpose made weep tubes or weep slots can be built in as work proceeds; check what is specified.
- Ensure that the dpc and the cavity tray project beyond the face of the wall, Figure 4.
- Ensure that cavity trays are kept free of mortar droppings;
  - — trays are easily damaged when mortar droppings are removed; the use of cavity battens to collect these droppings is amply justified when building cavity parapets.
- Ensure that any roof cover flashings are well lapped by the cavity tray, Figure 4.

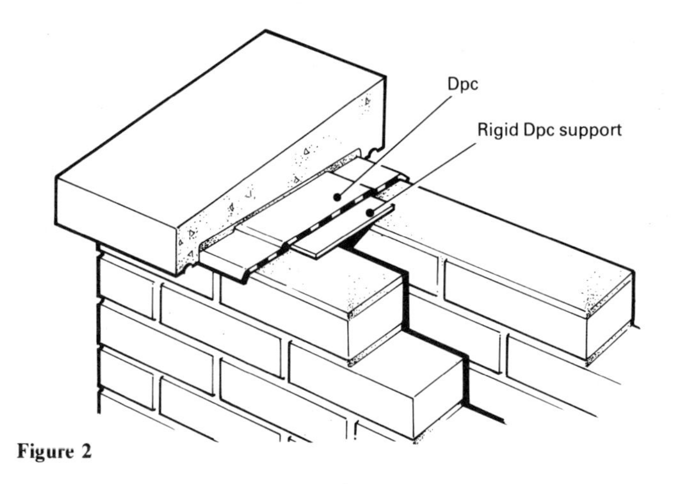

Figure 2

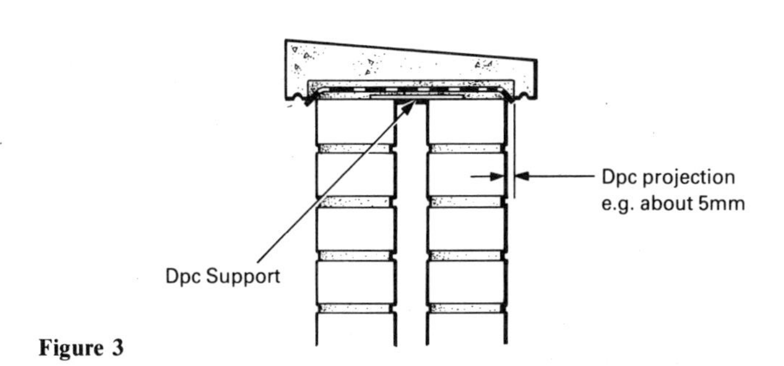

Figure 3

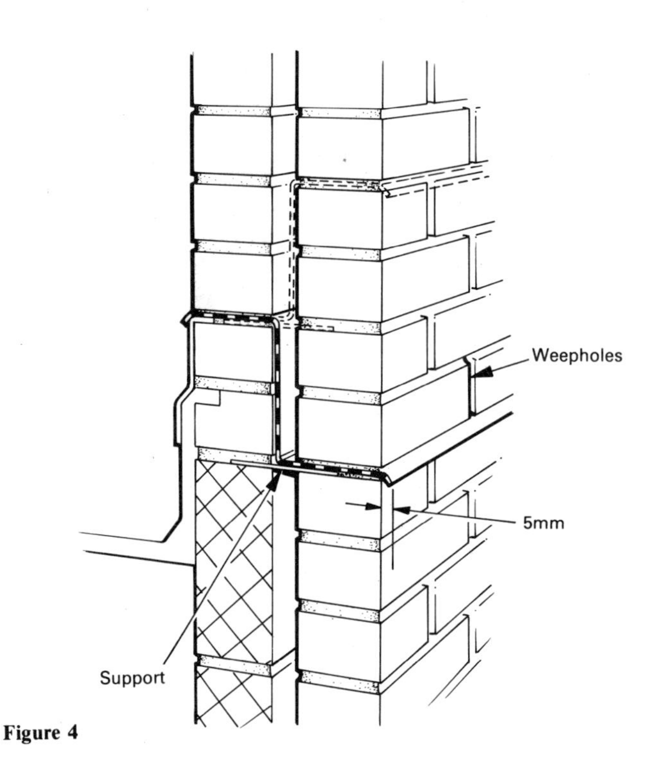

Figure 4

Walls and ceilings: remedying recurrent mould growth

FAILURE: **Recurrent mould growth on walls and ceilings**

DEFECTS: **Inadequate thermal insulation, ventilation, or both, in relation to heating and usage of the dwelling; thermal capacity of walls too great for use with intermittent heating; possibly also thermal insulation reduced by wetting from rain penetration, rising damp, leaking water services, or interstitial condensation; thermal insulation discontinuous**

Figure 1 Mould growth

Mould growth on walls, ceilings and wooden window frames is a problem often reported by Local Authorities. Moulds grow on organic materials, such as paints, wallpaper and wallpaper adhesives and also even on traces of household dust. Mould may also grow within a paint film, causing a stain, usually pink or purple, but the mould itself may not be visible.

Mould growth requires sustained high relative humidity but is unlikely to start unless surface water is present. Persistent surface condensation is the commonest though not the only, cause of such conditions; this DAS therefore deals with that case. (Other DASs deal with other causes of dampness). The occurrence of surface condensation depends on the relationship between, on the one hand, heating, ventilation, insulation etc, and on the other, the pattern of occupants activities. ***The predominant cause can therefore either be in the design provisions or in the occupants' usage.*** There is no point in treating the symptoms unless the predominant causes have been identified and cured, since mould growth will otherwise recur.

PREVENTION

Principle: Kill the mould and change the environment to one which does not encourage re-growth.

Practice: Getting rid of the mould

Specify the following steps:

- Clean off the mould using clean water
- Apply household bleach of the sodium hypochlorite type, diluted one volume of bleach to four of water, (it may bleach some colours)
- Alternatively — or if the household bleach treatment proves unsuccessful — use a toxic wash cleared under the Pesticide Safety precautions Scheme (PSPS); see BRE Digest 139, 1982[1].
- Where repainting is necessary, use a mould resistant paint (Note that the toxic ingredient protects only the paint film — it does not protect the substrate, nor does it sterilise any subsequent dirt deposits which can support mould growth).

Curing the cause:

- Check that condensation is indeed the cause.
- Identify probable major causes of condensation and select relevant actions from those below.
- Improve thermal insulation (or its continuity if mould growth is associated with localised construction features forming cold bridges — eg uninsulated lintels).
- Provide some insulation at or near indoor surfaces of walls if thermal capacity on the indoor side of the insulation is large (eg dense concrete inner leaf) and heating intermittent. Consider applying insulating linings (these will need a vapour check). See Figure 2.
- Provide for extraction of water vapour, preferably at source, and possibly automatic (eg humidistat controlled extractor fans in kitchens and bathrooms) — but note the need for maintenance and that running costs and noise may discourage their use.
- Ventilate unheated internal stores on external walls to ***outside*** not to inside. Venting to the inside can make matters worse. See Figure 3.
- Provide more, or more acceptable, ventilation — eg finely adjustable wall or window ventilators in unheated bedrooms.
- Advise occupants about activities that promote condensation and mould growth — eg flue-less heaters, tumble driers not vented to outdoors, excessive indoor clothes drying, failure to use (or blocking off) ventilators.
- Consider whether the provision of continuous background heating may be more economical than remedial measures which would entail substantial alteration to the building.
- Record actions taken and check subsequently that they have been effective.

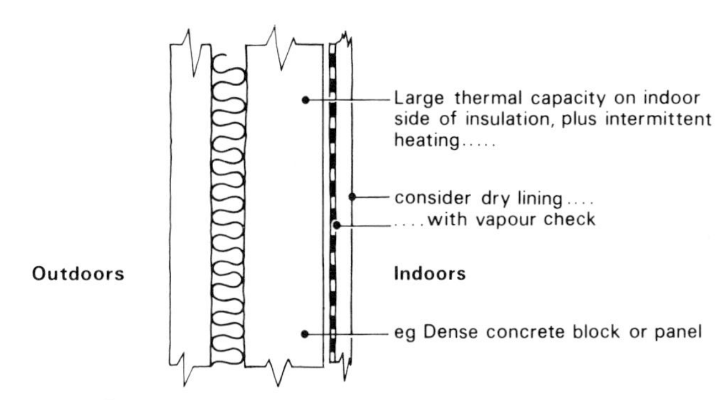

Figure 2

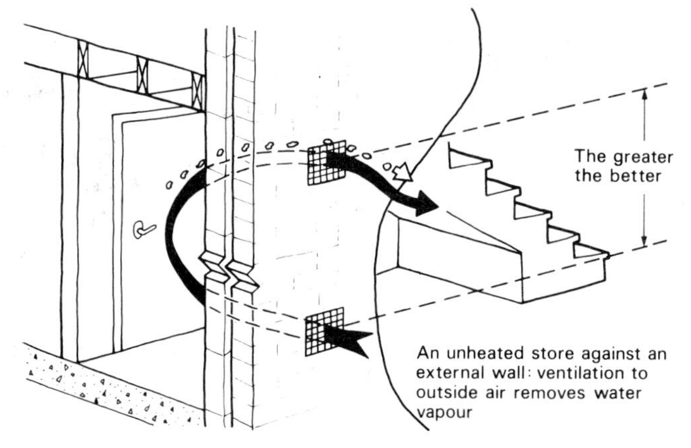

Figure 3

REFERENCES AND FURTHER READING

1 *BRE Digest* 139 (1982 Edition) 'Control of lichens, moulds and similar growths'.

BRE Digest 198 'Painting walls, Part 2; failures and remedies'.

PART SEVEN

Walls 4: Fire Barriers and Sound Insulation

External and separating walls: cavity barriers against fire — location*

FAILURE: **Spread of fire within wall cavities and between wall cavities of adjoining houses**

DEFECTS: **Cavity barriers omitted in design; not designed so that they fully close the cavity; not designed so that they and their fixings will be durable over time and in a fire**

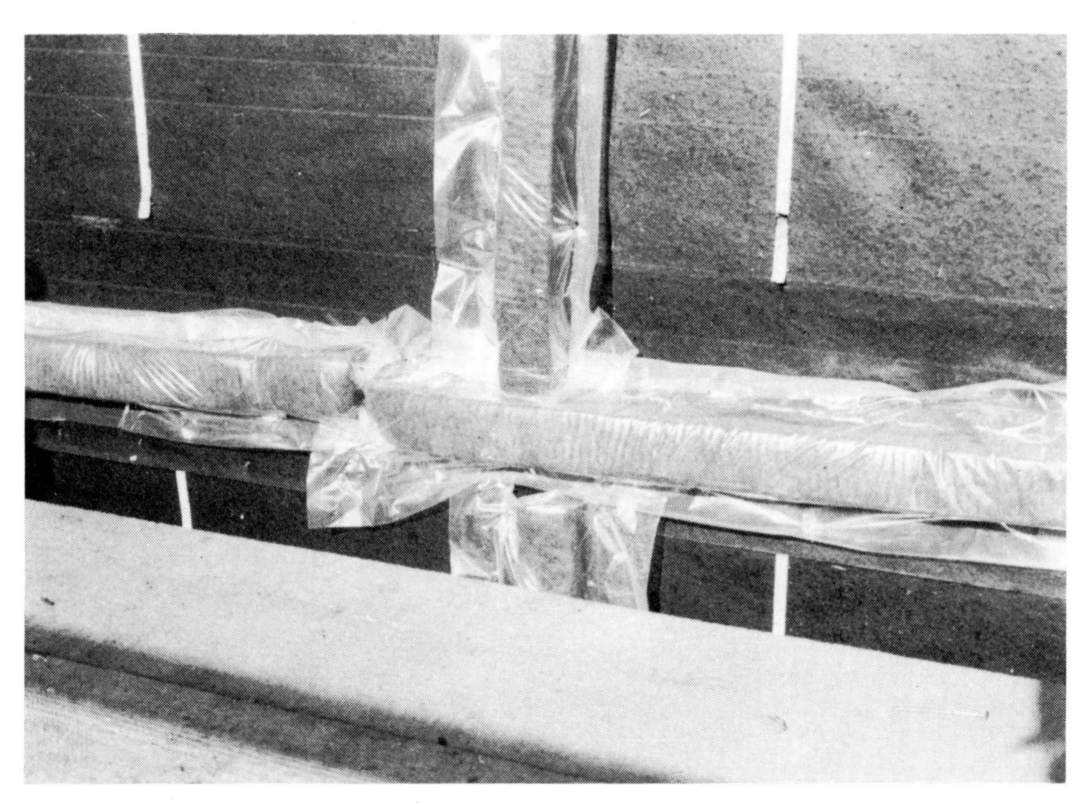

Figure 1 Continuity of barriers is important

This *Defect Action Sheet* deals only with houses, not with flats and maisonettes. Requirements are given in Reference 1 (in Scotland Reference 2).

Where both leaves of the wall are non-combustible, cavity barriers are only needed if either leaf is less than 75 mm thick or if the cavity is more than 100 mm wide (provided that cavities are closed at their tops and at the tops of openings).

*Requirements in Scotland and in Inner London may differ from those described in this *Defect Action Sheet.*

Where either leaf is combustible, as in timber-framed houses, special measures are needed to ***close the cavities. 'Cavity barriers'*** are used for this purpose and may include construction which, though primarily for other purposes, will also ***prevent fire spread within cavities.*** If a cavity barrier cannot fully close a cavity, the ***imperfections of fit*** must be ***made good by fire stops.***

External and separating walls: cavity barriers against fire — location*

DAS 29
June 1983

PREVENTION

Principle — Fire must not spread within or between cavities nor must it by-pass elements required to have fire resistance.

Practice

- Check where cavity barriers are needed.
- Check whether construction fulfilling another purpose will also provide a satisfactory barrier where needed or whether an additional one is necessary.
- Check that any barrier will, in practice, close the cavity fully (despite, for example, unavoidable inaccuracies[3] in construction).
- Provide fire-stopping for any gaps where imperfection of fit is inevitable.
- Specify materials and construction, for cavity barriers, that will be durable over time and in a fire.
- Specify fire-stopping materials that are non-combustible and capable of filling irregular spaces to prevent the passage of smoke and flame.
- Specify vertical cavity barriers at:
  - external wall/separating wall junctions located to separate their cavities (see Figure 2);
  - eight metre intervals in external and separating walls (see Figure 2), or at 20 m intervals if surfaces exposed within the cavity are Class 0*;
  - sides of every opening through walls (except those for meter boxes).
- Specify horizontal cavity barriers at:
  - upper floors in external and separating walls (see Figure 3);
  - top floor ceiling level and roof level in external and separating walls (see Figure 3);
  - top and bottom of every opening through walls.
- Check that the foregoing will provide:
  - a continuous cavity barrier around every opening and at all junctions between vertical and horizontal barriers: ensure that continuity, which is of critical importance, will be practicable on site at all joints within the length, abutments and junctions of barriers (see Fig 4).
- Select durable fixings for barriers or ensure that barriers will be tight enough to stay in place for the life of the building and in a fire.
- Check the design for potential effects on other performance — for example, rain penetration across the cavity.
- Instruct site staff to ensure that each barrier fully closes the cavity or that fire-stopping makes good every imperfection, and to report for action any difficulty in achieving satisfactory closing of the cavity.

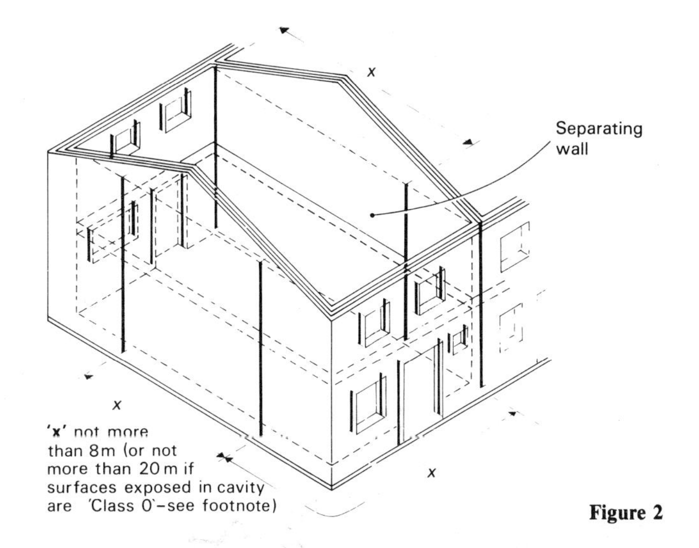

Figure 2

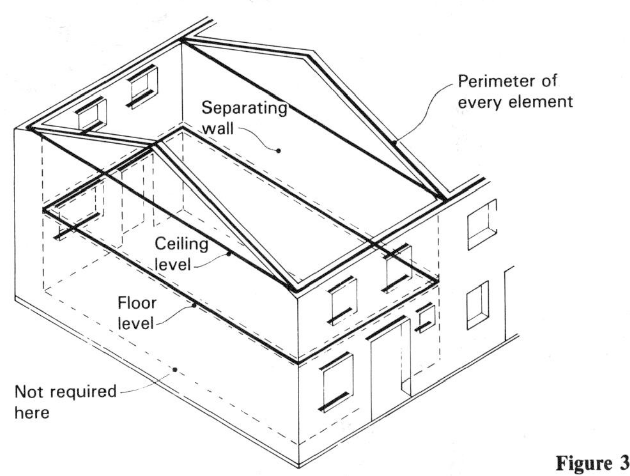

Figure 3

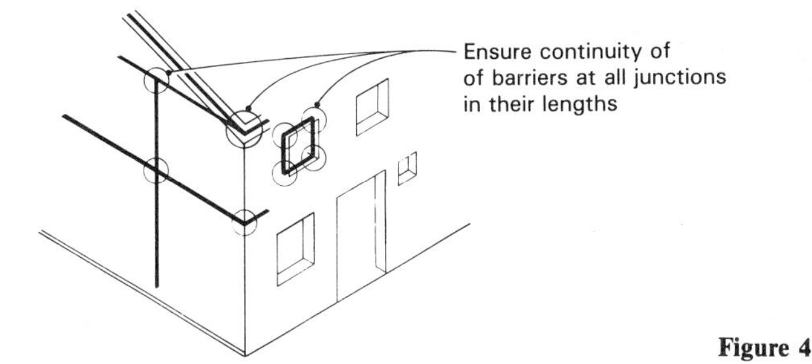

Figure 4

REFERENCES AND FURTHER READING

1 The Building Regulations (England and Wales) 1976, Regulation E 14.

2 The Building Standards (Scotland) Regulations, 1981 (as amended), Regulation D 14.

3 *British standard* 5606:1978 'Code of practice for accuracy in building'.

Defect Action Sheet DAS 30 'External and separating walls: cavity barriers against fire — installation'.

*Class 0 is not a BS classification. The Building Regulations 1976 (England and Wales) state that Class 0 shall be construed as a requirement that: (i) the material of which the wall or ceiling is constructed shall be non-combustible throughout; or (ii) the surface material (or, if it is bonded to a substrate, the surface material in conjunction with the substrate) shall have a surface of Class 1 and, if tested in accordance with BS 476:Part 6:1968, shall have an index of performance (I) not exceeding 12 and a sub-index (i_1) not exceeding 6.

External and separating walls: cavity barriers against fire — installation

FAILURE: **Spread of fire within wall cavities and between wall cavities of adjoining houses**

DEFECTS: **Cavity barriers omitted or removed during construction; not positioned or not fixed properly; not fully closing the cavity; not tightly lapped or butted; imperfections of fit not made good by fire stopping**

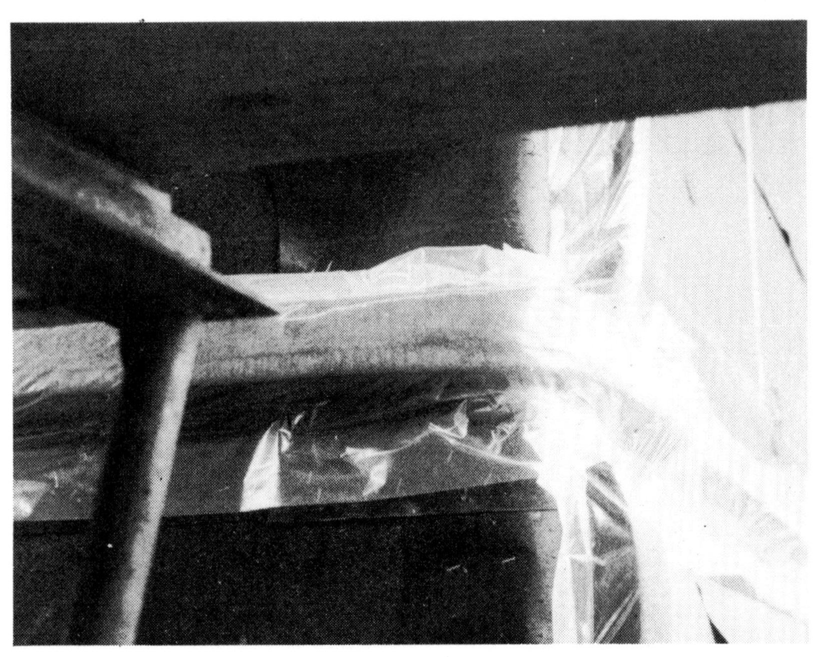

Figure 1 Barrier flattened at corner — gaps at joints

Defect Action Sheet DAS 29 describes circumstances in which cavity barriers will be needed. Cavity barriers prevent fire spreading within cavities, with the intention of reducing the risk to life and property in a fire. To achieve this, each barrier must be in the right place and must close the cavity completely, if necessary by fire-stopping to close any gaps. A barrier must stay in place for the life of a building and during a fire.

If ***cavity barriers*** are ***not correctly sited*** or ***not firmly fixed*** or compressed in place, or are ***not fully closing the cavity,*** or if gaps in the closure are ***not fire-stopped, fire may spread*** unseen from one part of the house to another, or to adjoining houses.

PREVENTION

Principle — Fire must not spread within or between cavities nor must it by-pass elements required to have fire resistance.

Practice

- Check where cavity barriers are needed[1].
- Ensure that no barrier is omitted or removed during construction.
- Ensure that any barrier which does not fully close the cavity has its gaps fire-stopped. **Note:** in some cases construction serving another purpose may also be intended as a cavity barrier, whose gaps must be fire-stopped too.

- Check that continuity is achieved at all joints within and between cavity barriers (see Figures 2a and 2b). At intersections the horizontal barrier should preferably be the continuous one (see Figure 2b).
- Report, for design action, any cases where, even with fire-stopping, barriers will not in practice fully close the cavity.

If the barriers are made up as 'sausages' of mineral fibre encased in polythene tube:

- Check that no gaps and no twists have developed between lengths of mineral fibre in the tube. **Note:** a clear polythene casing will facilitate visual check.
- Check that sausages are thick enough to fill the built cavity at all points. **Note:** a 50 mm thick sausage is unlikely to close a 50 mm nominal width cavity, since the cavity will exceed this width in practice at various points.
- Check that sausages are well fixed (not by corrodible staples or nails) and/or tight enough to stay in place permanently within the cavity.
- Check that sausages are not fixed by stapling or nailing through the mineral fibre as this produces gaps at every fixing (see Figure 3).
- Check that lengths of sausages do not overlap in such a way as to create gaps in the barriers (see Figures 4a and 4b).
- Check that horizontal barriers are not taken around corners in such a way as to flatten them and so produce gaps.

If barriers are of wire-reinforced mineral wool:

- Check that barrier is thick enough to fill the built cavity at all points.
- Check that barriers are well fixed (not by corrodible staples or nails) and/or tight enough to stay in place permanently within the cavity.
- Check that lengths of mineral wool do not overlap in such a way as to create gaps in the barriers. **Note:** overlaps are particularly vulnerable at junctions of vertical and horizontal barriers and/or at local widening of cavities.
- Check that wire mesh has not been buckled prior to installation or during construction in such a way as to produce a gap in the barrier.
- Check that barriers are not fixed via the mineral wool but via the expanded metal to avoid local gaps at each fixing.

If barriers are timber battens or rigid sheet materials:

- Check that all gaps around barriers are fire-stopped to close the built cavity fully at all points. **Note:** a timber batten or rigid sheet material cannot be expected to close tightly against a masonry leaf: all gaps must be fire-stopped by mortar or other suitable material (see Figure 5).

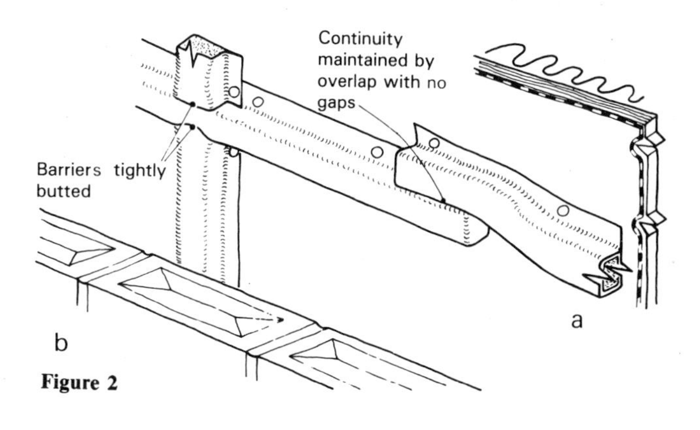

Figure 2

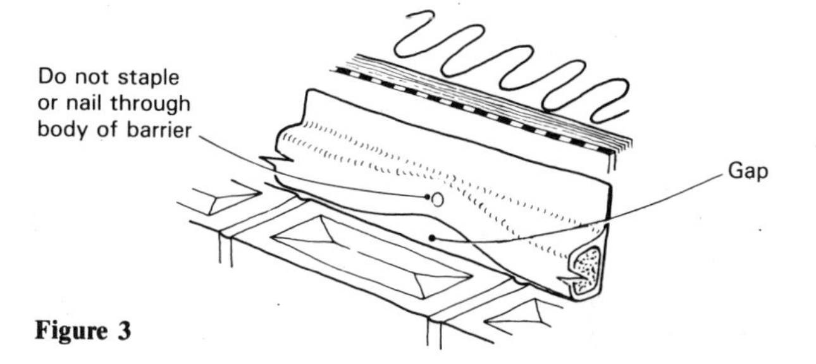

Figure 3

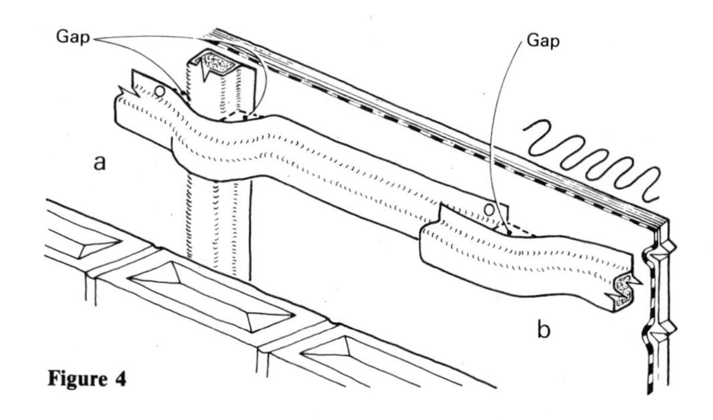

Figure 4

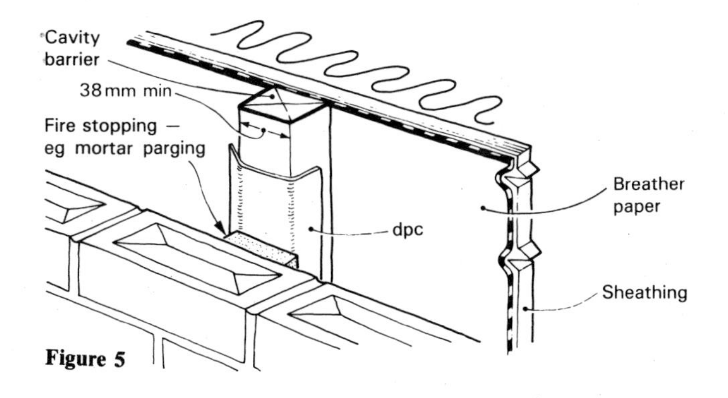

Figure 5

REFERENCES AND FURTHER READING

1 *Defect Action Sheet* DAS 29 'External and separating walls: cavity barriers against fire — location'.

External walls: combustible external plastics insulation: horizontal fire barriers

FAILURE: **Extension of fire spread and damage**

DEFECTS: **No horizontal fire barrier**

Figure 1

It is normally unnecessary to provide horizontal fire barriers within combustible thermoplastic external insulants for two storey housing. However, for more than two storeys there is a risk that, in fire, large areas of thermoplastic insulants can melt and fire can be transmitted up the face of a building even where there is no cavity. This can happen irrespective of the material of the outer finish of the insulant.

If ***horizontal fire barriers*** are ***not incorporated*** within external plastics insulation systems for walls over two storeys, ***fire*** can be ***spread upwards.***

External walls: combustible external plastics insulation: horizontal fire barriers

DAS 131
May 1989

PREVENTION

Principle — Barriers are needed to prevent progressive involvement of insulants in fire by restricting hot gases or flames from travelling upwards within the insulation layer, or within any cavity.

Practice

Note it may be necessary to specify movement joints to control cracking of the render or surface finish, and these may be conveniently incorporated with fire barriers.

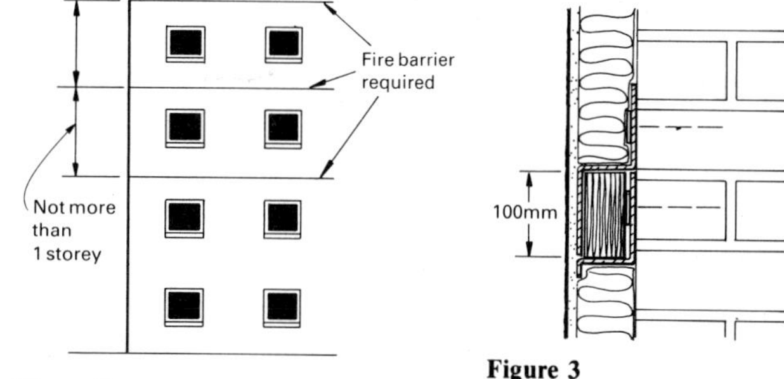

Figure 2

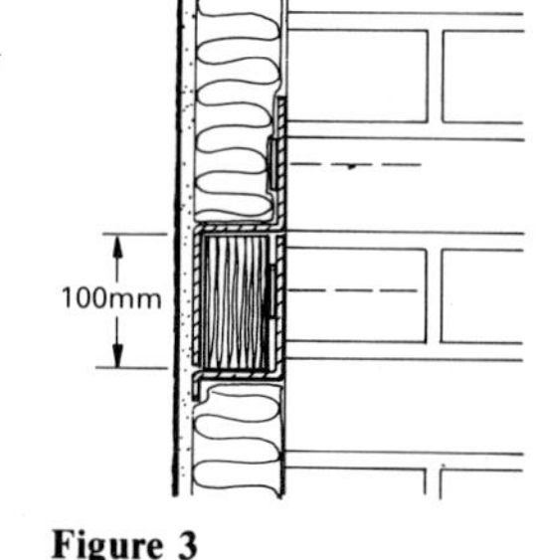

Figure 3

For adhesive fixed systems: thermoplastic insulant

- Specify horizontal fire barriers at intervals of every storey from the second storey Figure 2[1].
- Specify strip of non-combustible insulation eg mineral fibre interrupting insulation, Figure 3;
  - — strip should be 100 mm high and full thickness of insulation.
- Specify support for the protective finish;
  - — support should be bonded across the non-combustible strip to both wall and finish.

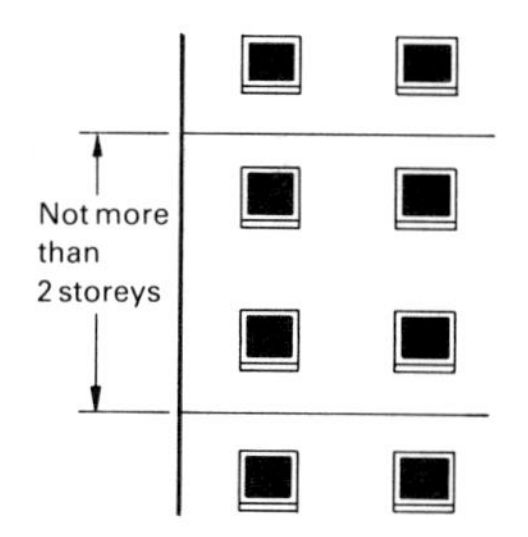

Figure 4

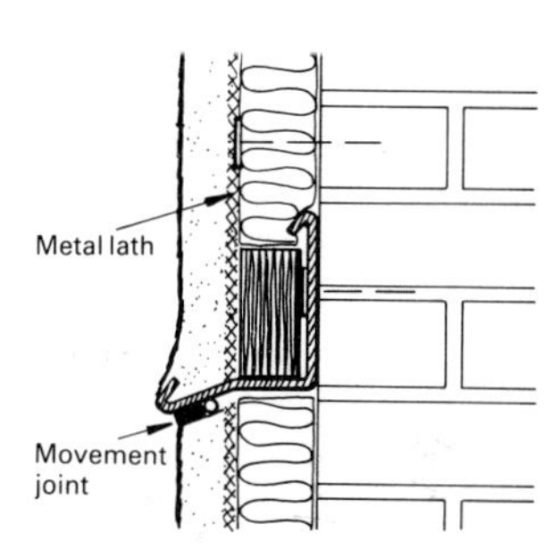

Figure 5

For rendered metal lath systems over thermoplastic insulants

- Specify horizontal fire barriers at intervals of not more than 2 storeys, Figure 4[1].
- Specify render stop filled with non-combustible insulation, Figure 5.
- Specify sealant to outer face of movement joint under render stop[2].

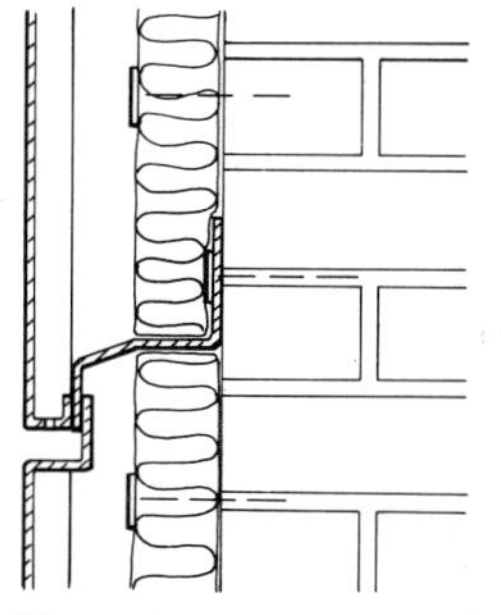
Figure 6

For sheet overcladding incorporating a cavity for thermosetting or thermoplastic insulants

- Specify horizontal fire barriers at intervals of not more than 2 storeys, Figure 4.
- Specify protected 3 mm steel cavity barrier through full thickness of cavity and insulation, Figure 6.

REFERENCES AND FURTHER READING

1 **Rogowski B F W, Ramaprasad R and Southern J R.** *Building Research Establishment BRE Report.* Fire performances of external thermal insulation for walls of multi storey buildings. *In preparation.*

2 **Building Research Establishment** *Defect Action Sheets* DAS 68 and 69. Detailing and application of sealants. Garston 1985.

External walls: external combustible plastics insulation: fixings

FAILURE: **Protective finish bowing outwards and exposing insulants to fire**

DEFECTS: **No metal fixings, insufficient metal fixings**

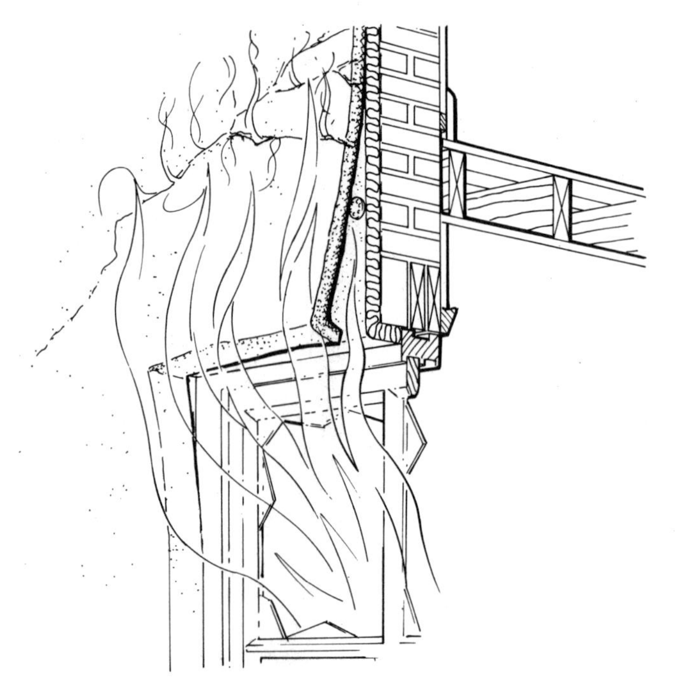

Figure 1

Plastics materials used, for example, for fixing rendered lath external insulation systems to walls are likely to melt or burn when heated by a fire, Figure 1. This can lead to rendered lath bowing away from the wall, allowing direct exposure of the insulation to the effects of a fire, and to the fire spreading extensively. Plastics fixings must be supplemented by some metal fixings, usually of stainless steel.

If some ***metal fixings*** are ***not used,*** the ***cladding*** may ***deform*** in fire, ***and damage*** will be ***extensive.***

PREVENTION

Principle — Plastics fixings must be supplemented with metal fixings.

Practice

- Check specification to see whether metal fixings are specified alone or in addition to plastics whether of polypropylene or of nylon;
  — If not, alert designer
- Use not less than 1 metal fixing (in addition to those of plastics) per square metre of insulation, Figure 2.
- Always use metal fixings for fire barriers, Figure 3.

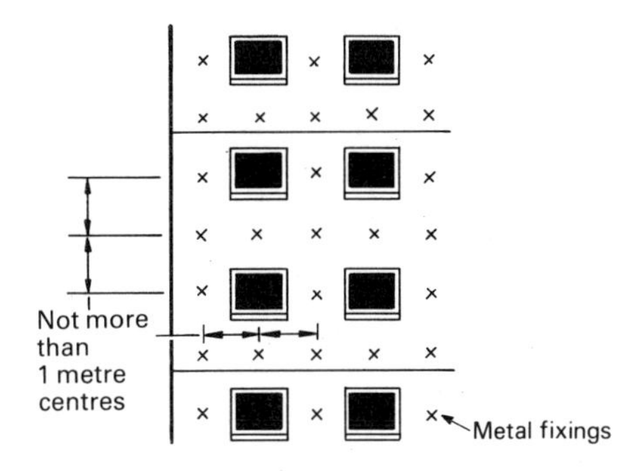

Figure 2

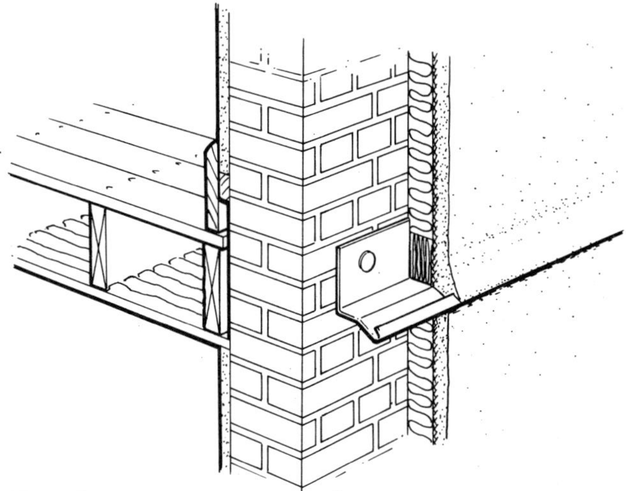

Figure 3

Masonry separating walls: airborne sound insulation in new-build housing

FAILURE: Excessive airborne sound transmission between adjacent dwellings

DEFECTS: Unacceptable airborne sound paths through or around separating walls; insufficient mass of separating and flanking walls

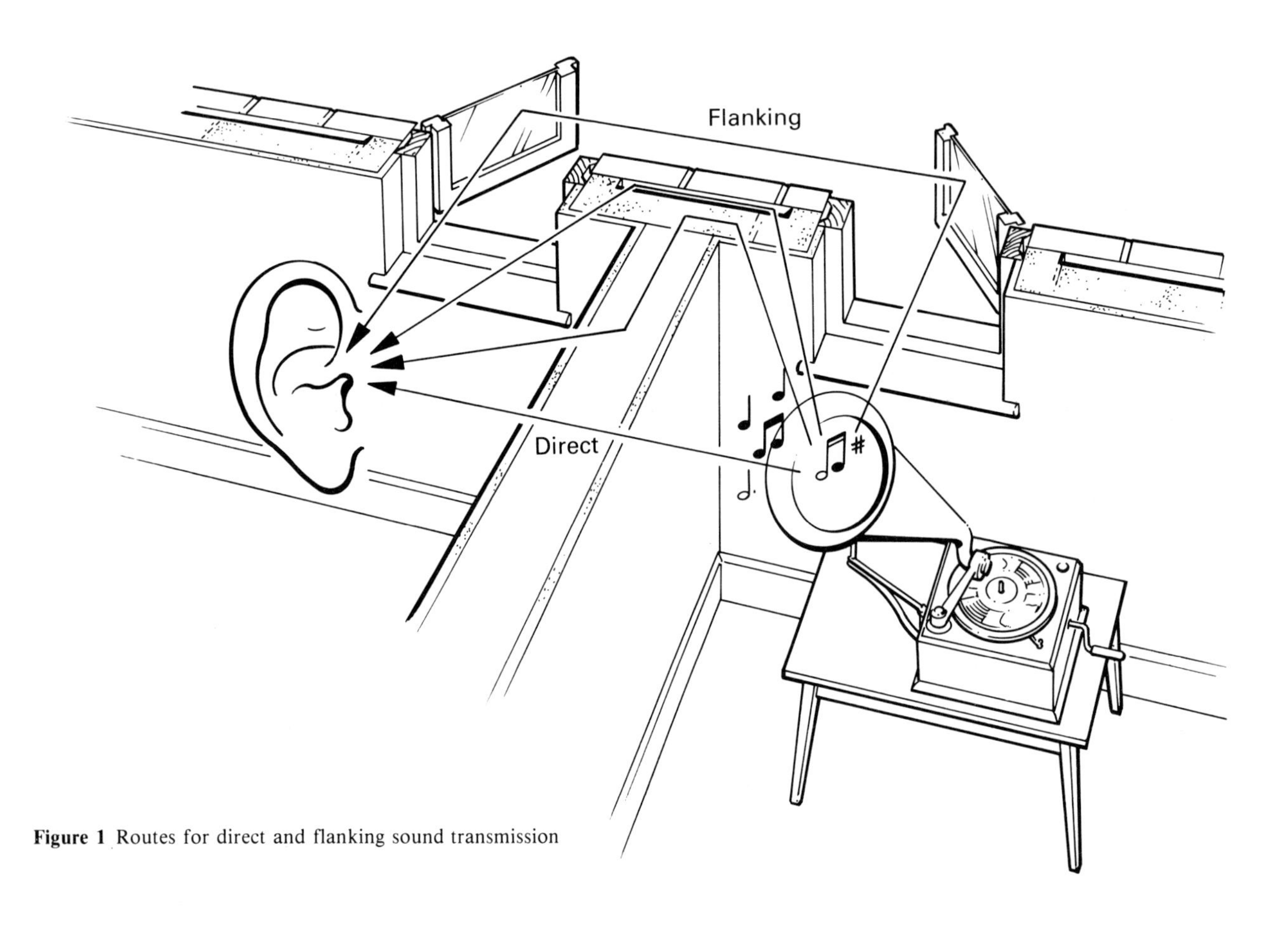

Figure 1 Routes for direct and flanking sound transmission

A BRE survey[1] of post-1970 terrace and semi-detached houses showed that poor sound insulation was of major concern.

This DAS is concerned with ways of reducing airborne sound transmission such as speech or noise from loud speakers. It does not deal with impact noise such as footsteps and banging doors. Airborne sound can be transmitted between dwellings both directly through walls and indirectly through adjoining elements — the latter is called flanking transmission, Figure 1. Both direct and flanking transmission need to be considered.

If ***materials*** of ***insufficient mass***/m^2 have been used or if ***air paths*** between dwellings are ***not sealed*** or if ***openings*** through external walls in adjacent dwellings are ***too close together,*** there can be ***weak links*** in the sound insulation of the structure which could ***reduce*** the overall ***sound insulation*** between dwellings to an ***unaccaptable level.***

Masonry separating walls: airborne sound insulation in new-build housing

DAS 104
July 1987

PREVENTION

Principle — Airborne sound paths through and around separating walls must be minimised.

Practice

- Specify walls of sufficient average mass;
  - — solid walls : brickwork at least 375 kg/m² including finishes
    : blockwork or concrete at least 415 kg/m² including finishes
  - — cavity walls: 50 mm cavity — at least 415 kg/m² including finishes
    : 75 mm cavity — at least 250 kg/m² including finishes*.
- Ensure that lightweight blocks intended for the inner leaf are not incorporated in any part of the separating wall;
  - — where blocks of different mass/m² are to be used for inner leaf and separating wall, specify a means of distinguishing on site (eg colour coding);
  - — indicate block weights on drawings so that dry sample blocks can be check-weighed.
- Ensure minimum connections are used between the leaves of cavity separating walls consistent with structural stability;
  - — specify that any wall ties should be butterfly wire type.
- Specify that bricks are laid 'frog up' and that frogs and perpends are filled with mortar.
- Specify joist hangers (plus lateral restraint ties where necessary) to avoid the need to perforate the separating wall.
- Specify a rough render coat on at least one side of the separating wall in roof space and in intermediate floor depth, Figure 2.
- Specify, for preference, that cavity separating wall construction be continued unmodified right through the roof space, Figure 3;
  - — reference 2 permits a reduction in mass to not less than 150 kg/m².
- Ensure external flanking wall has an average mass of 120 kg/m². Where the mass is less, a different basis for design will be needed[2] to provide discontinuity to control flanking transmission, Figure 4;
  - — check that the thermal insulation of the external walls satisfies the Building Regulations (0.6 W/m²k); increase the thermal insulation if necessary (eg specify thicker insulating blocks, use added insulation).
- Ensure that any permanent openings, such as air bricks, in the external wall on either side of the separating wall are at least 650 mm apart, Figure 5.
  - — windows, without permanent openings, in a cavity wall may be less than 650 mm apart provided the cavity is closed.

*Certain lighter constructions have been demonstrated to comply.

REFERENCES AND FURTHER READING

1 *BRE Information Paper* IP1/82. 'Noise from neighbours and the sound insulation of party walls in houses'.

2 *Approved Document* E/1/2/3 Building Regulations 1985 for England and Wales.

☐ *BRE Digest* 252 'Sound insulation of party walls'.

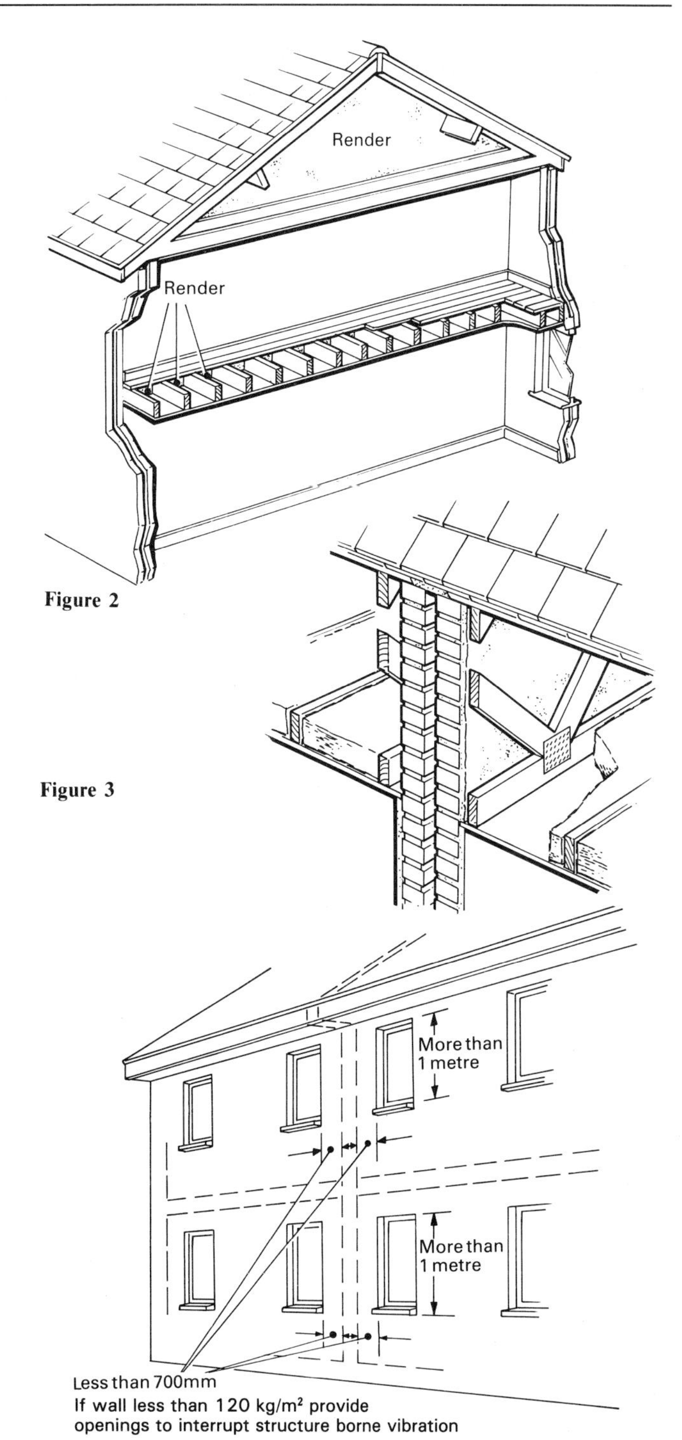

Figure 2

Figure 3

Figure 4

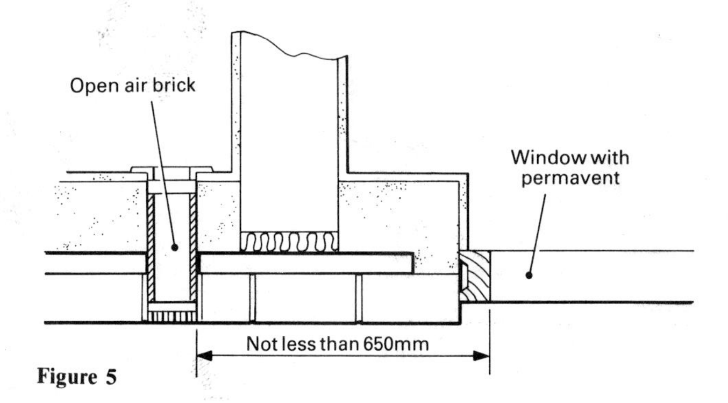

Figure 5

DAS 105

July 1987

Design

CI/SfB 8(22.8)(W7)(P2)

Masonry separating walls: improving airborne sound insulation between existing dwellings

FAILURE: **Unacceptable sound transmission through separating walls**

DEFECTS: **Airborne sound paths, or insufficient mass in separating walls; flanking sound transmission**

Figure 1

BRE inspections of rehabilitation work in progress have revealed instances of wholly inadequate separating walls between dwellings, including half brick or brick on edge, large gaps or absence of walls in roof spaces, and holes alongside floor joists or other timbers built into the walls, Figure 1. Conversion and subdivision of large dwellings may require existing partition walls to become separating walls.

Unless direct air paths are ***blocked***, and ***unless*** the separating wall is upgraded to give ***sufficient mass*** over all its area, or isolation is provided by independent leaves, the ***sound insulation*** through separating walls will be ***unacceptable.***

REMEDIAL WORK

Principle — Increase resistance of separating wall by blocking all holes or by construction of an independent leaf.

Practice

- Specify that all air paths through the separating wall are sealed;
  - — construction will need to be opened up to gain access eg to interfloor spaces;
  - — rough rendering can be used to make good unfilled perpends and gaps around joists etc;
  - — small gaps, difficult to reach, can be filled with, eg injected foamed polyurethane;
  - — do not retain back to back services, Figure 2; use surface mounting, or, remove from separating wall altogether.

- Specify an independent new leaf or leaves for all separating walls of less than one brick thick.

- Specify an independent leaf sealed at the perimeter and having absorbent material hung loosely (not squashed) in the cavity;
  - — leaf may be of proprietary plasterboard partitioning or 2 layers of 13 mm plasterboard fixed to break joint on studs at 600 mm centres (studs must be clear of wall), Figure 3;
  - — cavity width, for optimum performance, to be not less than 125 mm;
  - — absorbent material may be mineral wool quilt 25 mm or more thick, Figure 3;
  - — leaf should be fixed only at its perimeter (top and bottom for preference), NOT to the separating wall and sealed using eg, coving, scrim or sealant, as appropriate.

- Specify that any gaps in an existing separating wall* in the roof space are made good to reduce flanking transmission;
  - — for walls less than 1 brick thick, rough render one side;
  - — flanking sound transmission through the separating wall in the roof space may be further reduced by applying a layer of plasterboard to the existing ceiling, Figure 4.

*To meet fire requirements there must be a separating wall in the roof space.

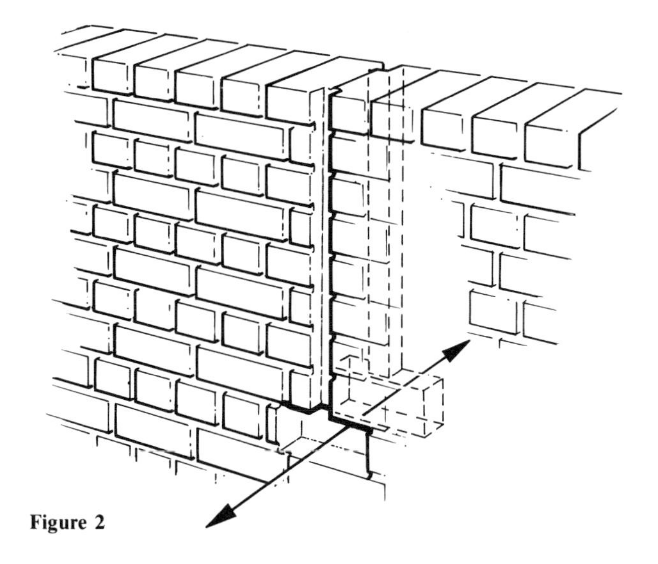
Figure 2

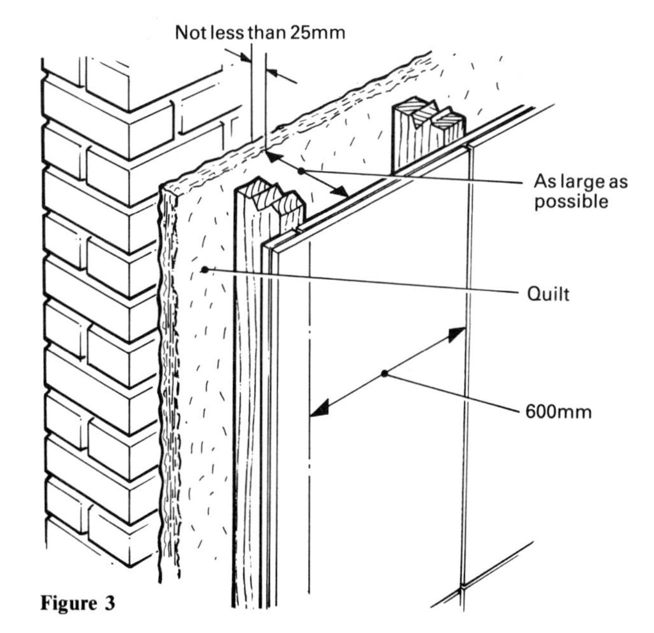

Figure 3

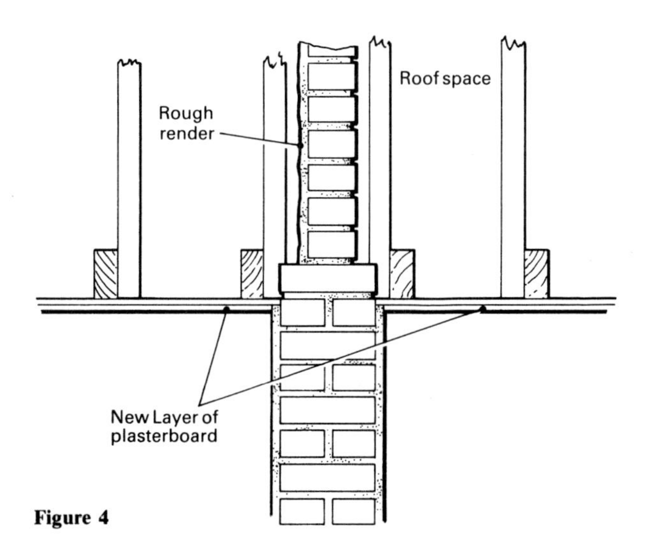

Figure 4

REFERENCES AND FURTHER READING

- ☐ *Defect Action Sheet* DAS 45 Intermediate timber floors in converted dwellings — sound insulation.
- ☐ *British Standard* BS 5628:Part 1:1978 (1985) Structural use of unreinforced masonry.
- ☐ *BRE Digest* 293. Improving the sound insulation of separating walls and floors.

PART EIGHT

Flat Roofs

Flat roofs: built-up bitumen felt — remedying rain penetration

FAILURE: Rain penetration through roof membrane

DEFECT: Inadequate provision to isolate felt from building movements

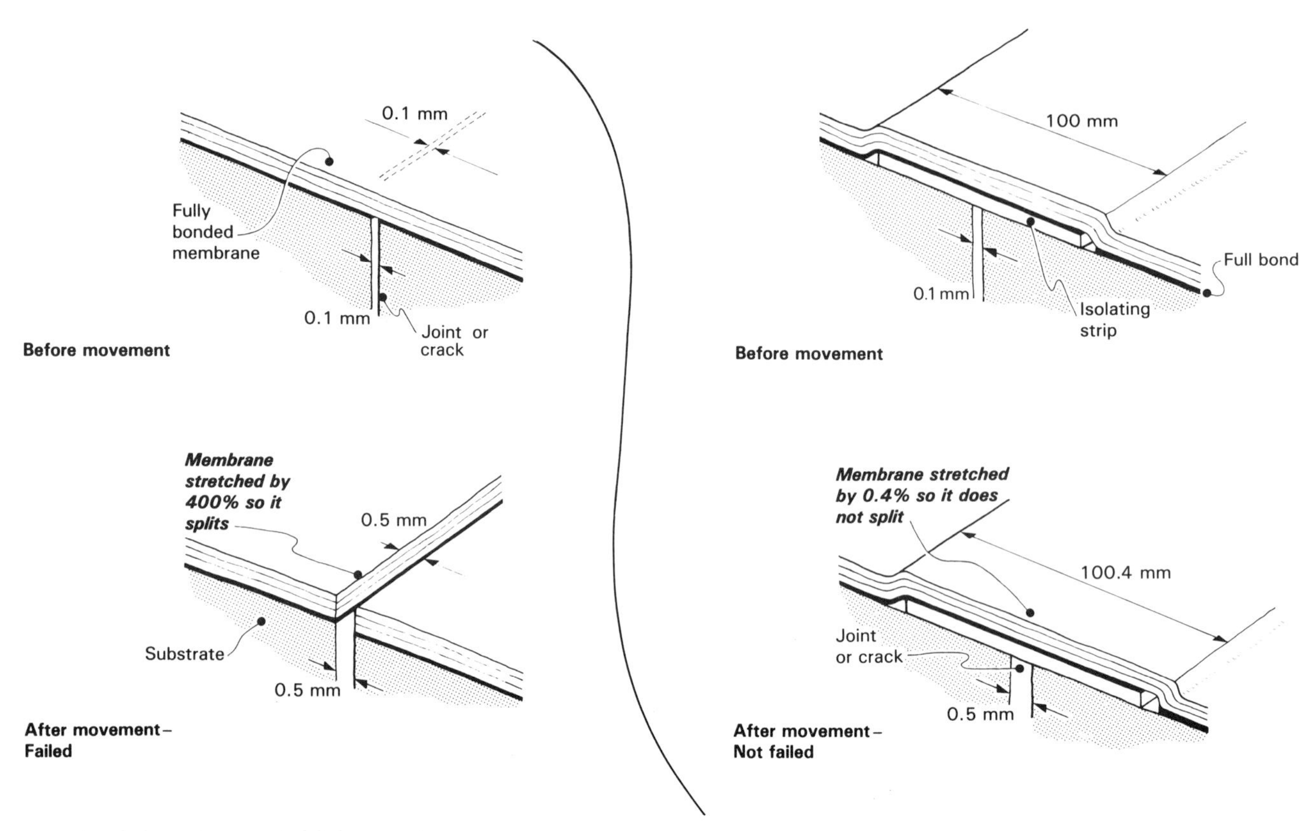

Figure 1 Isolating strip protects felt from strain

This *Defect Action Sheet* deals with remedying localised rain penetration in existing 'cold deck'[1] felted roofs, insufficient to justify replacement of the felt.

In a BRE survey of bituminous felt roofs about half of the failures were attributed to splits produced by locally concentrated movements of the substrate. Very few were attributed to ageing of the felt. Blisters and ridges in the felt are rarely directly responsible for leakage, but they may be vulnerable to damage and impede drainage.

If a bituminous felt ***membrane is not isolated from*** the effects of ***localised movements*** in the substrate it is ***likely to split and leak*** and ***rainwater*** may penetrate into the building and ***into walls and decks.*** If decks are of material that will degrade when wet, leaks should be repaired urgently.

REMEDIAL WORK

Principle — Isolate the roof membrane from localised building movements.

Practice

- Check that dampness is due to roof leakage.
  - Other possible sources of water include condensation within the roof, leakage from water supply pipes, roof drainage pipes and gutters. Check whether leakage is occurring at upstands and flashings. If so see DAS 34[2] and this DAS does not apply.
- Check for obvious signs of mechanical damage and make good.
- Check past records of leakage and repairs.
  - Note particularly any repeated, unsuccessful, repairs: the underlying cause probably has not been correctly diagnosed.
- Relate obvious potential leakage points in membrane to details of construction; a pattern may become apparent and indicate causes.

If splits are found to correspond with positions of minor substrate joints (such as those between screed bays, woodwool slabs, insulation panels) then:

- Where splits are parallel to direction of falls, lay a strip of bituminous felt about 100 mm wide over the entire length corresponding to the substrate joint, unbonded or bonded to one edge only of the existing felt. Follow with two further, progressively wider strips, fully bonded (Figures 2a, b and c) (remove reflective chippings, etc., where repair is to be bonded; repair with felt of the same type as existing material; reinstate reflective treatment).
- Where splits are at right angles to falls (so that a repair as in Figure 2 would impede drainage) cut a 300 mm wide slot centrally over and through to the substrate joint, lay an isolating strip of felt, unbonded or bonded along one edge only, and make good (Figure 3). Note that the 'through joint' in the membrane, so produced is not located over a substrate joint and thus not subjected to movement; it must be carefully sealed and it should be capped by a bonded felt strip even though this may slightly impede drainage.
- If splits are numerous, re-felting may be preferable to patching — but maintain the principle of isolating the felt at substrate joints.
- Check that substrate is sound and dry before repairs are started.
- Check that falls are sufficient to ensure complete drainage.
- Check roof condition regularly:
  - check that gutters and outlets are not blocked;
  - inspect covering, especially all previous repairs.
- Record all repairs so that inadequate remedies are not repeated.

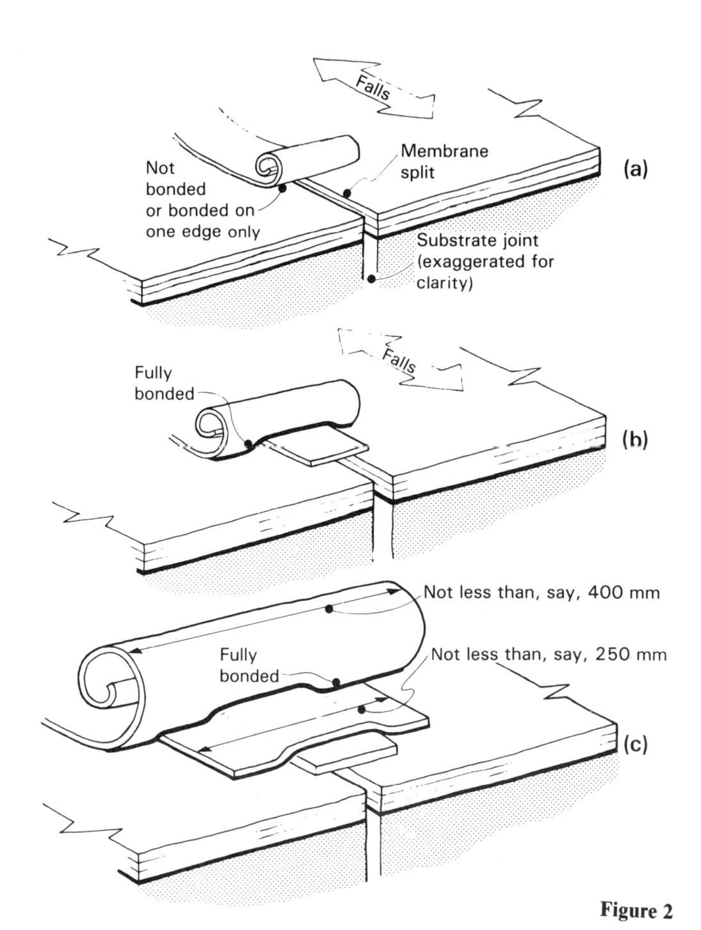

Figure 2

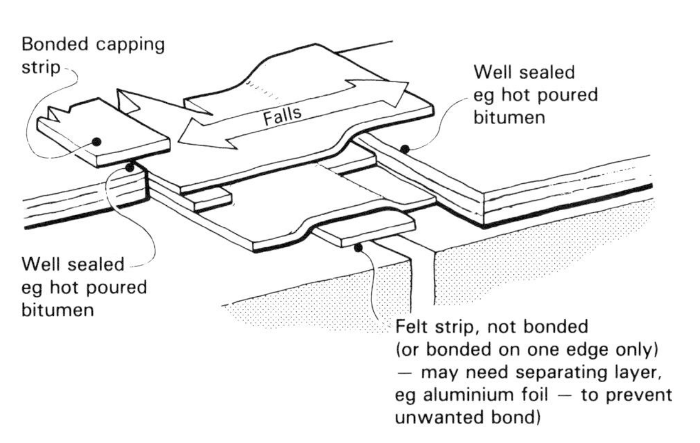

Figure 3

REFERENCES AND FURTHER READING

1 *BRE Digest* 221 'Flat roof design: the technical options'.
2 *Defect Action Sheet* DAS 34 'Flat roofs: built-up bitumen felt — remedying rain penetration at abutments and upstands'.

Flat roofs: built-up bitumen felt — remedying rain penetration at abutments and upstands

FAILURE: Rain penetration at abutments and upstands

DEFECTS: Inadequate detailing at abutments; inadequate provision to isolate felt from building movements

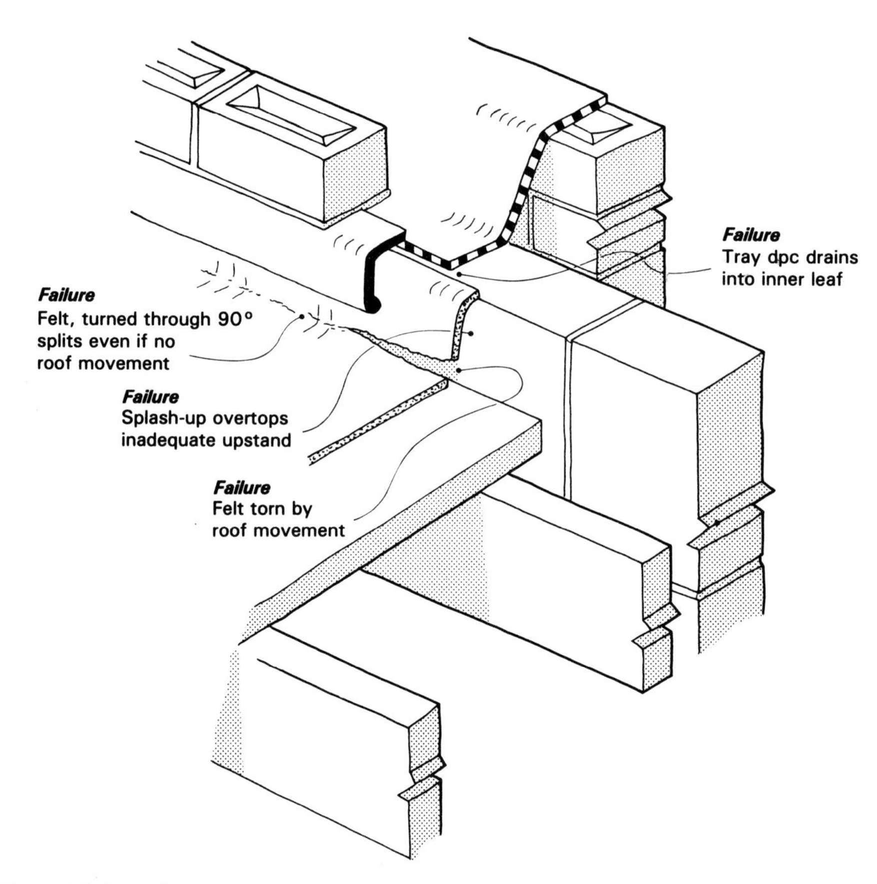

Figure 1 Failures due to poor detailing

This *Defect Action Sheet* deals with remedying localised rain penetration in existing 'cold deck'[1] felted roofs, insufficient to justify replacement of the felt.

In a BRE survey of bituminous felt roofs, about one third of the failures were attributed to poor detailing at abutments.

If a bituminous felt ***membrane is not isolated from*** the effects of ***localised movements*** at abutments it is ***likely to split and leak***; if ***upstands and flashings*** are ***not well detailed, rainwater*** may penetrate into the building and ***into walls and decks***. If decks are of material that will degrade when wet, leaks should be repaired urgently.

REMEDIAL WORK

Principle — Isolate the roof membrane from localised building movements; correct inadequate detailing and upstands.

Practice

- Check that dampness is due to roof leakage.
  — Other possible sources of water include condensation within the roof, leakage from water supply pipes, roof drainage pipes and gutters. Leaks may be occurring in areas away from abutments. If so see DAS 33[2].
- Check for obvious signs of mechanical damage and make good.
- Check past records of leakage and repairs.
  — Note particularly any repeated, unsuccessful, repairs: the underlying cause probably has not been correctly diagnosed.

If splits are found at upstands against abutments such as parapets:

- Check whether deck spans parallel to, or at right angles to, the abutment.

If deck spans parallel to the abutment:

- Check cause of the split: if due to deflection of the deck (or downward movement of its supporting walls), provide an upstand curb, fixed to the deck not the wall, and a new flashing if necessary (Figure 2). Make good.

If deck spans at right angles to the wall and split is at upstand:

- Check whether an angle fillet has been incorporated (Figure 3). If not, provide angle fillet and make good.

If rain penetration of the roof has been identified as the source of dampness, but careful inspection reveals no potential leakage points in the membrane, then:

- Check that all upstands are at least 150 mm high and well lapped by secure flashings.
- Check that flashings lap under DPCs in abutment walls, not over them, especially in a cavity parapet wall where the DPC is a cavity tray draining water towards the roof (Figure 2)
- Ensure that inadequate remedies are not repeated: keep records of repairs.
- Inspect at regular intervals.

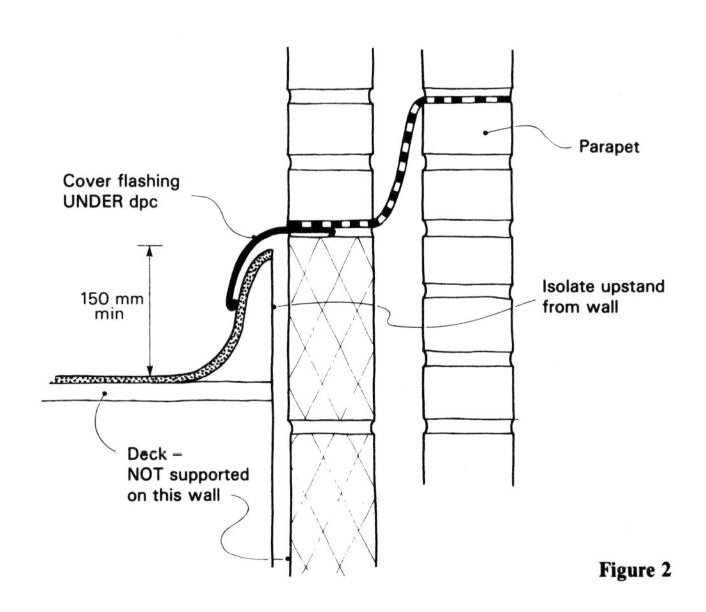

Figure 2

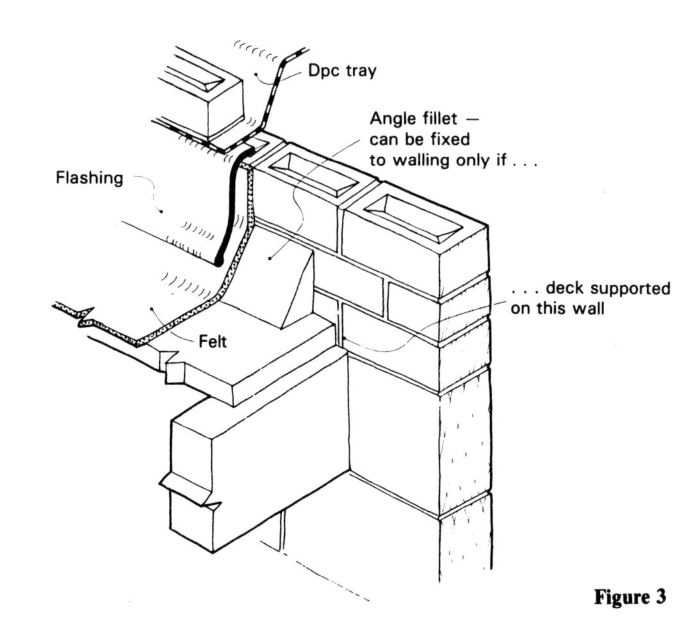

Figure 3

REFERENCES AND FURTHER READING

1 *BRE Digest* 221 'Flat roof design: the technical options'
2 *Defect Action Sheet* DAS 33 'Flat roofs: built-up bitumen felt — remedying rain penetration'

Felted cold deck flat roofs: remedying condensation by converting to warm deck

FAILURES: Wetting of insulation and deterioration of roof structure, due to condensation within roofs; excessive heat loss

DEFECTS: Inadequately ventilated cold deck roof; inadequately insulated cold deck roof

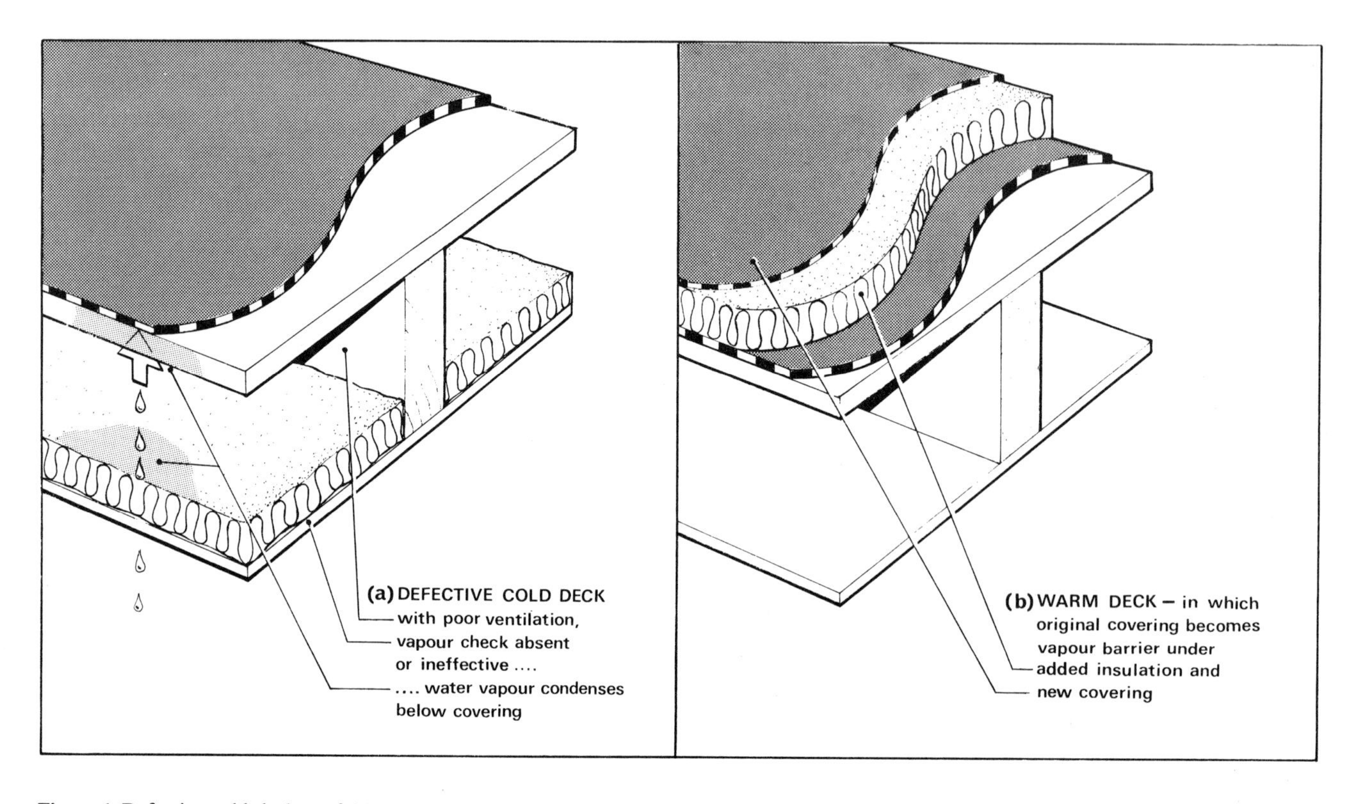

Figure 1 Defective cold deck roof (a) converted to warm deck (b)

In a cold deck roof, Figure 1a, a bitumen-felt roof covering can act as an unwanted vapour barrier on the cold side of the roof. ***If ventilation of roof voids is poor and a vapour check is absent or deficient, condensation occurs within the roof*** — sometimes in amounts sufficient to drain out at light fittings. ***Often*** the construction is such that ***ventilation cannot be improved.*** Usually insulation is absent or inadequate. A remedy sometimes adopted is to ***convert the roof to a 'warm deck' type,*** Figure 1b, by ***adding insulation above*** the existing covering ***and a further covering*** above that. BRE surveys have revealed a number of bad practices in such work (and also in new-build warm deck roofs). If the ***insulation added*** above the deck is ***insufficient,*** becomes ***wet,*** or is ***not continuous, condensation*** in the roof may ***not*** be ***cured.***

Felted cold deck flat roofs: remedying condensation by converting to warm deck

DAS 59
September 1984

REMEDIAL WORK

Principle — Insulation installed above the deck must be sufficient to make condensation below the original roof covering unlikely; the effectiveness of the added insulation must not be reduced by discontinuity or by ingress of water or water vapour.

Practice

Before deciding to convert:

- Check cause of dampness. Ensure it is due to interstitial condensation, not to rain penetration.
- Check that roof structure and deck are sound.
- Confirm that there is no existing insulation below the deck:
  - — if insulation is present, ideally it should be removed so that all insulation in the converted roof is above the vapour barrier formed by the original roof covering;
  - — if its removal is impracticable, insulation installed above the deck will reduce the incidence of condensation but may not do so acceptably. Acceptable reduction of the risk may require an impracticable thickness of insulation. A dew-point calculation[1] is necessary if the risk of condensation within the deck is to be fully evaluated.
- Check that existing covering is sufficiently free from irregularities for added insulation boards to lie flat;
  - — note that chippings may be difficult to remove without damage to the covering.
- Check that insulation can be laid as a continuous layer so that there will be no cold bridges, eg at drainage channels or at changes of roof level, Figure 2.

If, after the above checks, it is decided to convert:

- Check product manufacturer's recommendations for laying insulation and roof covering, as requirements vary.
- Specify that insulation boards are to be kept dry at all stages:
  - — fibrous or open-cell materials can become saturated and even closed-cell materials must be kept dry if bonding problems are to be avoided;
  - — instruct site supervisors to check that no more insulation is laid than can be quickly protected from rain and that insulation board edges are protected from water if work is left uncompleted, Figure 3.

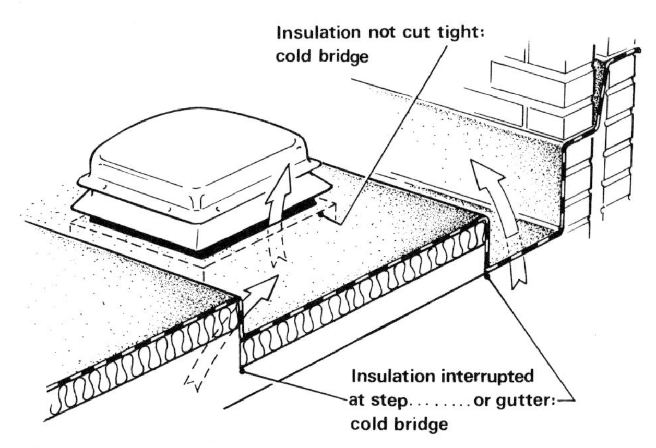

Figure 2

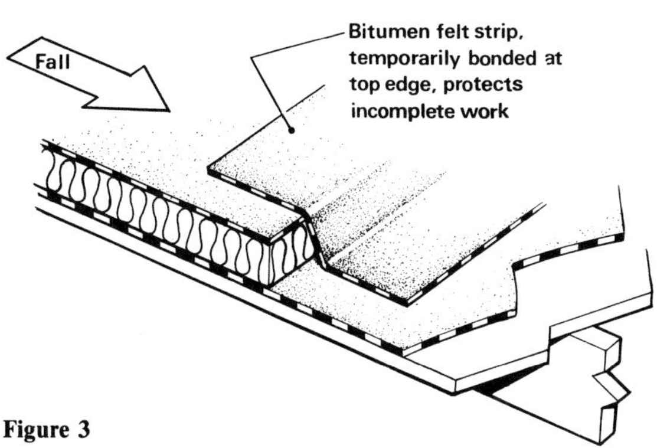

Figure 3

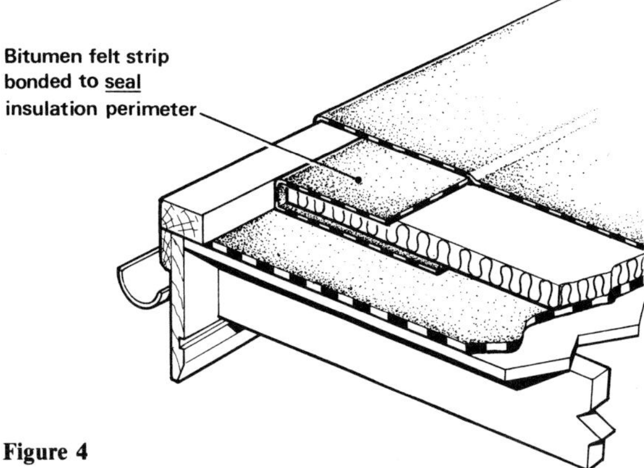

Figure 4

- Specify that the vapour barrier (ie the former roof covering) is to be sealed to the new covering at the perimeter of the insulation, Figure 4.

The last practice points apply equally to new-build warm deck roofs.

REFERENCES AND FURTHER READING

1 *BRE Digest* 110 'Condensation'.
Flat Roofs Technical Guide. Obtainable from the Property Service Agency, Library Sales Office, Whitgift Centre, Croydon CR9 3LY. Price £7.00 net.

DAS 63
August 1985

Site
CI/SfB 8 (27)(H121)

Flat or low-pitched roofs: laying flexible membranes when weather may be bad

FAILURES: Cracked membranes; loss of adhesion

DEFECTS: Laying membranes in cold, wet, or windy weather or on damp decks

Figure 1 Working in bad weather may produce a permanently bad roof

A flexible roof membrane, of any type, will not perform well *if laid on* a *wet or frosted deck. Adhesion* will be *poor* and *trapped moisture* may later cause *blisters.* It is *vitally important* that *deck* surfaces are kept *dry up to* the point *when membranes are laid.*

The *risk of cracking* membranes as they are unrolled *increases with falling temperature. Low temperature* may also cause *lack of adhesion* even when adhesives are hot-applied. *High winds* increase the *risk of damage* to membranes during laying and later if adhesion is poor.

Flat or low-pitched roofs: laying flexible membranes when weather may be bad

DAS 63
August 1985

PREVENTION

Principle — Checks and precautions become critically important if bad weather may occur during or immediately before laying roof membranes; in some conditions work should not proceed.

Practice

- Do not handle or lay sheet materials in strong winds (Force 5) since membranes may be damaged:
  - in Force 5, force of wind is felt on the body;
  - on some roofs work may have to stop at lower wind speeds than this for safety reasons.

- Do not handle or lay sheet membranes when the air temperature is low enough to reduce their workability (generally less than 5°C for heavier types or 0°C for lighter types, Figure 2):
  - some membranes are more tolerant than others of low temperatures: check manufacturers' recommendations for 'conditioning', unrolling in advance of laying, etc.

- Do not lay membranes on damp or frosted surfaces, or when ***any*** rain, sleet or snow is falling.

- Do not lay a greater area of hot-applied adhesive than will stay fluid until membrane is laid.

- Protect, immediately, day-work joints in warm deck roofs with a fully bonded lap, (see Figure 3 and DAS 59).

- Note that the provision of shelter and heat may enable work to proceed.

- Remember that pre-recorded forecasts are available — 'Weather line' numbers in your dialling code booklet. Forecasts specific to contractors' needs can be obtained, for a fee, from your local Weather Centre (see directory).

- For your locality, the Meteorological Office can compute the proportion of working hours in which conditions will be unsuitable for the operations covered by this DAS – a charge is made – England and Wales, telephone 0344 420242, Extension 2278, Scotland, 031 334 9721, Extension 524, Northern Ireland, 0232 228457. A list of threshold conditions for many building operations (together with local data for Plymouth) is given in Reference 1(in preparation).

REFERENCES AND FURTHER READING

1 *Building Research Establishment Report,* 'Climate and Construction Operations in the Plymouth area'; Keeble and Prior. (In preparation.)
British Standard BS 6229 : 1982 : 'Code of Practice for flat roofs with continuously supported coverings'.
National Federation of Roofing Contractors, 'Roofing and cladding in windy conditions'.

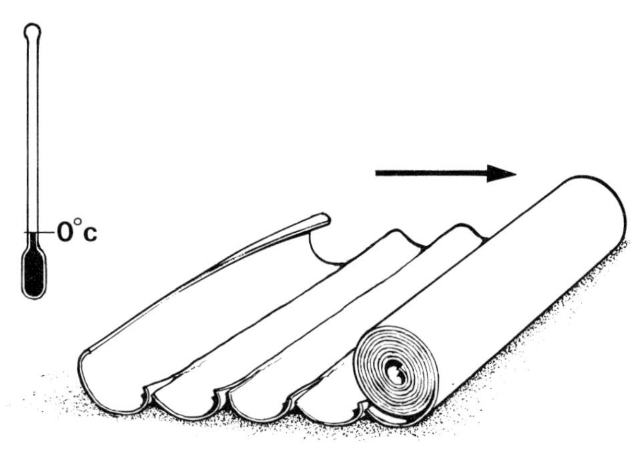

Figure 2

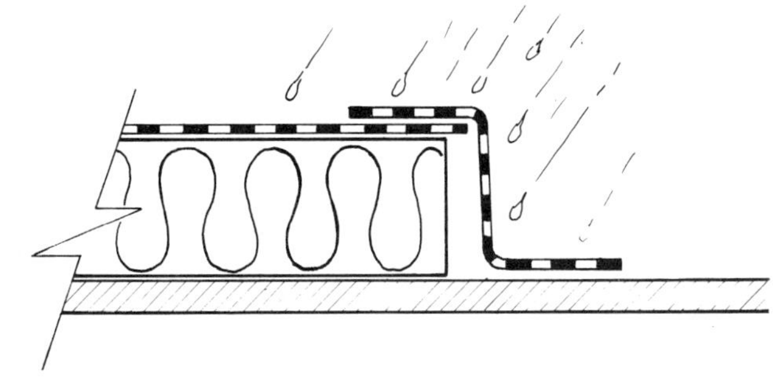

Figure 3

PART NINE

Pitched Roofs

Slated or tiled pitched roofs: ventilation to outside air

FAILURE: Condensation, mould growth, corrosion and rot in roof space

DEFECT: Inadequate provision for ventilating roof space to outside air

In recent years there has been an increasing number of reports of serious condensation in the roof spaces of slated or tiled pitched roofs. The increase has almost certainly been due to a combination of factors — colder roof spaces due to higher levels of thermal insulation, and higher moisture content in the air within the dwellings caused by changed living patterns.

Avoidance of this condensation problem requires two actions — restricting the amount of water vapour which gets into the roof space from the house[1] — and providing ventilation of the roof space to the outside air to a greater degree than has been customary of late. The ventilation hole shown in Figure 1 was inadequate by a factor of 20. A recent survey by BRE revealed that in none of the houses on the building sites visited had the designers made provisions for ventilating the roof spaces to the outside air to the levels required in Clauses 22.8 to 22.16 of BS 5250:1975 'The control of condensation in dwellings'[2]. These levels are now incorporated as an acceptable provision in the Building (Second Amendments) Regulations 1981.

Unless ventilation, as described later, is achieved over the whole of the roof space, ***water vapour leaking into it from the dwelling below is likely to condense*** on the cold underside of the roof covering. If excessive, this can ***wet the structural timbers*** (or prevent their drying out if they were installed wet) with increased risk of ***rot, and corrosion of fixings.*** It may also wet the over-ceiling insulation, reducing its effectiveness, and may ***stain or damage the ceiling*** and ***short-circuit electrical wiring.***

Figure 1 Ventilation hole inadequate

PREVENTION

Principle — Provide ventilation gaps which will give adequate ***ventilation from the outside air*** over the whole roof space (Figure 2).

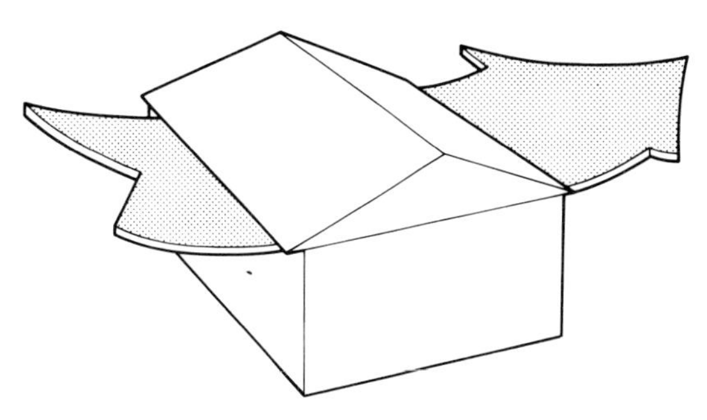

Figure 2

Practice

- Any roof void above an insulated ceiling shall be arranged so as to be cross-ventilated by means of permanent vents not less than 0.3% of the roof plan area; or if associated with a pitched roof of square or rectangular plan, so as to be cross-ventilated by means of permanent vents situated on two opposite sides of the roof, such vents having an area equivalent to that of a continuous gap along those sides of width not less than:
  (i) 10 mm where the roof pitch exceeds 15°; or
  (ii) 25 mm where the roof pitch is 15° or less.

- Dimension the gaps on the drawing. Do not rely on a vague instruction or show only holes of unspecified size.

- Design the ventilation openings so as to prevent the ingress of large insects and birds (for example by using suitable mesh).

- Show a stop or spacer in the eaves so that the ceiling insulation can be installed well into the eaves to prevent a cold bridge without blocking the ventilation[3] (Figure 3).

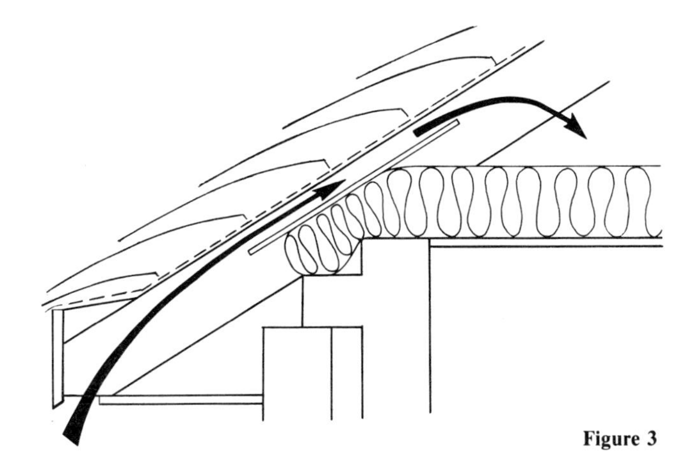

Figure 3

- If it is possible to arrange for ventilation at eaves level on only one side of a roof — for example in mono-pitch construction — it is necessary to provide further ventilators (air bricks in gables, ventilating tiles, ridge ventilators) at high level to promote a through-flow of air. ***High level ventilators must not be used on their own,*** without vents also at low level in the roof. (Failure to observe this may result in an increase in vapour transfer from the dwelling due to negative pressures being created in the roof space.)

REFERENCES AND FURTHER READING

1 *Defect Action Sheet* DAS 3 'Slated or tiled pitched roofs: restricting the entry of water vapour from the house'.

2 *British Standard* BS 5250:1975 'The control of condensation in dwellings'.

3 *Defect Action Sheet* DAS 4 'Pitched roofs: thermal insulation near the eaves'.

BRE Digest 180 'Condensation in roofs' (revised 1978).

Slated or tiled pitched roofs: restricting the entry of water vapour from the house

FAILURE: Condensation, mould growth, corrosion and rot in the roof space (Figure 1)

DEFECT: Passages into the roof space for water vapour from the house

In recent years there have been an increasing number of reports of serious condensation in the roof spaces of slated or tiled pitched roofs. The increase has almost certainly been due to a combination of factors — colder roof spaces due to higher levels of thermal insulation, and higher moisture content in the air within the dwellings caused by changing living patterns.

Avoidance of this condensation problem requires two actions — providing adequate ventilation of the roof space to the outside air (dealt with in Defect Action Sheet 1[1]), and restricting the amount of water vapour which gets into the roof space from the house.

Figure 1 Condensation and mould growth in the roof space

Large amounts of water vapour are produced in houses, particularly from bathrooms and kitchens. Some of this ***water vapour can find its way into the roof space,*** mostly by being carried by air currents ***through holes in the ceilings.*** The most significant holes are gaps ***around loft access hatch lids,*** (Figure 2) or ***through the hatch itself*** if the lid becomes displaced, and gaps around pipework, especially in bathroom airing cupboards (Figure 3). Lesser amounts of water

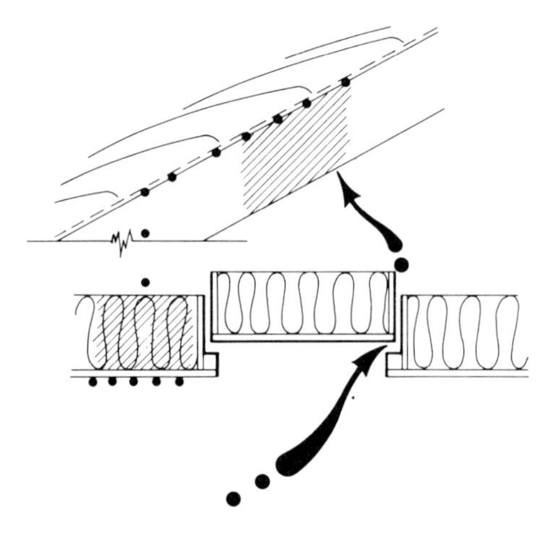

Figure 2

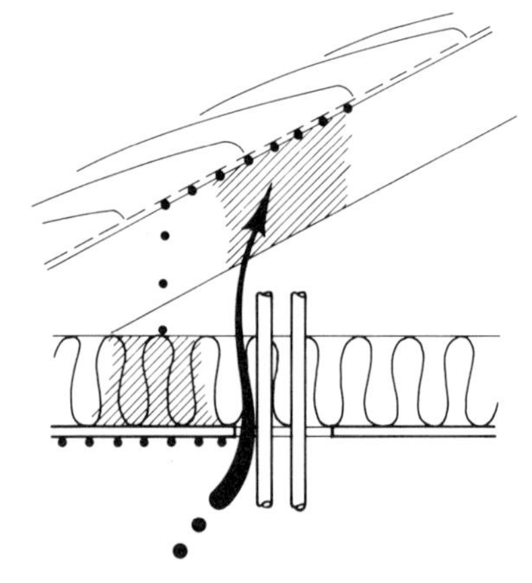

Figure 3

vapour pass through gaps around light fittings. Some water vapour also penetrates ceilings by diffusion.

Water vapour reaching the roof space and not removed by ventilation may cause ***mould growth*** in the roof and a high moisture content in roof timbers, which can lead to ***rot.*** It can also cause ***corrosion*** of steel fixings, nails and connections in the roof space.

Condensation in the roof can lead to ***deterioration of insulation, damage to ceilings*** and decorations and ***electrical risks*** with wiring.

PREVENTION

Principle — Design to restrict passage of air and water vapour from the house into the roof space to amounts that roof ventilation can remove.

Practice

- Avoid gross perforations of the ceiling, or show how they can be sealed (Figure 4).
- Provide a compression seal for loft access hatches (Figure 5).
- Provide a weight or a latch to loft access hatch lids (Figure 5).
- Above bathrooms and kitchens and for areas beneath roofs with a pitch below 15° ceilings should include a substantial vapour check, for example not less that 250 gauge polyethylene with sealed joints.

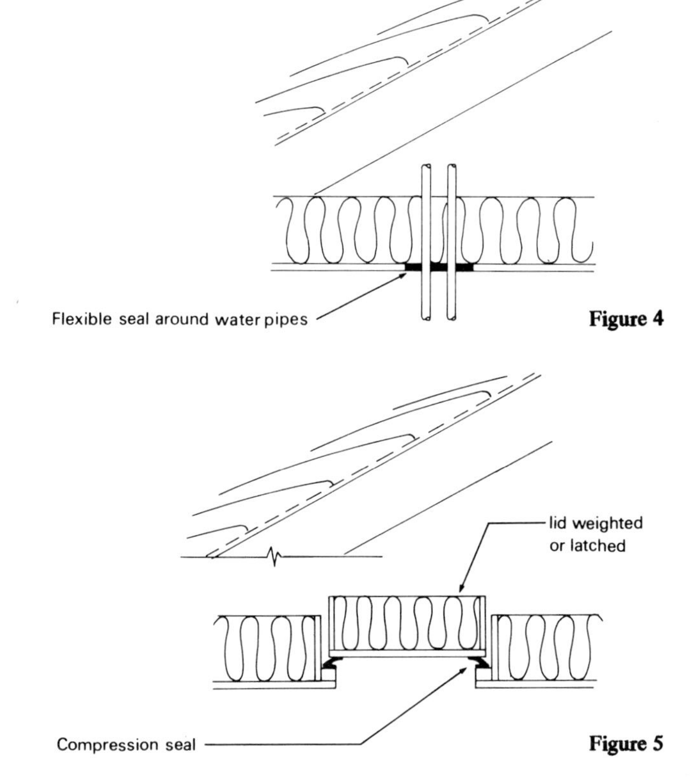

Figure 4

Figure 5

REFERENCE AND FURTHER READING

1 *Defect Action Sheet DAS 1.* 'Slated or tiled pitched roofs: ventilation to the outside air'.

British Standard 5250:1975 'The control of condensation in dwellings'.

BRE Digest 110 'Condensation'.

BRE Digest 180 'Condensation in roofs'.

Pitched roofs: thermal insulation near the eaves

FAILURES: Condensation and mould growth in the house and the roof space; rot in timber and corrosion of steel fixings in the roof space

DEFECTS: Insulation not covering the ceiling at the edges, or pushed into the eaves so that eaves ventilation is blocked

On a recent BRE survey of new low-rise traditional housing, ceiling insulation was frequently found to be missing at the edges or to be pushed into the eaves so that the eaves ventilation was blocked (Figure 1).

Lack of care in placing the insulation in the eaves can have serious consequences.

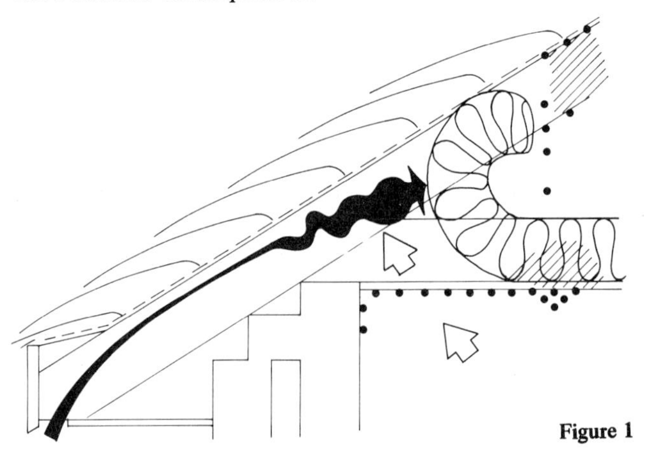

Figure 1

If the ***insulation is not carried into the eaves,*** (Figure 2) the edges of the ceiling will not be insulated and a ***'cold bridge'*** will result which may lead ***to condensation and mould growth*** in the room below at the junction of wall and ceiling.

If the ***insulation is pushed or blown too far into the eaves,*** (Figure 3) ***or left partially unrolled*** (Figure 1) ***cross ventilation*** provided at the eaves may be ***reduced or totally blocked.*** Cross ventilation is vital. Without it water vapour from within the house can cause ***mould growth in the roof,*** high moisture content in roof timbers which can lead to ***rot***, and ***corrosion*** of steel fixings, nails and connectors in the roof space.

It can also cause ***condensation*** which may wet the insulation, reducing its effectiveness and ***damaging ceilings and decoration.***

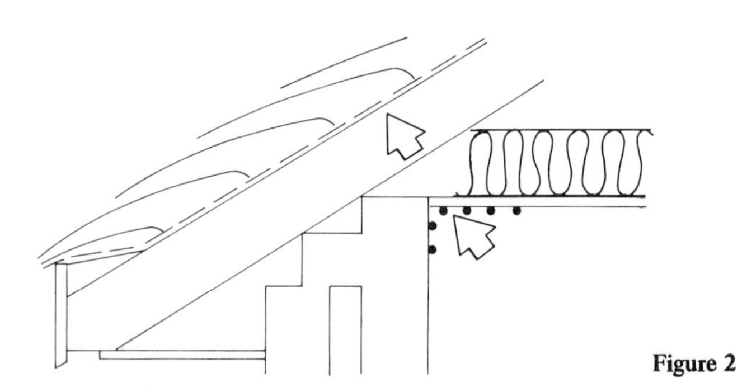

Figure 2

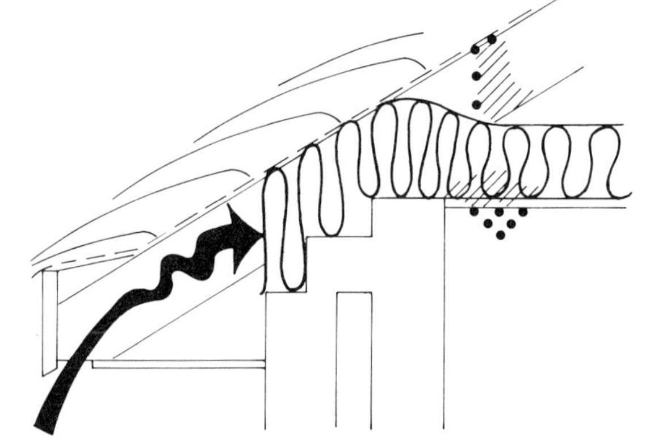

Figure 3

PREVENTION
Principle — Full coverage of insulation with unobstructed ventilation

Practice
- Take care at the eaves that the insulation covers the whole ceiling (A, Figure 4).
- Ensure that the insulation is layed up to stops or spacers which should have been provided for roof ventilation (B, Figure 4).
- When the insulation has been laid, check that the cross-ventilation provided for in the design is not blocked in any way (C, Figure 4).

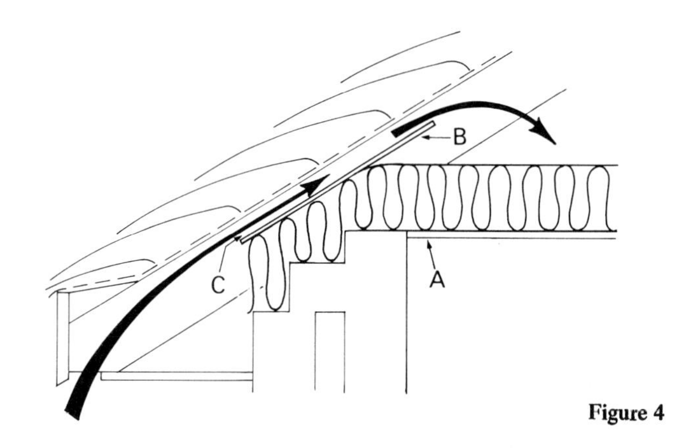

Figure 4

Slated or tiled pitched roofs — conversion to accommodate rooms: ventilation of voids to the outside air

FAILURES: **Condensation; rot in timbers; corrosion of steel fixings**

DEFECTS: **Insufficient area of ventilation openings to the outside air; air flow across roof voids restricted; discontinuous or absent vapour control layer**

Figure 1

Adequate cross ventilation of roof voids is particularly difficult to achieve when rooms intrude into them. Thermal insulation around rooms in the roof may obstruct ventilation of the roof void, Figure 1.

If adequate ***cross ventilation is not achieved*** everywhere in the roof void, persistent ***condensation may drip onto the ceiling, may reduce insulation*** effectiveness and ***promote rot*** and ***corrosion.***

PREVENTION

Principle - All roof voids must be adequately ventilated to the outside air.

Practice

- Specify ventilation openings to the roof void, at low (eaves) level, equivalent in nett area to a continuous 25 mm slot along the full length of the eaves, Figure 2.

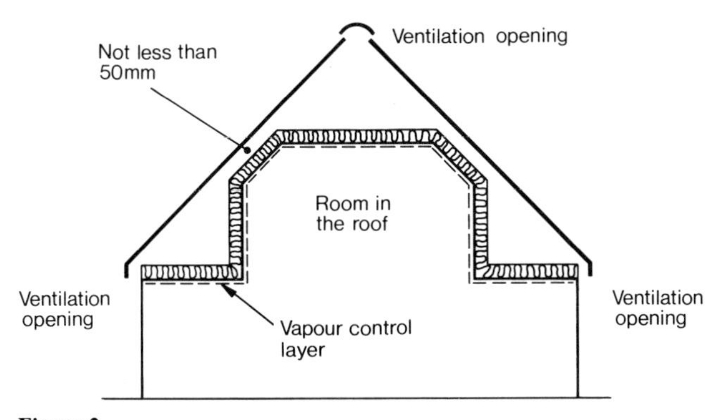

Figure 2

- Specify ventilation openings at high (ridge) level;
  - these should be equivalent in nett area to a continuous opening of 5 mm, Figure 2;
  - proprietary products such as ventilating ridge tiles are available;
  - high level ventilation must only be provided in addition to low level ventilation. High level vents used alone increase the risk of vapour transfer from the dwelling.

- Provide ventilation openings, so that cross ventilation is achieved in every part of the roof void;
  - where ventilation openings are intermittent the openings must be positioned to ensure that every roof void is ventilated;
  - check that dormer constructions will not obstruct cross ventilation of the main roof.

- Specify screens for ventilation openings to resist entry of large insects, birds or small mammals.

- Specify, where insulation is to be positioned at rafter level, an unobstructed air-path above the insulation of not less than 50 mm depth;
  - where insulation is installed between the rafters, take into account the effect of noggings and similar intrusions;
  - rafter depth may be increased by fixing battens to their underside;
  - the use of certain board insulants can reduce the total thickness of insulation needed;
  - insulated linings fixed to the underside of the rafters can provide scope to accommodate an air-space between rafters.

NOTE: The fire risk of materials used either in original ceiling level insulation or in new board insulants and insulated linings may require special consideration[1].

- Consider, where quilted insulation is to be used, specifying a paper or polythene backed quilt for rafter sections — the backing can be tacked to the underside of the rafters to hold the quilt in position until the ceiling is fixed. Specify that the quilt must be installed with its backing on the underside (nearest the habitable room).

- Specify that holes (eg for service entries) between roof void and room are to be sealed;

- Specify 'draught stripping' of access hatches.

- Specify a vapour control layer to the entire habitable area intruding into the roof space (this is essential if permeable insulants are used)
  - for example 60 μm polythene sealed at joints.

REFERENCES AND FURTHER READING

1 *BRE Digest* 233 Fire hazard from insulating materials.

☐ *Defect Action Sheet* DAS 3 Slated and tiled pitched roofs: restricting the entry of water vapour from the house.

☐ *British Standard* BS 5250 Basic data for the design of buildings: the control of condensation in dwellings.

Slated or tiled pitched roofs — conversion to accommodate rooms: installing quilted insulation at rafter level

FAILURE: **Condensation; rot in timbers; corrosion of steel fixings**

DEFECTS: **Insulation obstructing air-paths through roofs; obstructed ventilation openings; incomplete vapour control layer**

Figure 1

Where roofs contain rooms it is usually necessary to insulate the roofs, or parts of the roof, at rafter level — where space for installing the insulation is often comparatively limited. Such roofs may not be adequately ventilated, because the air-path can be obstructed by poorly installed thermal insulation, Figure 1.

If ***insulation*** is installed in a way that ***obstructs*** the ***air-paths above*** it, the ***roof*** void may ***not be ventilated*** adequately; ***condensation, rot*** in timbers, ***corrosion*** of steel fixings and damaged ceilings and decorations may follow.

Slated or tiled pitched roofs — conversion to accommodate rooms: installing quilted insulation at rafter level

DAS 119
August 1988

PREVENTION

Principle — Full coverage of insulation must be achieved to habitable spaces without obstructing the ventilation to the roof void.

Practice

- Ensure that insulation is installed before fixing ceiling of room in roof;
  - — provide support — eg fibrous insulation without backing should be supported by netting;
  - — where polythene or paper backed quilt is specified ensure that it is installed with the backing below the quilt and tacked to the underside of the rafters, Figure 2. The backing must be on the room side of the insulation;
  - — do not permit insulation to be installed after ceiling boards are in place since air-paths cannot be assured.

NOTE: a separate vapour control layer (eg polythene) is required where fibrous insulation is used.

- Ensure that a space of not less than 50 mm is maintained above the insulation, Figure 3.

- Check that the insulation is installed without gaps;
  - — insulation should be dressed over noggings — providing an acceptable space above the insulation can be achieved, Figure 4.

- Check that excess quilt is trimmed away; do not permit 'overpacking';
  - — spaces alongside walls and adjacent to roof lights are often narrower than normal rafter spacing and insulation should be trimmed accordingly.

- Check that ventilation openings at eaves are not obstructed.

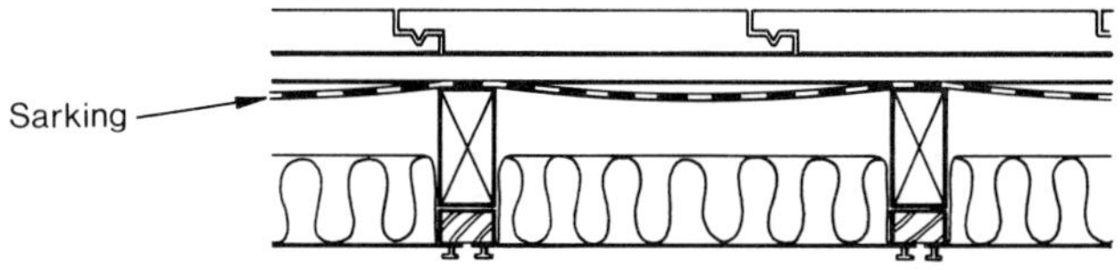

Sarking

Figure 2

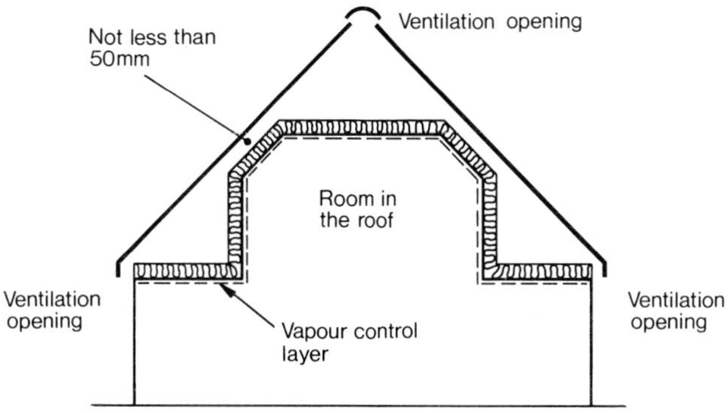

Not less than 50mm
Ventilation opening
Room in the roof
Ventilation opening
Vapour control layer
Ventilation opening

Figure 3

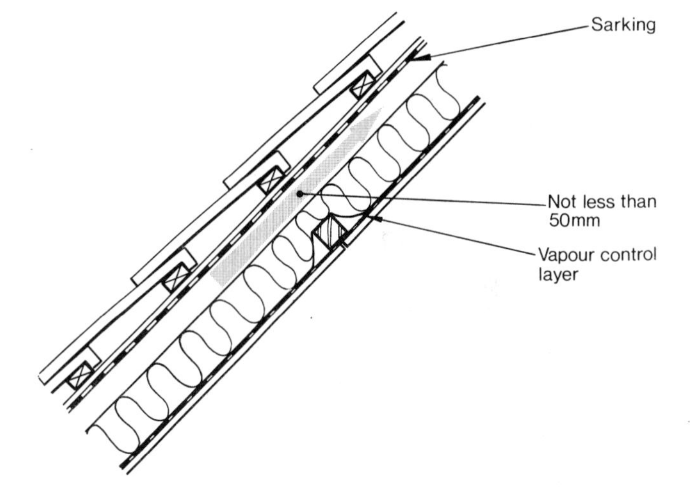

Figure 4

REFERENCES AND FURTHER READING

□ *Defect Action Sheet* DAS 4 Pitched roofs: thermal insulation near the eaves.

Pitched roofs: boxed eaves — preventing fire spread between dwellings*

FAILURE: **Spread of fire, within boxed eaves, between adjoining dwellings**

DEFECTS: **No cavity barrier to close voids in boxed eaves at separating walls; incomplete separating walls**

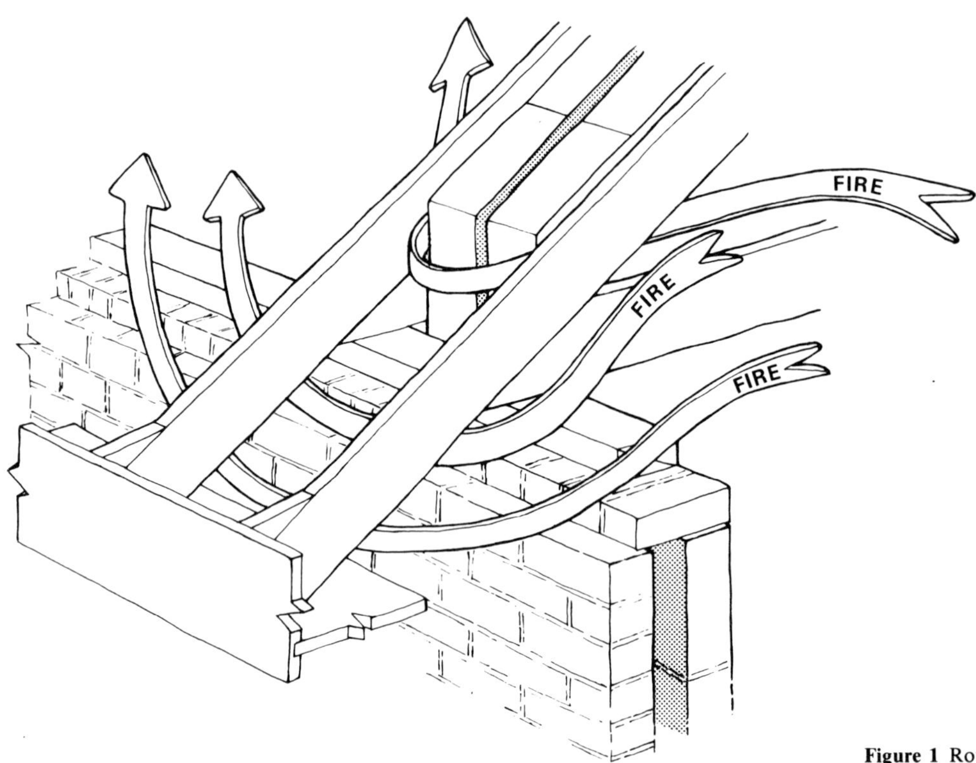

Figure 1 Routes for spread of fire at eaves

In a BRE Survey of house design and construction, there was found to be a potential route for fire between adjoining dwellings at every continuous 'boxed eaves' detail. Most also had other routes for fire past the end of separating wall (Figure 1) and some of these routes were wider than they might otherwise have been as a result of incomplete blockwork which had been concealed by the adjacent rafters from subsequent inspection. Fires are known to have spread between dwellings through these routes.

Since these defects are so common in boxed eaves, it is clear that designers and builders are generally unaware of their significance. The Building Regulations 1976, E8 (1) require a separating wall to be imperforate and to form a complete vertical separation between dwellings, including their roof spaces. There will be routes by which ***fire can spread between dwellings if the separating wall stops at the inner edge of the wall plate*** and if the ***voids within the boxed eaves are not closed.*** These defects are significant in all cases, and particularly so where there are habitable rooms in either roof space.

*Requirements in Inner London may differ

Pitched roofs: boxed eaves — preventing fire spread between dwellings*

DAS 7

October 1982

PREVENTION

Principle — there must be a complete separation in the plane of the separating wall between dwellings, which cannot be by-passed by fire.

Practice

- Design to simplify the shape of the void to be closed, (A, Figure 2.) Consider extending the separating wall to the outer face of the external wall within the eaves.
- Do not carry the separating wall over an uninterrupted wall plate: movements in the timber will disrupt the brickwork or blockwork.
- Select material for filling the eaves void that can be readily cut to profile — or, better, that is sufficiently resilient to achieve a tight fit without gaps.
- Ensure that the filling material can be securely fixed without support from the soffit board: it must remain in place if the soffit board is destroyed by fire.
- Consider using:
  - (a) Wire reinforced mineral wool, 50 mm thick, (B, Figure 2);
  - (b) Mineral wool, wired to expanded metal lath for support;
  - (c) Semi-rigid mineral wool batt, spiked or wedged in place;
  - (d) Compressed mineral board cut to close fit;
  - (e) Plywood, not less than 19 mm thick and treated with a flame retardant, cut to close fit;
  - (f) Sand:cement (or pre-mixed vermiculite: cement) render on expanded metal lath.

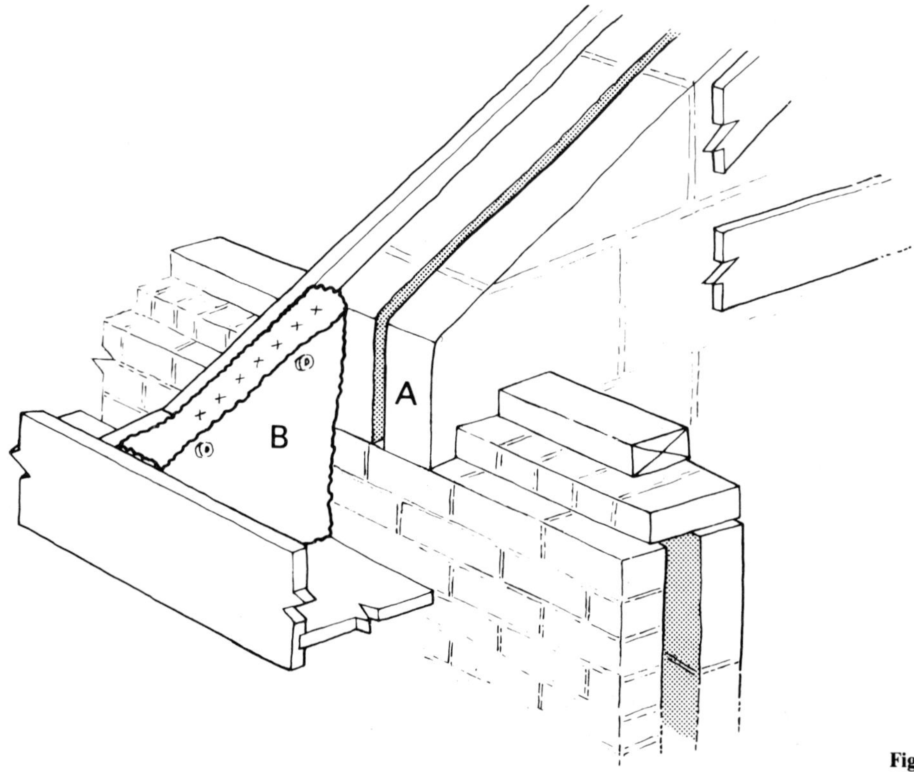

Figure 2

REFERENCES AND FURTHER READING

The Building Regulations 1976, E8(1).

BRE Digest 218 'Cavity barriers and ventilation in flat and low-pitched roofs.

Pitched roofs: separating wall/roof junction — preventing fire spread between dwellings*

FAILURE: **Spread of fire over top of separating wall between adjoining dwellings**

DEFECT: **Inadequate fire-stopping between the top of the separating wall and the underside of the roof covering**

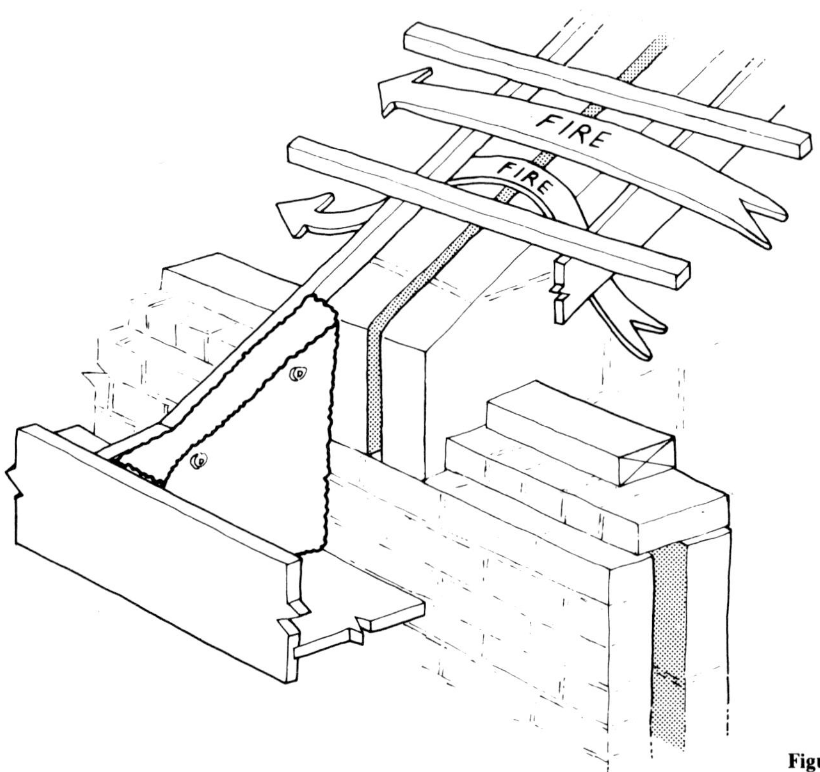

Figure 1 Routes for spread of fire over separating wall

A BRE survey of house design and construction revealed that fire-stopping at the top of separating walls is rarely effectively done. For adjacent small residential buildings, neither of which exceeds 12.5 m in height, the separating wall need not be carried up above the roof covering provided that the covering to the roof is non-combustible and the junction between the separating wall and the roof covering is fire-stopped. If the ***gap*** is ***not*** properly ***fire-stopped, fire can spread*** from one dwelling to the next.

There are three main ways in which defects occur. Sometimes no attempt is made to provide the necessary fire-stopping. In other cases the attempt is confined to inserting mineral wool between the top of the wall and the underside of the sarking felt after roofing has been completed, so leaving an unsealed void directly beneath the tiles. Alternatively, after battening, a mortar bed is trowelled onto the wall under and between the battens; this method also leaves unsealed voids beneath the tiles. Because the battens become supported by the mortar where they cross the wall, it also leads to subsequent hogging of the roof at the separating wall and possible displacement of the tiles.

*Requirements in Inner London may differ

Pitched roofs: separating wall/roof junction — preventing fire spread between dwellings

DAS 8
October 1982

PREVENTION

Principle — there must be a complete separation in the plane of the separating wall between dwellings, which cannot be by-passed by fire.

Practice

- Ensure that the top of the separating wall, when trimmed to the slope of the roof and mortared if necessary to achieve a fair line, will be about 25 mm below the top edges of the adjacent rafters. This will minimise the risk of hogging of the roof.
- Select for fire-stopping a rock-wool, slag-wool or glass fibre quilt that is resilient enough to fill irregular spaces, but not so resilient as to lift or dislodge tiles.
- Ensure that, before felting and battening, the quilt is laid down the top edge of the separating wall, with the edges tucked between faces of the wall and adjoining rafters to keep it in place initially, (A, Figure 2).
- Ensure that, after felting and battening, lengths of quilt are laid between battens as tiling proceeds or fixed by spot sticking in place before tiling, (B, Figure 2).
- Instruct the site to check, either as tiling proceeds or later by lifting tile ends, that quilt is in place.
- Do not use intumescent materials — they are not suitable for this application.

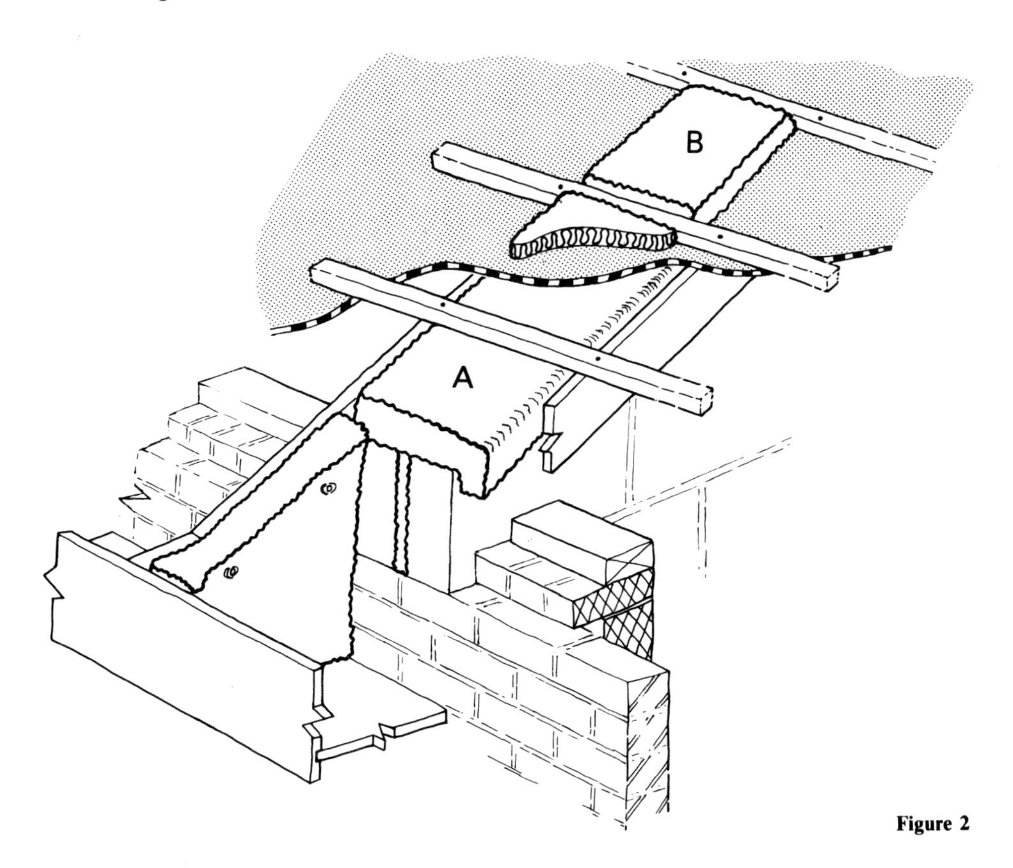

Figure 2

REFERENCES AND FURTHER READING

BRE Digest 218 'Cavity barriers and ventilation in flat and low-pitched roofs.

Pitched roofs: sarking felt underlay — drainage from roof

FAILURES: **Water penetration into eaves construction; degradation of eaves construction materials; dampness in internal walls under 'clipped' eaves; dampness in bargeboards**

DEFECTS: **Sarking felt not supported at eaves to enable water to drain out over fascia and into gutter; sarking felt not carried over bargeboards**

Figure 1 Rainwater ponding on sarking felt behind eaves fascia board

In BRE surveys of houses under construction, ponding of water was often seen on sarking felt underlays at the eaves. Wind-blown rain or snow can penetrate the slates or tiles. It is therefore important that water can drain safely from the sarking felt. BS 5534 : 1978 (Reference 1) says: 'The underlay should allow drainage of water and should extend over the tilting piece, fascia board and into the eaves gutter. Water traps behind the fascia board should be avoided'.

If the ***sarking felt is unsupported*** at the fascia board it ***sags***. ***Water*** collecting here cannot drain into the gutter but is likely to find ***a path into the eaves construction*** — putting soffit boards, timber framing and fastenings at risk. Where 'clipped' eaves are used, this water may find its way ***into the internal walls*** of the building. The omission of a tilting piece may seem a small oversight, but the consequences of omitting it may be serious. It is much cheaper to provide it in the first instance than to have to remedy the omission later. It may also seem relatively unimportant if the sarking felt does not extend, at verges, over any bargeboard. But the consequent persistent dampness in the board, due to water gaining access to its concealed face, can lead to early and repeated paint failure and eventual rot of the board.

PREVENTION

Principle — Water, such as from penetrating wind-blown rain or snow, must drain safely from the sarking felt and not get into the building.

Practice

- Use eaves sprockets — or, much better, a continuous tilting fillet or board — under the sarking felt at the eaves, (A, Figure 2).
- Ensure that the sarking felt projects at the eaves so as to drain to the gutter. (Note: there is evidence that the exposed projecting part of some sarking felts may disintegrate after a few years, and BS 747 states that type 1F felt is not recommended for external use: consider using a separate more durable flashing strip, such as bituminous dpc material, (B, Figure 2).
- Require the sarking felt to be carried over the top edge of any bargeboard, (A, Figure 3).

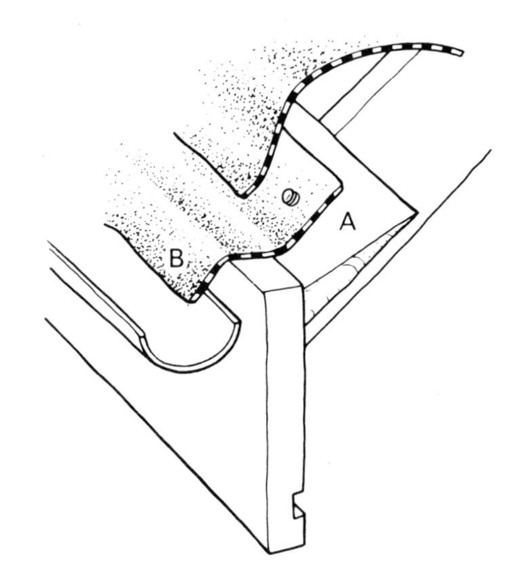

Figure 2

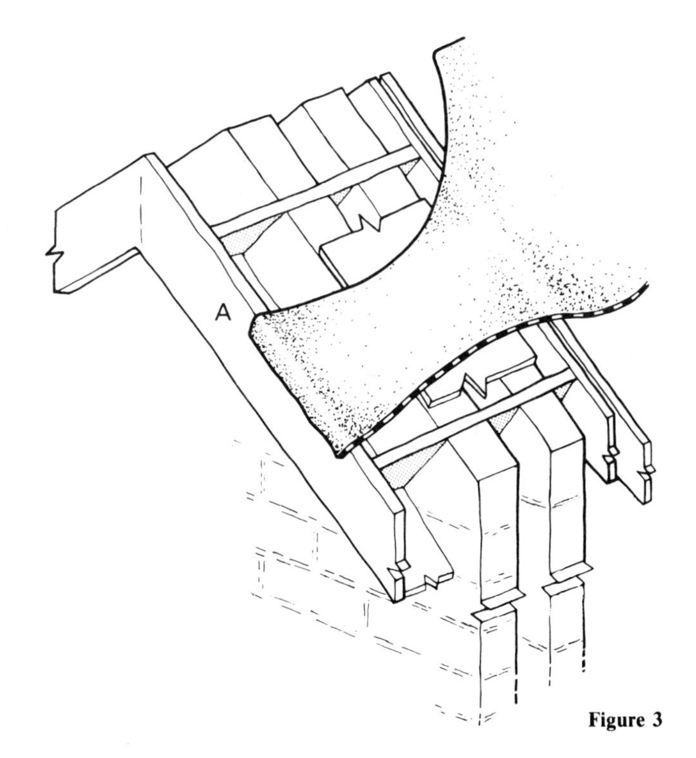

Figure 3

REFERENCES AND FURTHER READING

1 *British Standard* BS 5534 — 'Code of practice for slating and tiling' Part 1 : 1978.

2 *British Standard* BS 747 'Specification for roofing felts'.

3 *Defect Action Sheet* DAS 10 (Site) 'Pitched roofs: sarking felt underlay — watertightness'.

Pitched roofs: sarking felt underlay — watertightness

FAILURES: **Dampness and staining of ceilings, wetting of loft insulation; dampness in eaves and bargeboards; dampness in inner leaf of walls below 'clipped' eaves**

DEFECTS: **Sarking felt not fitted closely around SVP, not properly lapped, not dressed out over eaves gutters and bargeboards**

Figure 1 Sarking felt 'fitted' round soil and vent pipe

The installation of sarking felt to tiled roofs was faulty in various ways on a third of the building sites visited in a BRE survey. When wind-blown rain or snow reached the sarking felt, water would be led into the roof spaces and into construction materials.

Torn sarking felt round SVPs, as in Figure 1, can lead to ***wetting of the loft insulation*** and ***staining of the ceiling.*** Laps that are not big enough or that gape can allow water to enter the roof space generally. If sarking felt is not dressed out over bargeboards they, and the soffit and timbers of any gable ladder, will suffer from dampness. If sarking felt is ***not supported*** behind the fascia and ***dressed out into the eaves gutters*** water can ***drain into the eaves construction*** or, where 'clipped' eaves are used, ***into the top of the inner leaf.*** (The design should have provided support for the sarking felt so that it cannot 'trough' behind the fascia.) Because the exposed edge of the sarking felt where it projects into the gutter is likely to deteriorate, the design may have called for an additional flashing at this point. If so, the sarking felt should lap on top of it to ensure outward drainage.

PREVENTION

Principle — Water, such as from penetrating wind-blown rain or snow, must drain safely from the sarking felt and not get into the building.

Practice

- At SVPs cut a cross in the sarking felt with the cuts at about 45° to the roof slope, (A, Figure 2), and turn the tongues so formed upward around the pipe, (B, Figure 2). This will not be completely watertight though a great improvement on existing practice. For greater security wrap a sealant bandage around the upturned tongues. If the sarking felt is damaged or cut in the wrong place at the SVP, use an additional piece, cut as described, and tuck its top edge under the sarking felt next above.

- Ensure that vertical laps are 100 mm minimum, and occur only over rafters, and are securely fixed.

- Ensure that horizontal laps are:
  225 mm minimum for pitches of 12½° to 14°
  150 mm minimum for pitches of 15° to 34°
  100 mm minimum for pitches of 35° and over.

- Ensure that horizontal laps do not gape, (C, Figure 2).

- Install sarking felt to enable water to drain into eaves gutters, (D, Figure 2).

- Carry sarking felt over top edge of any bargeboard.

- Ensure that design provisions for eaves tilting piece (and any additional eaves flashing) are met.

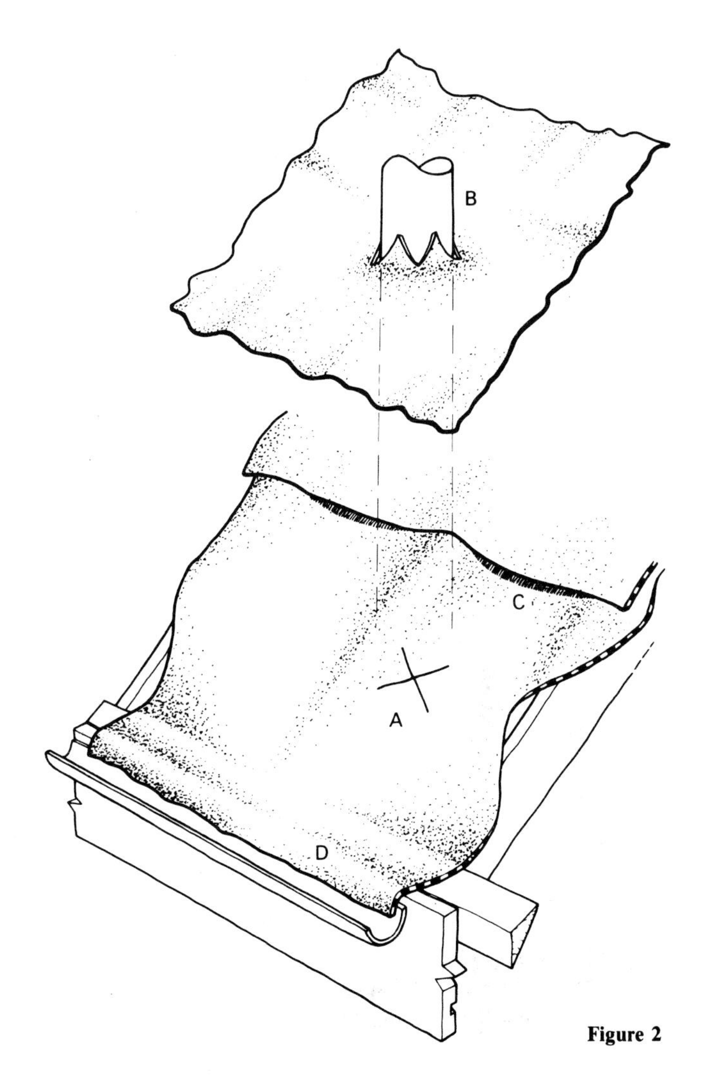

Figure 2

REFERENCE AND FURTHER READING

British Standard BS 5534 'Code of practice for slating and tiling' Part 1 : 1978.

Defect Action Sheet DAS 9 (Design)'Pitched roofs: Sarking felt underlay — drainage from roof'.

Slated and tiled pitched roofs: flashings and cavity trays for step and stagger layouts — specification

FAILURE: Rain penetration into a separating wall

DEFECTS: Cavity trays, soakers or flashings omitted, incorrectly formed or positioned

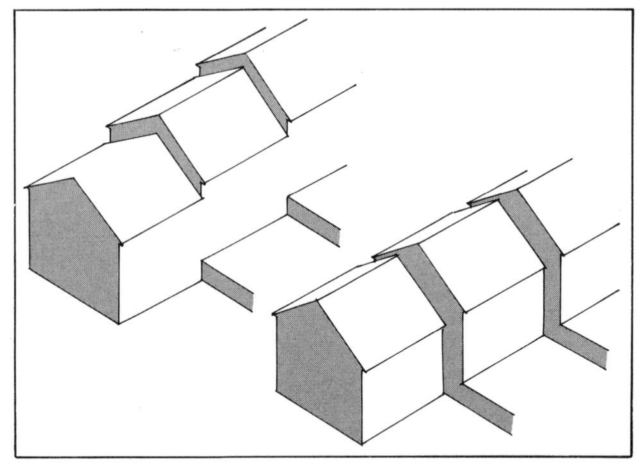

Figure 1a

Where a pitched roof abuts a separating wall, Figure 1a, the outer leaf of the wall becomes an inner leaf below roof level; particularly careful design is needed for steps and staggers to ensure that rainwater does not run down into the wall below the roof level, Figure 1b.

Figure 1b

If ***stepped cavity trays*** are ***omitted*** or if ***roof flashings*** and ***soakers*** are ***incorrectly designed,*** then ***rainwater may penetrate*** to the ***walling*** beneath.

PREVENTION

Principle — Rainwater in walls exposed to the weather must be prevented from draining to the interior of the building.

Practice

- Consider at the planning stage whether step and stagger relationships can be adjusted to avoid detailing difficulties, eg Figure 2.

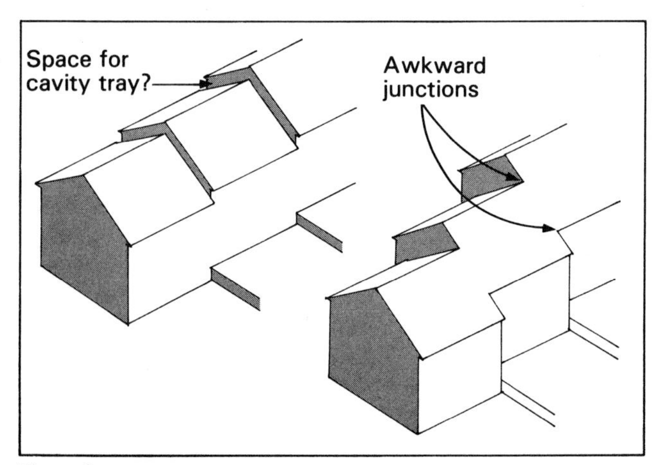

Figure 2

Slated and tiled pitched roofs: flashings and cavity trays for step and stagger layouts — specification

DAS 114
May 1988

- Specify a cavity tray to discharge water outwards, Figure 3;
  - — proprietary units are available that do not need to be built into the inner leaf, some are supplied with an attached lead flashing, Figure 3;
  - — where lead is specified the gussets should be lead-burned to form a watertight tray[1];
  - — where the wall above the abutment is to be tile hung or otherwise clad the cavity tray can be omitted provided that run-off from the cladding cannot drain into the wall. Note that during construction the wall will be unprotected until clad;
  - — where the step between roof levels is small (say 6 courses) it may not be possible to include a cavity tray.

- Specify a weep hole in the brickwork at the lowest tray, Figure 4.

- Specify, where adjacent dwellings are staggered as well as stepped, a stop end to the lowest tray as well as a weep hole, Figure 4.

- Specify, for plain tiles or slates, soakers at the end row against the wall, Figure 5;
  - — the width of the soaker should be a minimum of 175 mm to give an upstand of 75 mm against the wall and a lap of 100 mm under the tiles or slates;
  - — to prevent the soakers slipping out of position specify that their top edges are to be turned down over the tiles.

- Specify a stepped cover flashing to lap over the soakers between 50 and 65 mm, Figure 5;
  - — flashings should be set at least 25 mm into the mortar course and held with lead wedges at each step;
  - — flashing may be omitted if tile hanging or other cladding projects rain over the upstand of the soakers.

- Specify, for interlocking tiles that cannot accept soakers, a stepped flashing dressed over the tiles, Figure 6.

- Specify lead in accordance with BS 1178[2];
  - — for soakers, Code No 3 (1.25 mm thickness);
  - — for trays and cover flashings, Code No 4 (1.80 mm thickness);
  - — where wind is liable to lift flashings, eg where dressed over tiles, Code No 5 (2.24 mm thickness).

- Specify that cover flashings are not to exceed 1.8 m in length and that successive lengths are to be lapped at least 100 mm.

- Specify that lead trays built into masonry are to be painted on both sides with a bituminous paint, preferably solvent based.

- Specify, where the wall above is to be rendered, a render stop fixed to follow the line of the roof, Figure 7;
  - — as an alternative the render can be cut off to follow the line of the stepped flashing.

- Remember that pre-formed trays and flashings remove some of the risks of inadequate site work.

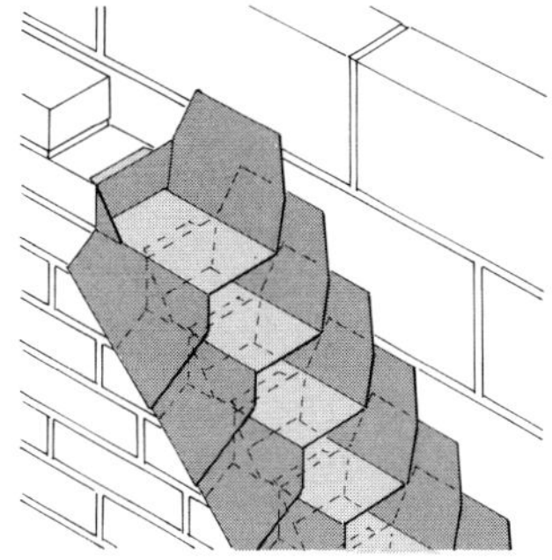
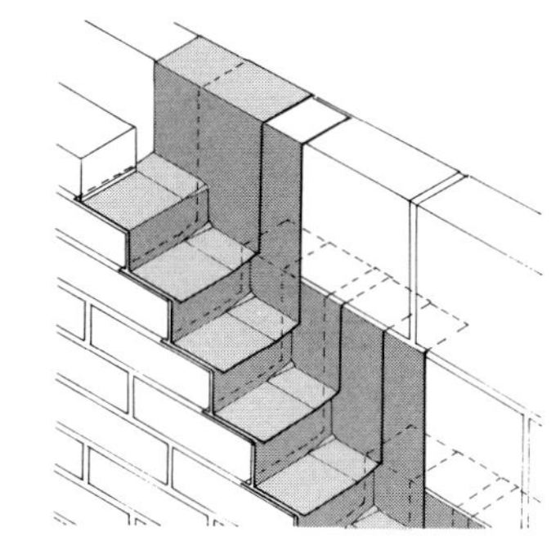

Figure 3

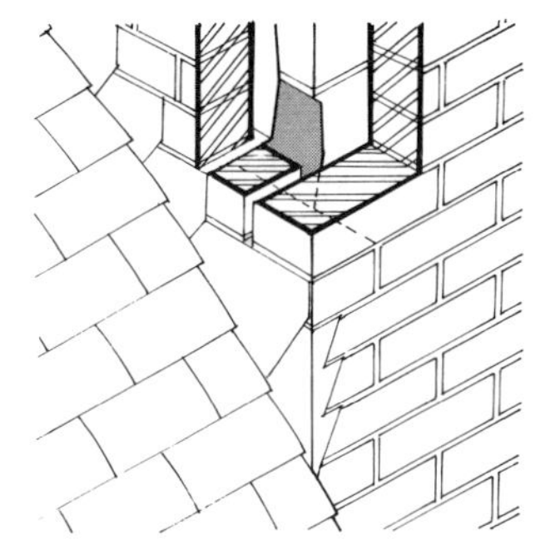
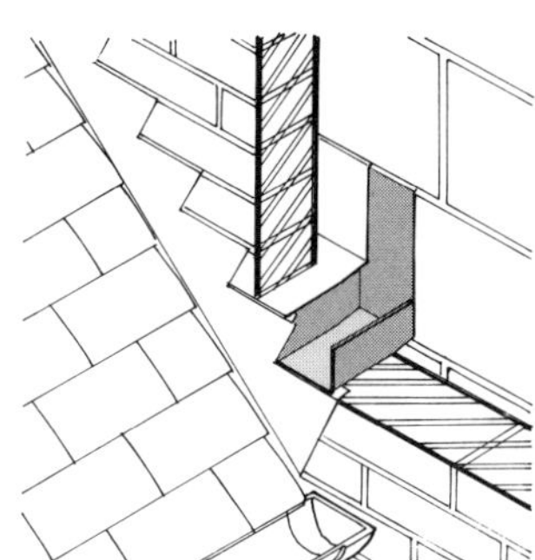

Figure 4

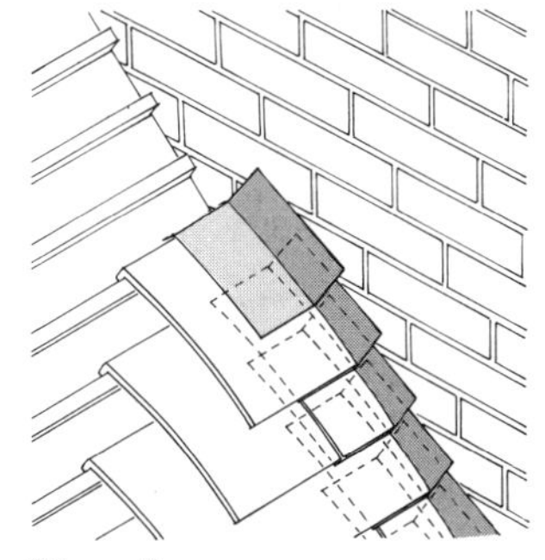
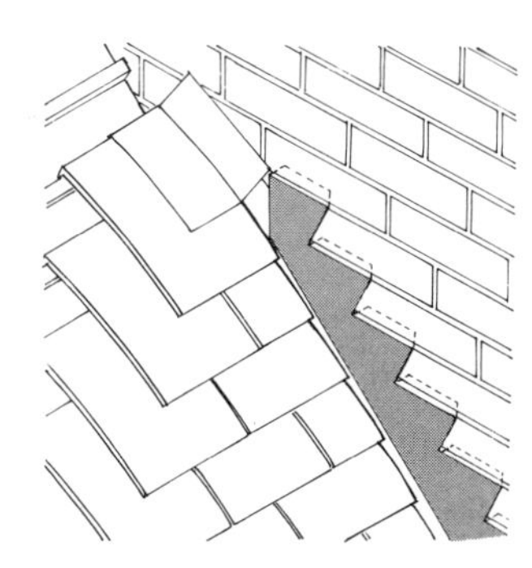

Figure 5

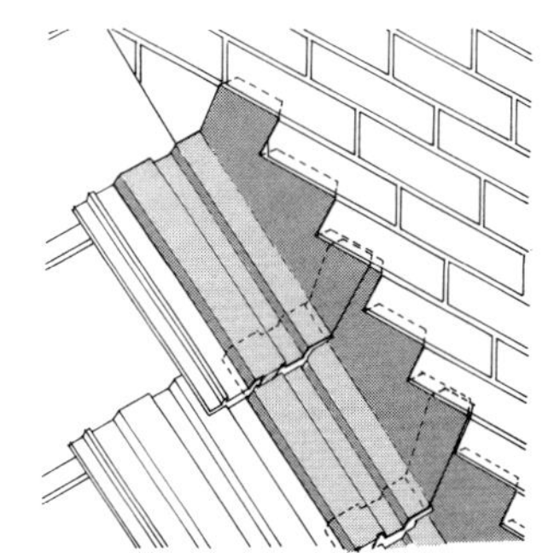

Figure 6

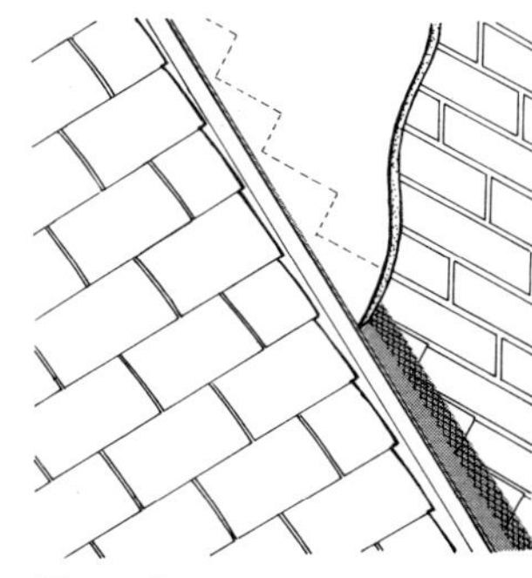

Figure 7

REFERENCES AND FURTHER READING

1 Lead Development Association 'Lead sheet in building'.

2 British Standard BS 1178:1982 'Milled lead sheet for building purposes'.

British Standard BS 5534:Part 1:1978, 'Code of Practice for slating and tiling, Part 1, Design'. ☐

Pitched roofs: renovation of older type timber roofs — re-tiling or re-slating

FAILURE: **Visible sagging of roof structure; collapse of roof structure**

DEFECTS: **Roof structure does not have the strength, stiffness or bracing to withstand the new dead loads to which it is subjected**

Figure 1a and 1b

Roof structures fabricated from rafters, purlins, struts and ceiling joists may not have sufficient strength, stiffness or stability to accept recovering in any material other than one with similar properties to the original specification, unless specific measures are taken to limit both deflection and movement of roof members and to prevent collapse.

BRE site inspections have revealed a proportion of older re-tiled or re-slated roofs which show signs of distress and in some cases of collapse, Figure 1.

If the ***original*** roof ***tiles or slates*** are ***replaced by heavier*** and/or more water absorbent ***tiles or slates*** the roof structure may become ***unstable*** and may collapse.

PREVENTION

Principle — Roof structure, Figure 2, must have sufficient structural strength, stiffness and stability to sustain the new dead load.

Practice

- Ensure that a thorough structural survey is made of the existing roof. Examine members for signs of timber decay and corrosion of fasteners.

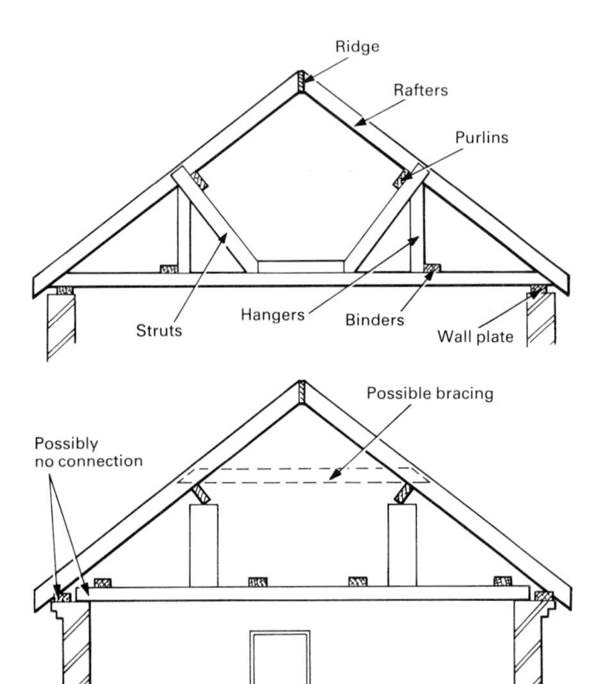

Figure 2 Two typical older type roof structures

Pitched roofs: renovation of older type timber roofs — re-tiling or re-slating

DAS 124
December 1988

- Check dry mass/unit area of the proposed tiles or slates and compare with the original tiles or slates;
  - — the mass/unit area of Welsh thins to Westmorland thick slates can range from 25 kg/m^2 to 78 kg/m^2;
  - — clay tiles can range from 39 kg/m^2 to 71 kg/m^2;
  - — concrete tiles can range from 28 kg/m^2 to 93 kg/m^2;
  - — for specific products obtain figures from manufacturer.

NOTE: The additional load or uplift due to the installation of an underlay may also need to be considered in the design calculations.

- Make allowances for the additional load due to water absorption of the proposed tiles or slates;
  - — slates have a maximum allowable water absorption value of 0.3% of their oven-dry weight[1], clay plain tiles 10.5%[2];
  - — allow 10% for concrete tiles[3];
  - — the increase in dead load, taking water absorption into account, can be as much as 100% when, for example, Welsh thin slates are replaced by average (48 kg/m^2) concrete tiles.

- Check that the roof structure is capable of accepting the new dead load.
  Either:
  Check timber sections against the Building Regulations[4,5]; Approved Document A1/2 for the appropriate dead loads after first establishing the species and grade of timber. The roof structure shown in Figure 3, if sized to the requirements of A1/2, will need no further check.
  Or:
  Calculate timber sections required to accept the new dead loads;
  - — inspect the existing structure for a frame system and analyse the forces in the members;
  - — check the rafters and purlins structurally as continuous beams, Figure 4;
  - — check the struts;
  - — check the bracing and other members connecting the rafters.

- Strengthen roof structure where necessary, eg specifying additional rafters, purlins, struts etc, or additional trusses between existing members. Replace defective timbers found in initial survey.

- Check the frame detail at the wall plate;
  - — the new forces must not cause distress to the roof or wall structures
  - — where necessary specify additional ties and fixings to the wall, vertical restraint must provide a resisting force of not less than 1.4 × total wind uplift force, Figure 5.

- Check overall stability of roof structure and consider bracing according to BS 5268:Part 3:1985[6].

- Check the existing dimensional accuracy.
  - — specify the members or sections of the structure requiring to be replaced or packed-up to ensure tiles or slates will seat down properly. Precise alignment is more crucial with inter-locking tiles than with plain tiles.
  - — rafters distorted or out of alignment with dividing walls or gable ends need to be replaced or packed to aid lateral stability.

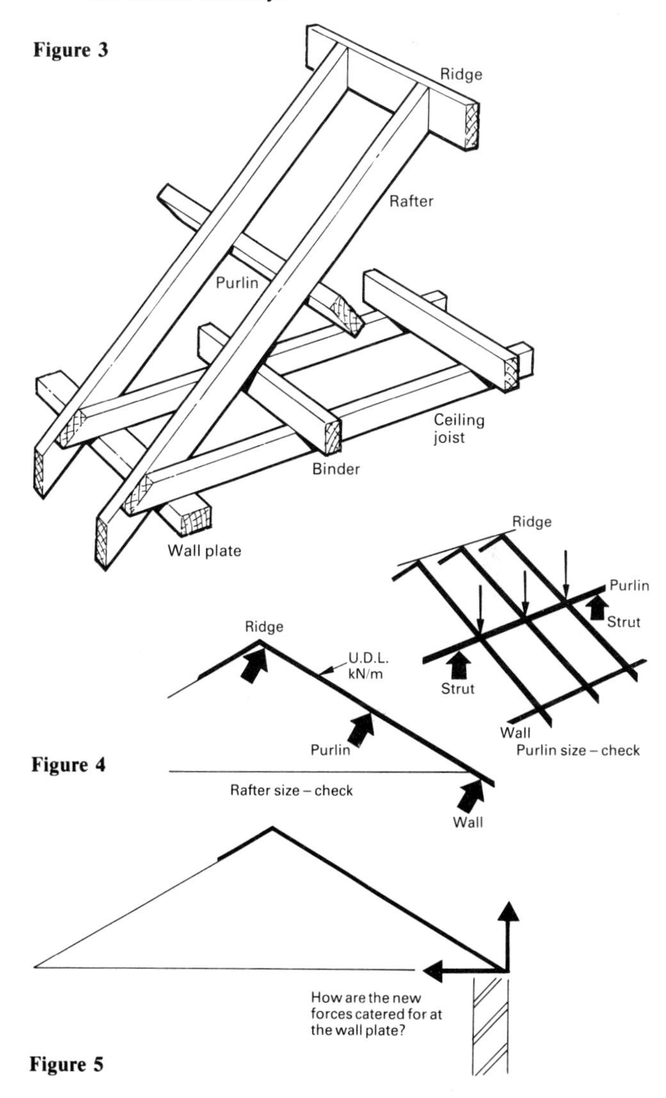

Figure 3

Figure 4

Figure 5

REFERENCES AND FURTHER READING

1 *British Standard* BS 680:Part 1:1944. 'Specification for roofing slates'.
2 *British Standard* BS 402:1979 'Specification for clay plain roofing tiles and fittings'.
3 *British Standard* BS 473, 550:1971 (1980). 'Specification for concrete roofing tiles and fittings'.
4 *The Building Regulations (England and Wales)* 1985: Approved Document A1/2.
5 *The Building Standards (Scotland)* Regulations 1981 to 1987 Part C.
6 *British Standard* BS 5268:Part 3:1985. 'Code of Practice for trussed rafter roofs'.
□ *British Standard* BS 5268:Part 2:1984. 'Structural use of timber. Code of practice for permissible stress design, materials and workmanship'.
□ *British Standard* BS 5534:Part 1:1978. 'Code of practice for slating and tiling — design'.
□ *British Standard* BS 8103:Part 1:1986 'Structural design of low-rise buildings. Code of Practice for stability, site investigation, foundations and ground floor slabs for housing'.
□ *Defect Action sheet* DAS 8 'Pitched roofs: separating wall/roof junction — preventing fire spread between dwellings'.

Pitched roofs: re-tiling or re-slating of older type timber roofs

FAILURE: **Visible sagging of roof structure; collapse of roof structure**

DEFECTS: **Roof structure does not have the strength, stiffness or bracing to withstand the new dead loads to which it is subjected**

Figure 1

New tiles or slates can impose substantially higher loads on the roof structure compared with the loads from the original tiles or slates. Older type timber roofs, comprising rafters, purlins, struts and ceiling joists, which have performed satisfactorily before re-tiling or re-slating may show signs of distress and may ultimately collapse, Figure 1.

The roof construction is designed to act as a complete structure. If ***all measures to strengthen*** the ***roof*** are not as specified, ***excessive movements*** and ***collapse may follow.***

Pitched roofs: re-tiling or re-slating of older type timber roofs

DAS 125
December 1988

PREVENTION

Principle — Roof structure must have sufficient structural strength, stiffness and stability to perform satisfactorily under the new dead load.

Practice

- Inspect roof members for signs of timber decay and corrosion of fasteners etc, which may reduce the strength. Replace items as necessary.
- Replace, plate or pack rafters, as specified, to rectify alignment with each other, with separating walls and with gable end walls to bring the roof structure to an even plane and to ensure tiles or slates will seat down properly, Figure 2;
  — check provision for fire stopping[1].
- Verify that the designer is aware of any new underlay as this may affect uplift and the dead load design considerations;
  — ensure correct fixing of underlay over rafters and secure by battens, Figure 3.
- Verify that the tiles to be used are the weight specified;
  — concrete tiles can range in dry mass/unit area from 28 kg/m² to 93 kg/m².
- Ensure all the measures specified to strengthen the roof are correctly placed and fixed;
  — galvanised fittings, nails etc, should be used.
- Ensure that any additional restraining straps to control vertical or lateral movement are fixed as specified, eg Figure 4;
  — vertical restraining straps should be fixed at no more than 1.2 m centres,
  — lateral restraint straps[2] should be fixed at no more than 2 m centres,
  — straps should be at least 30 × 5 mm cross-section and be galvanised.
- Ensure that any remedial bracing for stability of the roof structure is installed as specified.

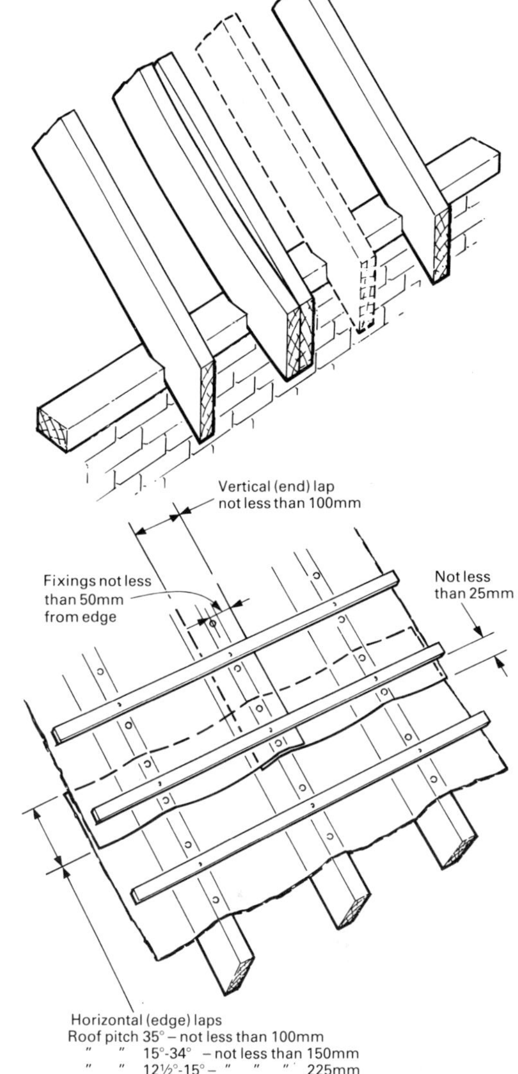

Figure 2

Figure 3

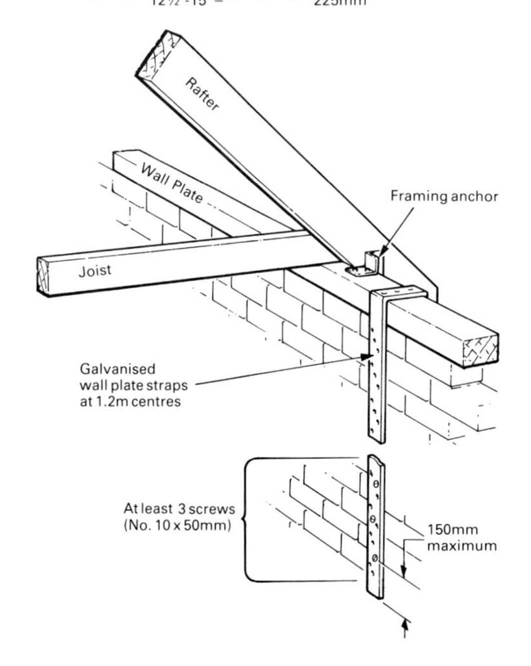

Figure 4

REFERENCES AND FURTHER READING

1 *Defect Action Sheet* DAS 8 'Pitched roofs: separating wall/roof junction — preventing fire spread between dwellings'.

2 *Defect Action Sheet* DAS 27 'External and separating walls: lateral restraint at pitched roof level — specification'.

Slate clad roofs: fixing of slates and battens

FAILURES: Slates lifting or rattling in strong winds; battens split and underlay penetrated by slate nails; water ingress to roof space.

DEFECTS: Slate or clout nails or clips, used to secure the slates to the battens, of incorrect size for the slate thickness and for prevention of uplift by wind action; battens of incorrect size for slate fixings used.

Figure 1

This *Defect Action Sheet* considers 'slates' as all types of flat nibless rigid material used for double-lap slating, eg natural slates, concrete slates, fibre cement slates etc. It does not deal with Scottish practice.

Slate or clout nails used for securing the slates should be of a durable material. Aluminium, copper or silicon-bronze nails to BS 1202 Parts 2 and 3 (Reference 1) are suitable for normal purposes. Austenitic stainless steel nails are also available. Clips should be aluminium alloy, austenitic stainless steel, or other alloys, non ferrous metals or plastics, provided the material is durable and fully tested.

Nails for securing battens may be cut or wire nails, smooth or ring shanked, and should be galvanised in coastal areas or aggressive atmospheres. Further selection of nail type may be necessary if battens are treated with wood preservative (see Note 1).

If the ***nails*** or ***clips*** are ***not specified*** to suit the slates and battens they will not hold the slates securely and the ***slates*** may ***rattle*** and ***work loose*** in strong winds. If the ***nails*** are too ***thick, splitting*** of the ***batten*** will occur; if they are too ***long, penetration*** of the ***underlay*** will occur which may give rise to ***water penetration*** problems in the ***roof space.***

PREVENTION

Principle — Nails for slates should be selected to suit the size and thickness of the slate. The length and thickness of the nail should also be checked against the batten size to ensure that the battens will not split and that no penetration of the underlay can occur. Clips should be shown by the manufacturer to be suitable for the conditions of service, and such that the weather-tightness of the slates will not be impaired.

Practice

BATTENS

- Specify battens in accordance with Table 1:
  — alternatively, battens should be in accordance with BS 5268: Part 2 (Reference 2).

- Specify battens to be treated with an appropriate wood preservative if they are likely to become wet during service (Reference 3).

- Specify battens to be not less than 1200 mm long and to span at least three support points (rafters, wall, trusses):
  — butt joints should be staggered, splay-cut at 45°, and over support points, Figure 2;
  — splicing between support points and cantilevering of battens is not permitted.

- Specify that battens detailed in Table 1 are nailed with 3.35 mm diameter 65 mm long nails in areas where conditions are as shown in Table 2:
  — for more severe conditions, the nailing requirements should be calculated in accordance with BS 5534: Parts 1 and 2 (Reference 4).

- Specify nails to be galvanised in coastal areas or areas of aggressive atmosphere. Some wood preservatives can react with metals under certain conditions (see Note 1).

Slate clad roofs: fixing of slates and battens

DAS 142
February 1990

SLATES

- Specify slate or clout nails to be aluminium, copper or silicon-bronze to BS 1202: Parts 2 and 3 (Reference 1):
  - — length of nails needs to suit the size and thickness of the slates: 65 mm long for thick heavy slates and 30 mm long for small thin slates;
  - — diameter of nails should not exceed 1/10 of the batten width to prevent splitting;
  - — nails should not penetrate completely through the batten as damage to the underlay may cause water ingress to the roof space;
  - — mechanical fixings, clips etc.; aluminium alloy, stainless steel and other non-ferrous metals or plastics may be available. Fixings are exposed to the weather. Evidence of the strength, durability and suitability of the fixing should be provided by the manufacturer.

NOTE 1 Some timber preservatives can increase corrosion under certain conditions. Aluminium/copper alloys should not be used with CCA preservatives. Both the batten nails and slate nails must be compatible with the preservative (References 3 and 5).

- Specify that all slates are holed twice, usually 20-25 mm from the edge. For head nailing to battens, the hole centres are usually 25 mm from the head of the slate:
  - — verge slates and narrow slates should be fixed with two nails to prevent the tails moving sideways;
  - — extra long nails will be required at the eaves for centre nailed slates.

NOTE 2 Man-made slates may need specific precautions to be taken eg. fibre-cement slates may need tail fixing to prevent curl. The manufacturer's instructions should be obtained.

- Specify the lap of slates (tail over head of slate in next but one course below) to be according to Table 3:
  - — side lap should be half the slate width but not close-butted. A small gap assists drainage;
  - — for vertical slates, minimum lap should be not less than 32 mm.

- Specify a double layer of slates at the eaves. The underlayer should consist of short slates laid on their backs and head-nailed to the first batten. The first course of full length slates can be either head-nailed or centre nailed Figure 3.

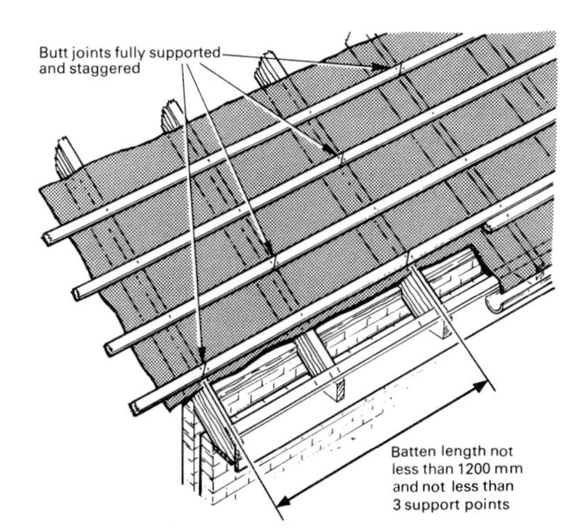

Figure 2

Table 1 Minimum batten sizes for pitched roofs and vertical work

Slates	450 mm span		600 mm span	
	width	depth	width	depth
sized (natural)	38	19	38	25
random (natural)	50	25	50	25
fibre-cement	38	19	38	25
concrete	38	19	38	25

Table 2 Conditions for use of 3.35 mm dia, 65 mm long nails (ridge height not exceeding 7.2 m)

Roof type	Nail type	Maximum basic wind speed (m/s) (4)
Dual pitch*	smooth round	48
	ring shank	56
Monopitch	smooth round	45
	ring shank	52

*Not less than 17½°

Table 3 Recommended lap for slates

Rafter pitch not less than (degrees)	Moderate Exposure lap in mm	Severe Exposure lap in mm
45	65	65
40	65	75
35	75	75
30	75	75
25	90	100
20	115	130

NOTE 3 Severe exposure conditions are experienced in coastal areas, elevated sites, etc. For further guidance refer to BRE Digest 346, 'The assessment of wind loads. Parts 1-7'.
NOTE 4 Fibre-cement slates should be confirmed by the manufacturer as suitable for roof pitches below 25° (moderate exposure), or below 30° (severe exposure).

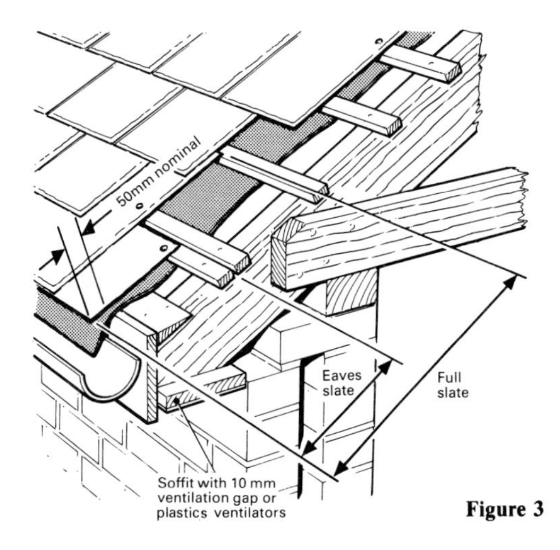

Figure 3

REFERENCES AND FURTHER READING

1. *British Standard* BS 1202: Part 2: 1974: Copper nails and BS 1202: Part 3: 1974: Aluminium nails
2. *British Standard* BS 5268: Part 2: 1988: Structural use of timber. Code of practice for permissible stress design, materials and workmanship.
3. *British Standard* BS 5268: Part 5: 1989: Structural use of timber. Preservative treatments for constructional timber.
4. *British Standard* BS 5534: Part 1: 1978: Code of Practice for slating and tiling: Design and *British Standard* BS 5534: Part 2: 1986: Slating and tiling. Design Charts for fixing roof slating and tiling against wind uplift.
5. *British Standard* BS 4072: Part 2: 1987: Wood preservation by means of copper/chromium/arsenic compounds: Appendix A.

Pitched roofs: trussed rafters — site storage

FAILURES: **Structural unsoundness of roof; degradation of timber and corrosion of connector plates in trussed rafters**

DEFECTS: **Distortion and damage to trussed rafters; dampness in timber**

Figure 1
Badly stored trusses

Poor site storage damages building components of many kinds. The damage is often obvious and repairable. But trussed rafters can be damaged in ways that are not obvious without careful inspection, and unauthorised site repairs of any kind are not acceptable[1].

A BRE survey revealed that, on many sites, timber trussed rafters were stored in such a way that they had become damaged and wet before being fixed in place (Figure 1). Trussed rafters are important structural members and it is essential to try to minimise the occurrence of defects.

Timber trussed rafters are usually stacked on site prior to being used. If the ***bearers*** for the stack are ***not level,*** or are in the ***wrong position,*** or if the trussed ***rafters are banded too tightly,*** the trussed rafters can become ***permanently distorted.*** Distortion of trussed rafters which has occurred during storage is especially undesirable structurally since it is likely to be common to all trussed rafters in the stack and hence all trusses in a roof. If accidental damage is caused to stacked trussed rafters, for example by site traffic, several trusses in the stack are likely to be affected.

If the stack is left uncovered the trussed rafters can become wet, or develop shakes from rapid drying in full sun. ***If the stack and its cover are not kept so as to shed rainwater and provide adequate ventilation the trussed rafters can become wet from rain seeping in or from condensation.*** High moisture contents reached in the timber trussed rafters during storage are likely to remain high for some time in service — particularly where good roof ventilation is not achieved. The moisture content of truss timbers must not exceed 22 per cent at any time[2]. Prolonged dampness may cause rot in the timber and corrosion in the connector plates.

Trussed rafters may be stored horizontally or vertically. Although horizontal storage may seem easier, vertical storage, apex upwards, simulates the support that trusses receive when installed and enables the cover over the stack to shed rainwater readily. It has been found in practice that horizontal storage is often less well done than vertical storage. It more often produces distortion and allows any cover on the stack to collect pools of rainwater. Ventilation is more difficult to achieve.

PREVENTION

Principle — Arrange for storage of trussed rafters so as to avoid distortion, wetting and accidental damage.

Practice

General:

- Phase deliveries to minimise site storage time.
- Check on delivery that trusses are dry and undamaged.
- Mark stack to distinguish between apparently similar types.
- Take particular care of 'specials' — replacement may delay completion.
- Cover to protect from sun and to shed rainwater, ensuring good ventilation.
- Do not permit storage that would distort trusses
- Do not store where site traffic or building operations might cause damage.
- Ensure that bearers for the stack are level and on a firm well-drained base.
- Remove any tight banding which is distorting trusses.
- Do not allow vegetation to grow close to the stack.
- Preferably use vertical storage.

Vertical storage:

- Support trusses on bearers at wall plate positions and high enough to keep rafter feet clear of the ground (A, Figure 2).
- Brace the stack to keep trusses vertical (B, Figure 2)
- Keep bracing timber out of contact with connector plates, it may be wet or contaminated (C, Figure 2).

Horizontal storage:

- Support trusses on bearers at every joint, and also at intermediate points for long spans (A, Figure 3).
- Ensure that all bearers are level and in the same plane.
- Before stacking, cover bearers with a material such as polythene so that wet bearers will not be in contact with the bottom truss in the stack (B, Figure 3).
- If a second batch of trusses is put on the stack ensure that intermediate bearers line through vertically with those at the base (C, Figure 3).

Inspection: When trusses are taken from the stack check that:

- They are not damaged or distorted and their moisture content is below 22 per cent (check initially with moisture meter),
- They have no unacceptable shakes,
- There are no damaged finger joints
- There are no damaged or displaced connector plates,
- Connector plates are free from corrosion.

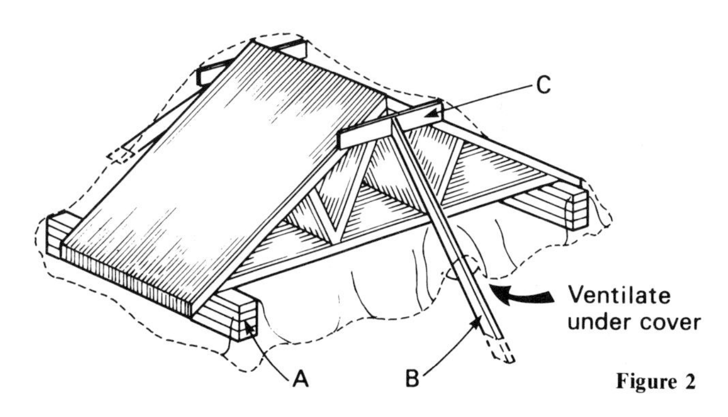

Figure 2

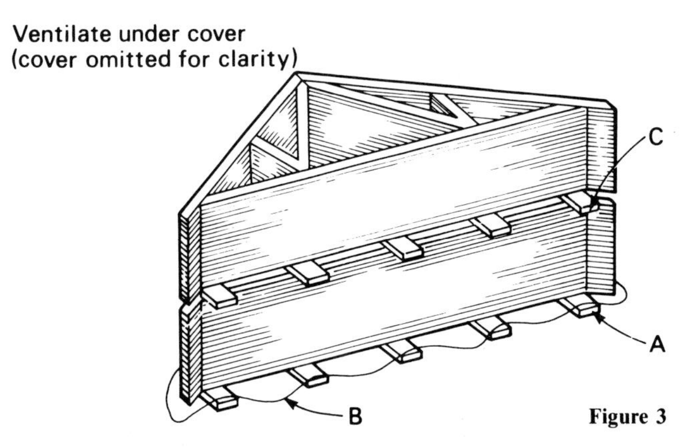

Figure 3

REFERENCES

1 *British Standards Code of Practice* CP 112:Part 3. The structural use of timber — Trussed rafters for roofs of dwellings.

2 Agrement Board certificates for connector plates. Obtainable from the British Board of Agrement, PO Box No 195, Bucknalls Lane, Garston, Watford, Herts, WD2 7NG.

Trussed rafter roofs: tank supports — specification

FAILURES: **Distortion of trussed rafter members; deterioration of chipboard tank platform; disturbance of plumbing joints**

DEFECTS: **Incorrect or incomplete specification for materials, and method of support, for tank platform**

Figure 1 Badly supported tank

A BRE survey revealed many instances of poorly designed support for cold water storage tanks in roof spaces. ***If loads*** from tanks are ***concentrated*** on too few trusses, or in the wrong places on trusses, truss members may deflect and ***trusses*** may be ***damaged; if bearers*** for tanks are ***not strong enough*** they may deflect and ***plumbing connections leak; if chipboard*** is used for tank platforms it will be ***wetted*** by condensation running off tank surfaces and may ***collapse.***

PREVENTION

Principle — Tanks must be supported on durable platforms so that loads are safely distributed on trusses.

Practice

- Ensure that the trussed rafter designer knows your requirements for tank size and location.
- Specify that tanks of not more than:
  330 litres actual capacity to marked waterline are to be carried on **four** trusses;
  230 litres actual capacity to marked waterline are to be carried on **three** trusses.
- Specify that tank bearers are to be positioned close to node points of trusses (Figure 2) and correctly spaced (see DAS 44).
- Specify cross-section of bearers A and C to be 50 × 75 mm and of bearer B to be as given in Table 1 below (Reference 1);
  — bearers may be of any species with a permissible bending stress not less than that of European redwood/whitewood of GS stress grade.
- Specify skew nailing of C to B and B to A to stop rotation of bearers;
  — where headroom is limited consider using joist hangers (Figure 3), but do not use joist hangers to carry any bearer on a truss tie.
- Do not specify chipboard for a tank support platform.
- Show all requirements clearly on drawings.

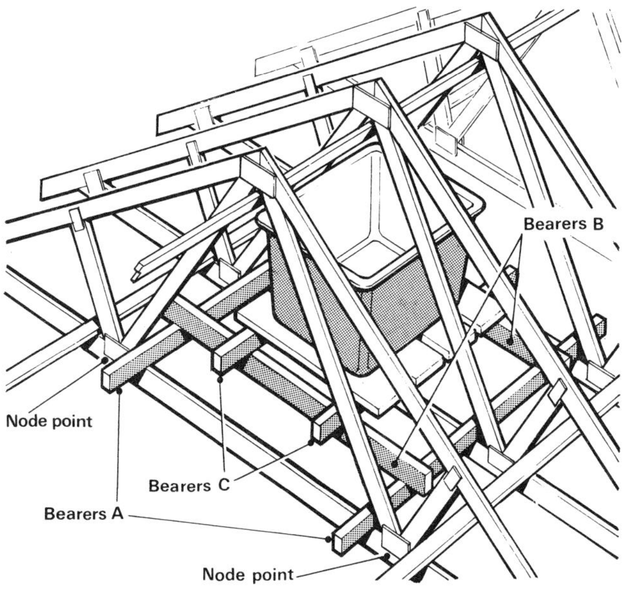

Figure 2

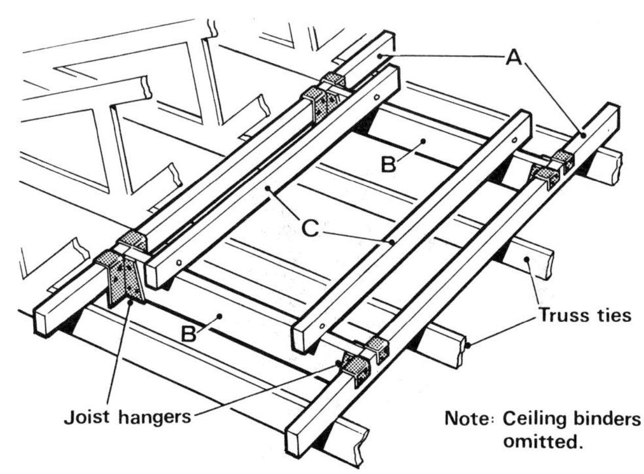

Figure 3

Table 1*

Size of bearers B

Truss span (metres)	330 litres actual capacity to marked water-line	230 litres actual capacity to marked water-line
6	50 × 125 mm	50 × 125 mm
8	50 × 150 mm	50 × 125 mm
11	75 × 150 mm	50 × 150 mm

*Taken from Reference 1. A draft British Standard (BS 5268:Part 3) is in preparation and, when published, its requirements may differ slightly from those in the Table

REFERENCES AND FURTHER READING

1 *Technical Bulletin* No 4 'Tank supports, details and limiting spans'. Obtainable from the International Truss Plate Association Ltd. Twinaplate Ltd, Threemilestone, Truro, Cornwall, TR4 9LD

Trussed rafter roofs: tank supports — installation

FAILURES: **Distortion of trussed rafter members; deterioration of chipboard tank platform; disturbance of plumbing joints**

DEFECTS: **Tank bearers incorrectly sized and positioned; chipboard used as tank platform**

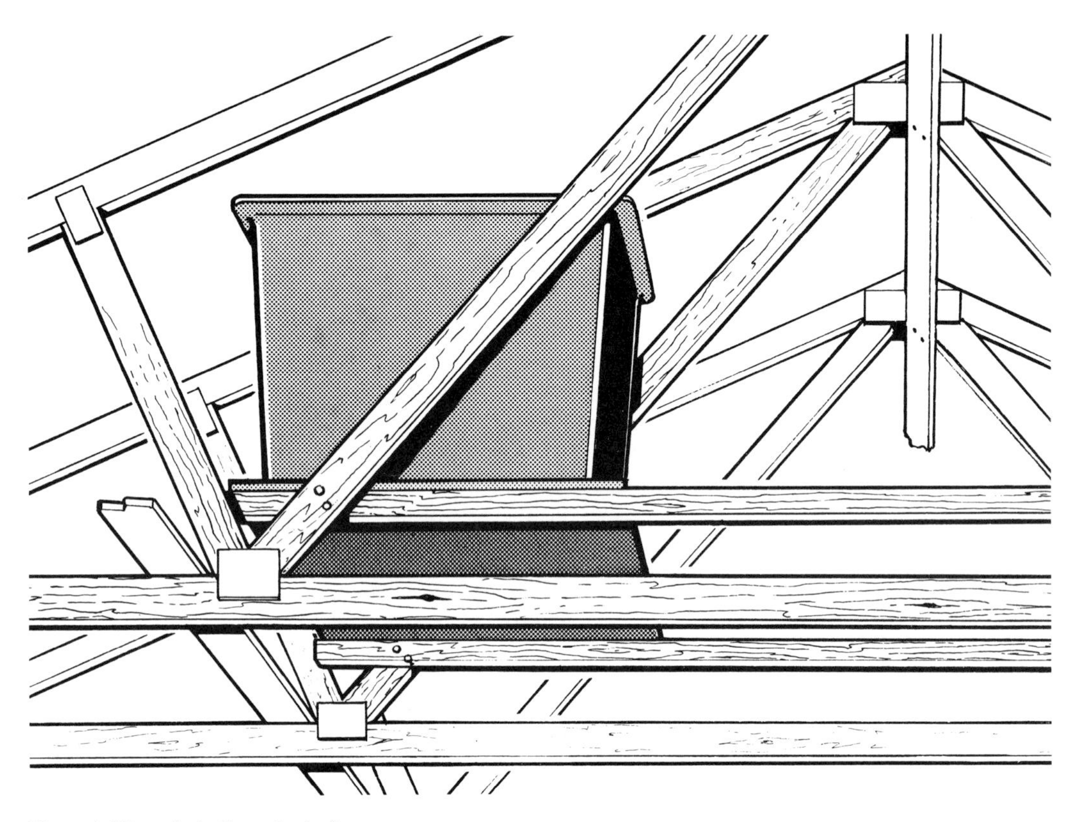

Figure 1 Wrongly-built tank platform

BRE site investigations have found many cases of badly built and unsafe supports for cold water storage tanks in roofs (Figure 1). A typical tank (230 litres nominal capacity to marked waterline) in service weighs about the same as three people — and a plumber changing a ball-valve would make a fourth. A typical larger tank (330 litres nominal capacity) is heavier still by about the weight of yet another person.

If ***tank bearers*** are ***undersized***, incorrectly positioned, or ***laid flat*** instead of on edge, the ***support*** for the tank will be ***insufficient or unsafe***; if ***chipboard*** is used for tank platforms it will be ***wetted*** by condensation running off the tank surfaces and may ***collapse.***

PREVENTION

Principle — Tanks must be supported so that loads are safely distributed on trusses; tank support platforms must be durable.

Practice

- Check whether tanks are to be carried by three trusses or four. (See DAS 43)
- Do not permit tanks or tank platforms to be carried directly on truss ties.
- Check that tanks and platforms are located where they are specified to be in the roof span.
- Ensure that tanks are supported on platforms and bearers as shown in Figure 2
  — alternatively, connection by joist hangers may have been specified.
- Check that bearers are of the specified sizes, species and stress grade and are laid on edge, not flat. (See DAS 43)
- Check that bearers A, B and C (Figure 2) are not interchanged: bearers B are the deepest.
- Check that bearers A are positioned in contact with the internal members in trussed rafters or as close to them as possible (they may be placed on the ceiling binders provided that support is continuous — Figure 2).
- Check that bearers B are spaced as in Figure 3 (3 trusses) or Figure 4 (4 trusses).
- Check that bearers are fixed against rotation, eg by skew nailing or by joist hangers.
- Do not permit chipboard to be used for tank platforms.

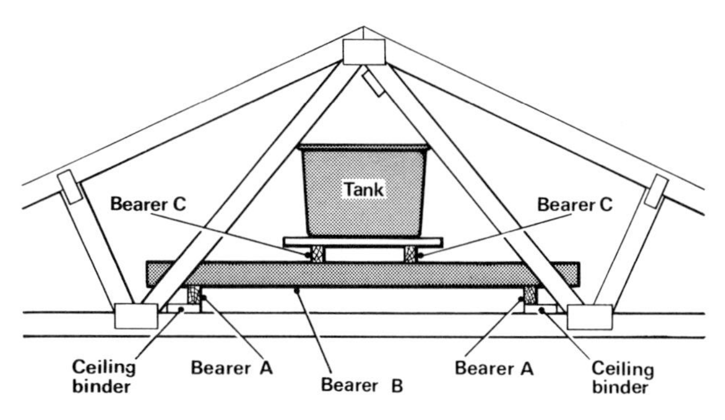

Figure 2

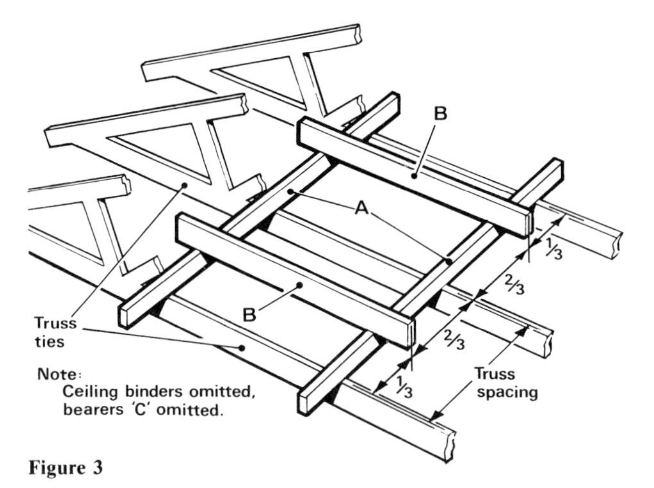

Figure 3

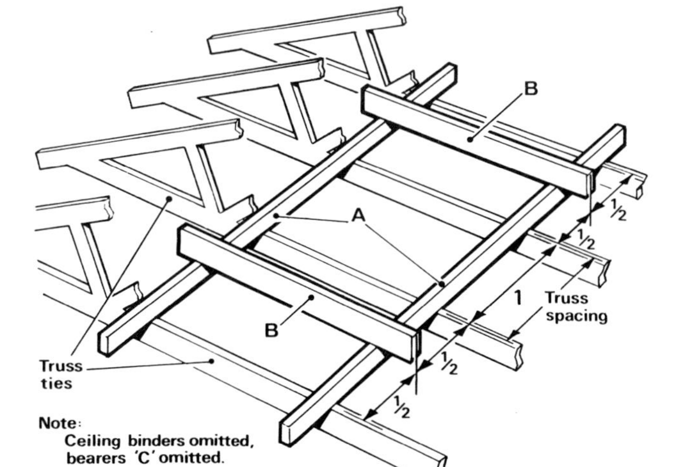

Figure 4

PART TEN

Dual Pitched Roofs

Dual-pitched roofs*: trussed rafters; bracing and binders — specification

FAILURE: **Movement of roofs**

DEFECTS: **Bracing and binders not shown on drawings, or position and fixing inadequately specified; no instructions on siting of items, such as tanks and flues, to avoid interrupting binders**

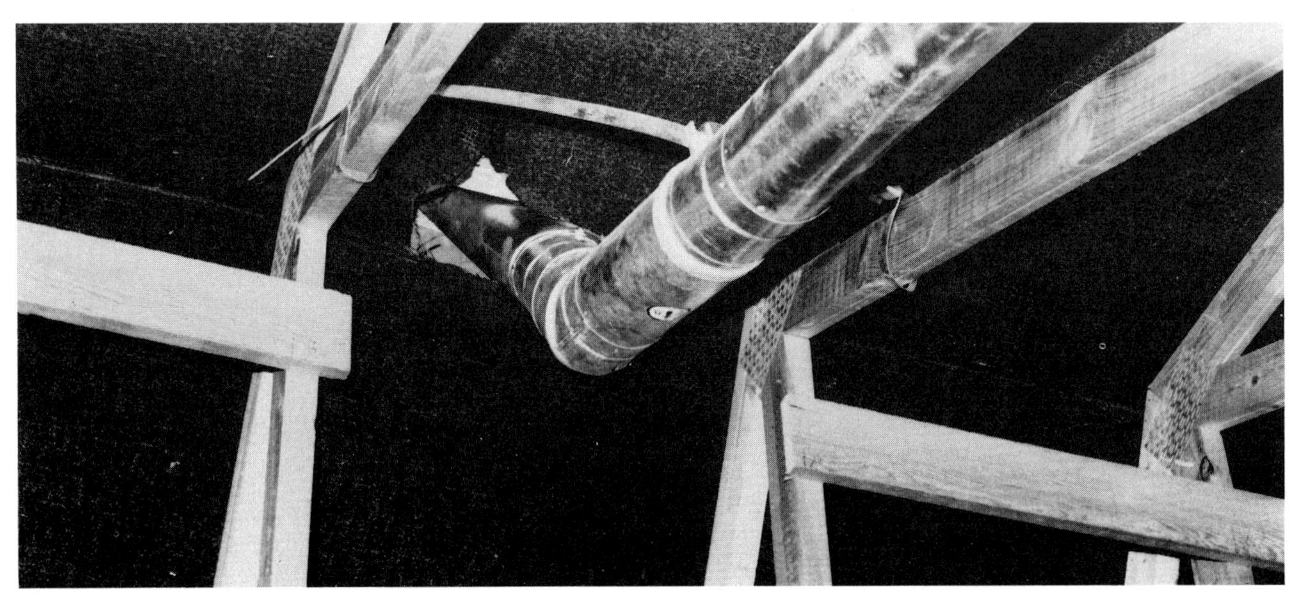

Figure 1 Flue interrupts ridge binder

In a survey of recent house construction, misplaced, interrupted or inadequately fixed raking braces and ridge and ceiling binders were relatively common, Figure 1. Whole roofs have been seen leaning sideways until the end truss touches the gable or separating wall. Displacements of 100 mm or more have been seen.

Building Regulations (for England and Wales) Approved document A, paragraph A2 states 'Trussed rafter roofs should be braced to the recommendations of BS 5268: Part 3 1985'**. The requirements of this Code may be unfamiliar to site staff. ***If specifications and drawings do not clearly set out requirements for bracing and binders, roofs may not comply with the Regulations.*** If the ***position of tanks, pipes and flues are not considered*** during design, ***they may interfere with*** positioning, continuity and access for nailing of ***bracing or binders.***

Note: Where lateral restraint connections are needed they should be as shown in Approved Document A (Building Regulations 1985, for England and Wales) and shown also in DAS 27.

* This DAS deals with roof pitches up to 30°, spans up to 11 m and 'basic wind speed' of 48 m/s; this 'basic wind speed' (Reference 1) applies to all of England and virtually all of Wales. For other conditions see Reference 2.

** This DAS is not applicable where roofs are constructed using rigid sarking boards, as in traditional Scottish practice.

Dual-pitched roofs*: trussed rafters; bracing and binders — specification

DAS 83
August 1986

PREVENTION

Principle — Bracing and binders are critically important: properly positioned and fixed they convert a collection of individual trusses into a single three-dimensional structural unit.

Practice

- Ensure that site supervisory staff understand that bracing and binders are crucial to the performance of the roof. Produce clear instructions — and 3-dimensional sketches where necessary — for the understanding of site workers.

- Specify, for every roof (or section of roof between cross walls) 100 x 25 mm raking braces, twice nailed to the underside of **rafters** of every truss. The braces should run at approximately 45° from ridge to eaves and be applied to both pitches, Figure 2.

- Specify, when the distance between centres of cross walls is not more than 1.2 x trussed rafter span, at least two 100 x 25 mm diagonal braces, twice nailed to every **ceiling tie** in every roof (or section of roof between cross walls) as shown in Figure 3.
  — where wall spacing exceeds 1.2 x span, specify at least four such diagonal braces in 'W' formation on each side.

- Specify (unless trusses are less than 5 m span), for every roof or section of roof between cross walls, 100 x 25 mm raking bracing twice nailed to every internal strut, Figure 4.

- Specify longitudinal binders, 100 x 25 mm, twice nailed and located as shown in Figure 5;
  — all binders should abut walls at both ends and for this purpose specify each binder to be in two overlapping lengths.
  — where binders cross raking braces the binders should be interrupted and plated, see Figure 4.

- Specify that all lap joints in braces and binders are to be lapped, and nailed over at least two rafters.

- Specify all nailing to be 3.35 x 75 mm galvanised round wire nails.

- Specify that no bracing or binders shall penetrate a separating wall.

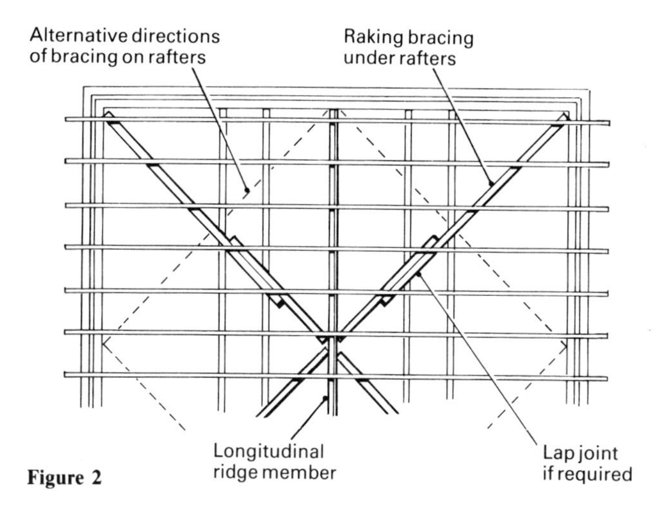

Figure 2

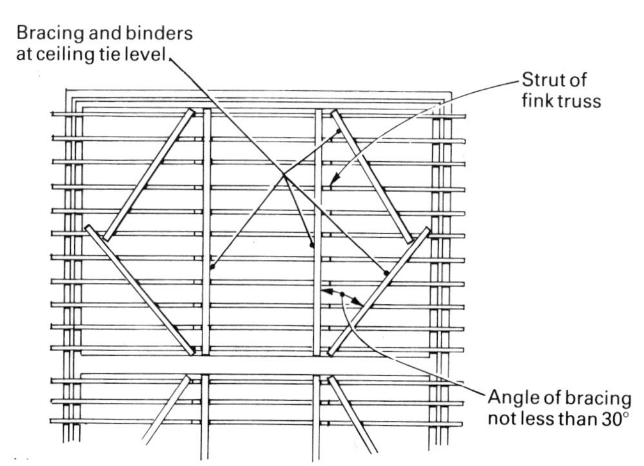

Figure 3

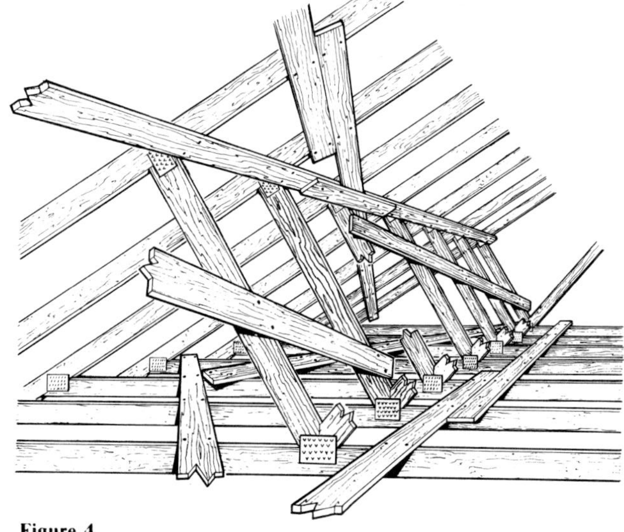
Figure 4

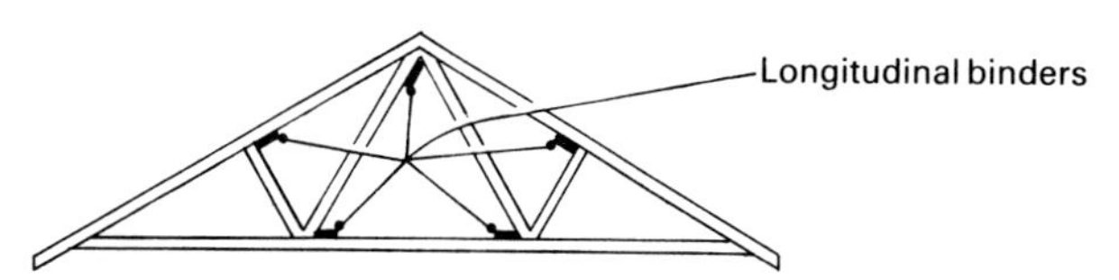

Figure 5

REFERENCES AND FURTHER READING

1 *BRE Digest* 119 The assessment of wind loads
2 *British Standard* BS 5268: Part 3 1985, Structural use of timber: Code of Practice for trussed rafter roofs.

Dual-pitched roofs: trussed rafters; bracing and binders — installation

FAILURE: Movement of roofs

DEFECTS: Bracing and binders missing or incorrectly positioned or fastened

Figure 1 Incomplete raking bracing

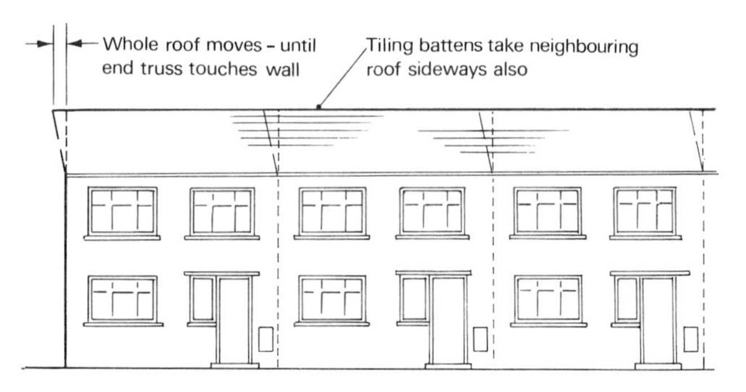

Figure 2 Roofs without diagonal bracing can move sideways

Inadequately braced roofs have been known to move, the trussed rafters leaning sideways. This commonly shows by oversailing of the verge tiles at gables, Figure 2. Tiling battens cannot by themselves prevent such movement. A new British Standard, (1985) calls for increased provision of braces and binders

If ***binders*** or ***bracing to rafters, struts*** and ***ceiling ties*** are ***omitted, wrongly positioned, badly fixed,*** or cut to make room for items such as pipes and flues, the ***roof will not comply with*** the British Standard nor with the guidance given in Approved Document A of the ***Building Regulations*** for England and Wales*.

* This DAS is not applicable where roofs are constructed using rigid sarking boards, as in traditional Scottish Practice.

PREVENTION

Principle — Bracing and binders are critically important: properly positioned and fixed they convert a collection of individual trusses into a single 3-dimensional structural unit.

Practice

- Check that trusses are plumb and not bowed.
- Check that, in every roof or section of roof between cross walls, 100 x 25 mm raking braces (at about 45° to the ridge) are fixed to the underside of rafters on each slope; any overlap in the length of a brace must be over at least two rafters, Figure 3; braces should be twice nailed to every rafter.
  — braces may run up to the ridge either toward or away from a flank wall.
- Check that at least two 100 x 25 mm diagonal braces are applied to ceiling ties, twice nailed to **every** tie in every roof or section of roof between cross walls. It is critically important that no trusses are unbraced at tie level.
  — if cross wall spacing exceeds 1.2 x truss span, at least four such diagonal braces are needed in every roof or section of roof between cross walls, laid in 'W' formation, to embrace every truss tie, Figure 4.
- Check that, unless trusses are less than 5 metre span, 100 x 25 mm raking bracing is applied, in every roof or section of roof between cross walls. Braces must be twice nailed to each of the struts that connect with the rafters at about mid-span of each rafter, Figure 3.
- Check that 100 x 25 mm longitudinal binders are applied to every roof or section of roof between cross walls, twice nailed to every truss;
  — there must be at least two binders on the ties, each at about one third span, plus one binder under each slope fixed close to where struts abut rafters, and one close to the apex of trusses.
- Check that all longitudinal binders are closely butted against cross wall faces, Figure 5.
  — any overlap in a binder must be over at least two trusses.
- Check that no brace or binder is lapped one on top of another, so producing a space that nails are forced to bridge.
  — where binders cross raking braces the binders should be interrupted and plated, Figure 3.
- Check that no brace or binder penetrates a separating wall.
- Check that site applied timbers are nailed with 3.35 (10 g) x 75 mm galvanised round wire nails.
- Check that nails are driven into the centre of truss members to avoid splitting.
- Do not permit modifications of trusses without reference to the truss designer.

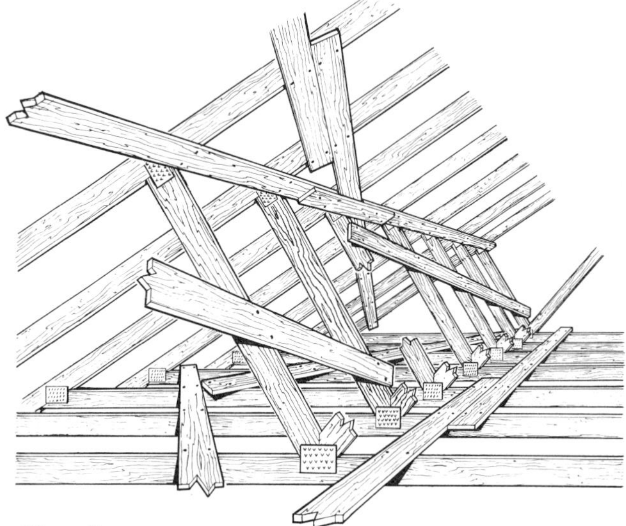

Figure 3

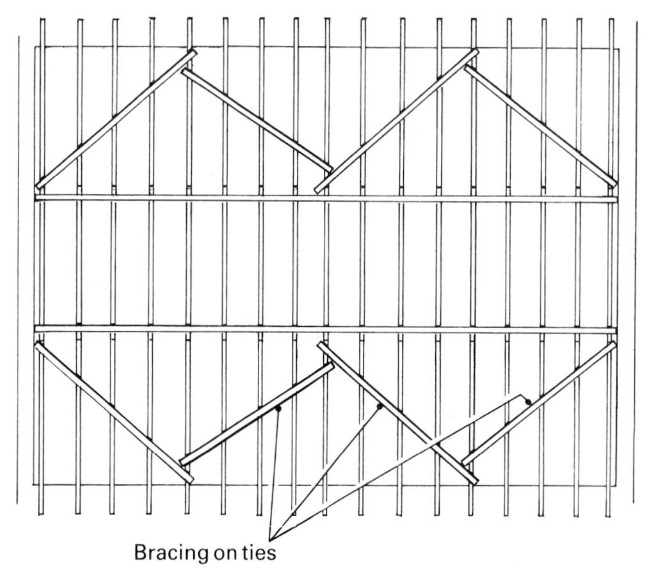

Figure 4

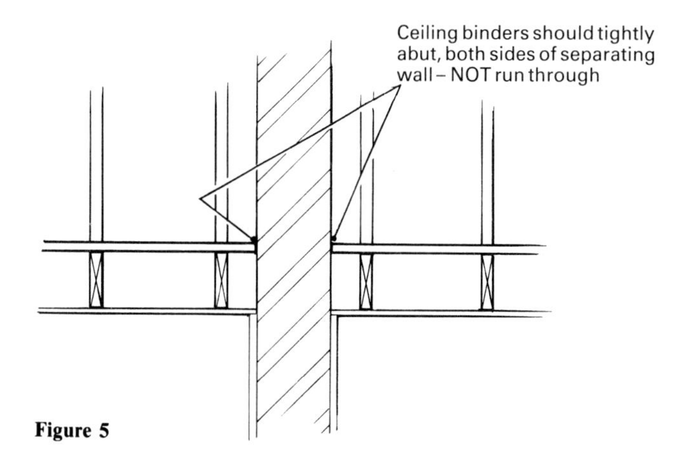

Figure 5

REFERENCES AND FURTHER READING

Defect Action Sheet DAS 83 (Design) 'Dual-pitched roofs: trussed rafters; bracing and binders — specification'.

Dual-pitched roofs: trussed rafters — specification of remedial bracing

FAILURE: Instability or collapse of roofs

DEFECTS: Bracing inadequate or absent

Figure 1

Many earlier trussed rafter roofs are inadequately braced. Those built before recommendations for bracing became widely known, in about the late seventies, must now be suspect. BRE surveys even as late as the mid eighties found inadequately braced roofs.

Inadequacy of bracing has resulted in distortion or leaning of trusses, Figure 1. When such movements have become appreciable they are revealed by disruption of tiling especially evident at verges.

If remedial action is not taken where ***roofs*** are ***inadequately braced,*** roofs will ***remain unstable*** and may be at ***risk*** of future ***collapse.***

Dual-pitched roofs: trussed rafters — specification of remedial bracing

DAS 110
December 1987

REMEDIAL WORK

Principle — Roof trusses must be braced so that they act together as a single three-dimensional structural unit.

Practice

- Inspect roofs suspected, by reason of age or structural performance, to be inadequately braced, to determine construction and condition:
  - — before 1976 the Building Regulations (England and Wales) did not specifically require trussed rafter roofs to be braced, though some may have been;
  - — in roofs built to the requirements of the 1976 Building Regulations consideration should have been given to the need for longitudinal braces (often called 'binders') and diagonal bracing to CP112 Part 3 1973;
  - — roofs built to the requirements of the 1985 Building Regulations should be braced to BS 5268 Part 3 1985.

Note: dates of relevant Scottish requirements may differ.

- Measure deviations from vertical in trusses and bow in truss members; if deviations exceed those in Table 1 or bow in truss members is greater than the 10 mm erection deviation permitted by BS 5268 Part 3 consult a structural engineer:
  - — where distortion or displacement of trussed rafters does not exceed 40 mm, Figure 2, remedial measures applied internally may be viable. To remedy inadequate stability longitudinal and diagonal bracing equivalent to that required in BS 5268 and additional internally applied plywood diaphragms will probably be the minimum necessary, Figure 3;
  - — where distortion or displacement exceeds 40 mm, immediate remedial action will be necessary: this may entail extensive re-building of the roof.

Where deviations are less than those in Table 1 and bow less than 10 mm:

- Check that construction of each roof satisfies the following:
  1. Diagonal bracing and longitudinal bracing to BS 5268 Part 3 (refer to DAS 83)[1];
  2. Lateral restraint strapping to DAS 27[2]
  - — if it does not, consider specifying bracing and binders as necessary to meet the requirements of BS 5268 Part 3;
  - — binders should be butted tight against gables and separating walls to provide restraint in compression;
  - — existing gable ladders may be providing effective restraint at verges.

  Note: it may not be practicable in an existing roof to emulate the construction of a new roof: alternative approaches, such as plywood diaphragms, may be structurally acceptable.

- Check the need to install holding-down straps.
- Where roofs are left unbraced consider periodical inspections.

Table 1 Maximum allowable deviation from vertical

Rise of trussed rafter (m)	1	2	3	4 or more
Deviation from vertical (mm)	10	15	20	25

REFERENCES AND FURTHER READING

1 **Defect Action Sheet** DAS 83 'Dual-pitched roofs: trussed rafters; bracing and binders — specification'.

2 **Defect Action Sheet** DAS 27 'External and separating walls: lateral restraint at pitched roof level — specification'.

British Standards Institution BS 5268:Part 3:1985 'Code of Practice for trussed rafter roofs'.

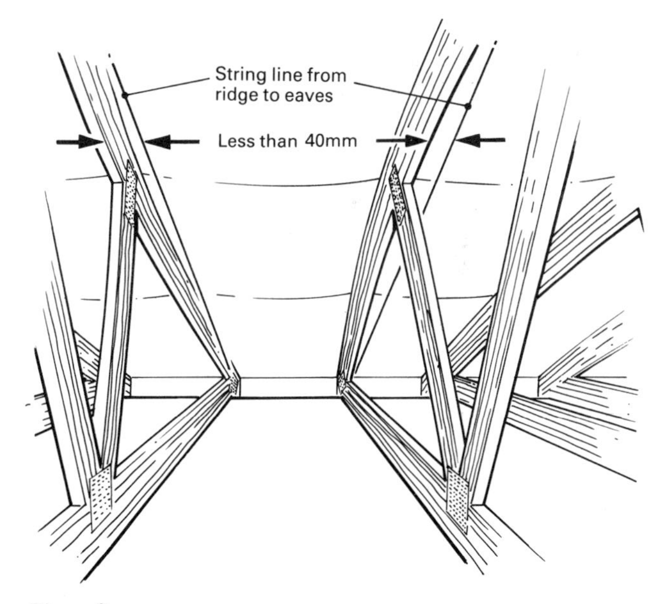

Figure 2

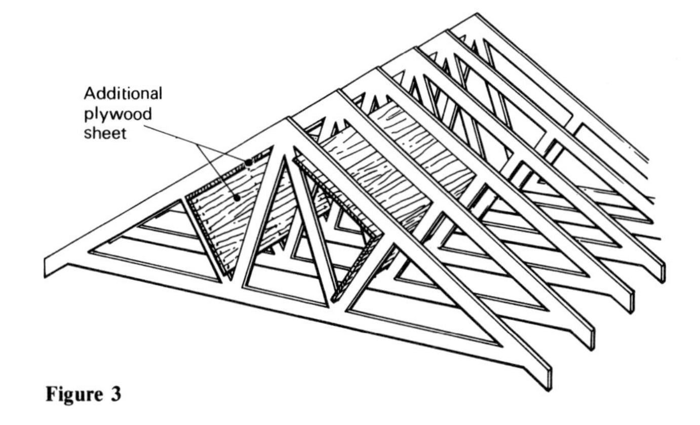

Figure 3

Dual-pitched roofs: trussed rafters — installation of remedial bracing

FAILURE: Instability or collapse of roofs

DEFECTS: Bracing inadequate or absent

Figure 1

Many earlier trussed rafter roofs were inadequately braced. In some cases this has led to distortion or leaning of trusses. Inadequately braced roofs may now need to be properly braced to prevent movement occurring. Where trusses are already distorted additional measures may be needed to restore stability. Where there is visible external damage, such as distortion of tiling, Figure 1, major repairs, or even rebuilding, will be necessary.

If ***remedial measures*** are ***not correctly carried out*** inadequately braced trussed rafter ***roofs*** may ***continue to move*** and ***may*** eventually ***collapse.***

REMEDIAL WORK

Principle — Roof trusses must be braced so that they act together as a single three-dimensional structural unit.

Practice

Where remedial bracing is specified:

- Ensure that all diagonal bracing and binders, where specified, are fitted as indicated in DAS 84[1].
- Ensure that any internally applied plywood diaphragms are of the type and thickness specified, and cover the specified number of rafters.
- Ensure that internally applied plywood diaphragms are nailed according to designers specification;
  — nailing will be required at relatively close centres, eg 3.35 mm × 75 mm galvanised nails at 100 mm centres, Figure 2.
- Ensure that, where lateral restraint straps are specified, they are fitted as indicated in DAS 28[2].
- Ensure that any blocking pieces, specified to straighten distorted trusses, are the full depth of the rafter and at least 38 mm wide and are skew-nailed into position.

Where major repairs are specified:

- Ensure that badly distorted or displaced trusses are rectified;
  — generally trusses more than 40 mm out of plumb will need attention, Figure 3; check with designer.
- Ensure, where complete re-building has been specified, that new roof construction conforms to the requirements of BS 5268 Part 3 (refer to DAS 84).
- Ensure that, where specified, holding down straps are installed.

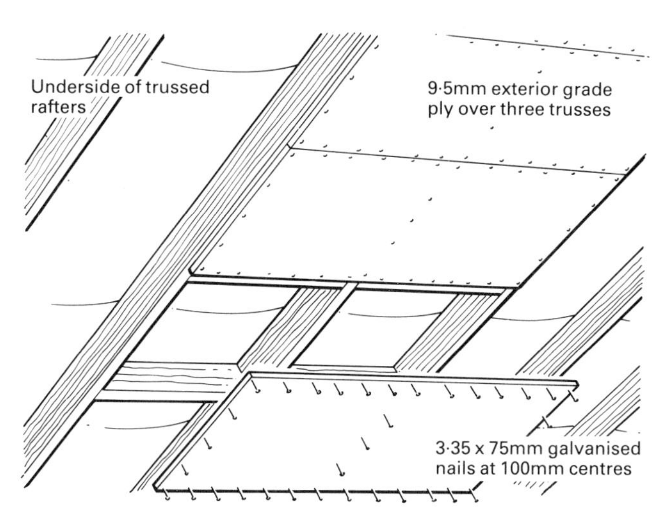

Figure 2

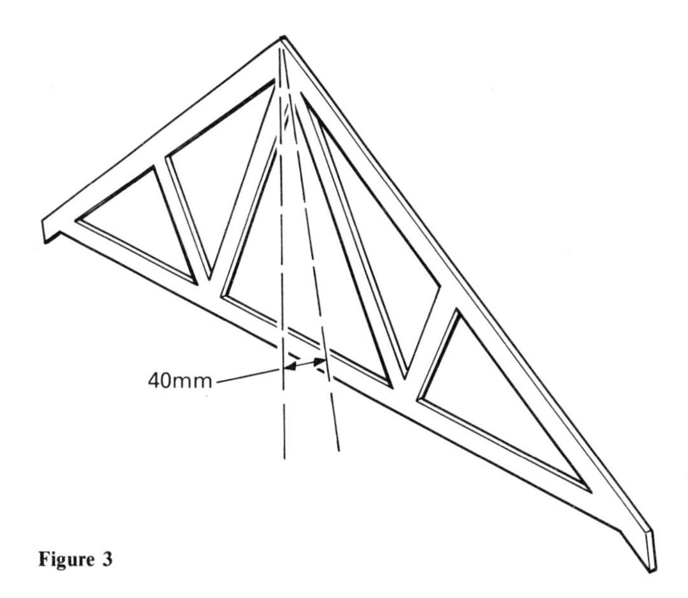

Figure 3

REFERENCES AND FURTHER READING

1 **Defect Action Sheet** DAS 84 'Dual-pitched roofs: trussed rafters; bracing and binders — installation'.

2 **Defect Action Sheet** DAS 28 'External and separating walls: lateral restraint at pitched roof level — installation'.

DAS 112
December 1987

Design
CI/SfB 8(27.9)Xi(A3u)(D7)

Dual-pitched roofs: trussed rafters — specification of remedial gussets

FAILURES: **Drawing of truss tension members from nail plate connectors; sagging of roofs**

DEFECTS: **Metal plate not correctly positioned during manufacture, of insufficient size; trussed rafter overloaded**

Figure 1

Photograph: TRADA

The structural integrity of a trussed rafter is critically dependent on the soundness of the jointing at each node point. If nail plates are inaccurately positioned in manufacture so that insufficient teeth engage securely in a truss member, joints, especially at the truss apex, can pull apart in service, allowing the truss to settle, Figure 1. Such joint failures can lead to local sagging of the roof sufficient to be apparent from outside.

Unless the ***defective trussed rafters*** are ***repaired,*** adjacent rafters will be overstressed, leading to further ***progressive distortion*** of the roof over time.

REMEDIAL WORK

Principle — Each trussed rafter must be capable of carrying its share of the roof load.

Practice

- Check integrity of joint at each truss node point; inspect nail plates on both sides of truss;
  - — give particular attention to each truss apex and to trusses which carry additional loads eg cold water storage cisterns.
- Consult a structural engineer if the integrity of the roof is in doubt.
- Specify that the ceiling under trusses with failed joints should be propped under the node points with a screw type prop and jacked to level;
  - — props/jacks must be properly supported on a spreader spanning several joists, Figure 2; Note: it may be necessary to prop, eg, the first floor back to the ground floor;
  - — web members should be jacked into tight contact with the underside of rafters, Figure 3.
- Specify in consultation with a structural engineer, remedial plywood gussets on the failed joints, Figure 4;
  - — size, thickness of plywood and number and size of nails will need to be calculated;
  - — no contribution from the failed metal plate to the strength of the joint should be assumed in the calculation.
- Specify, where necessary, remedial bracing to meet the requirements of BS 5268 Part 3, lateral restraint strapping and holding down straps.[1]
- Check that heavy loads such as cold water storage cisterns are adequately supported.[2]

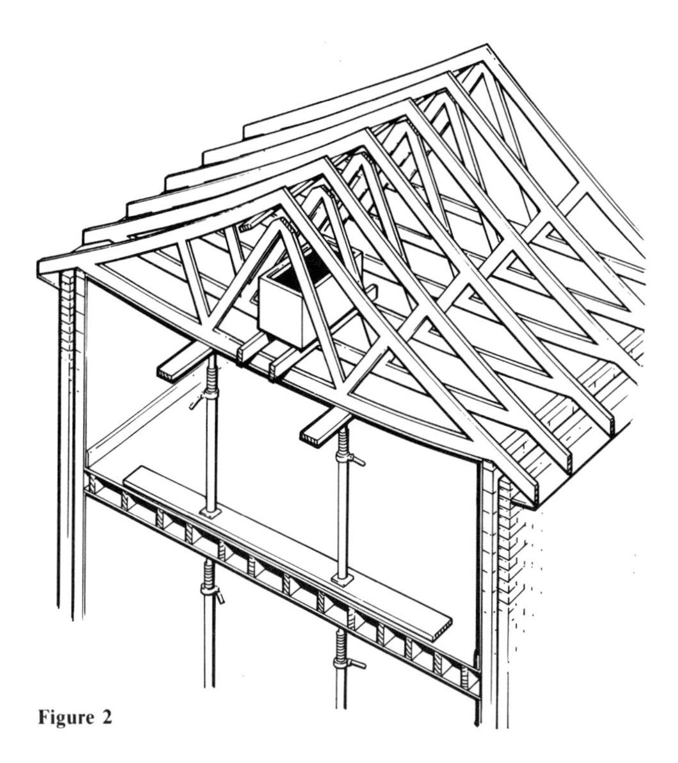

Figure 2

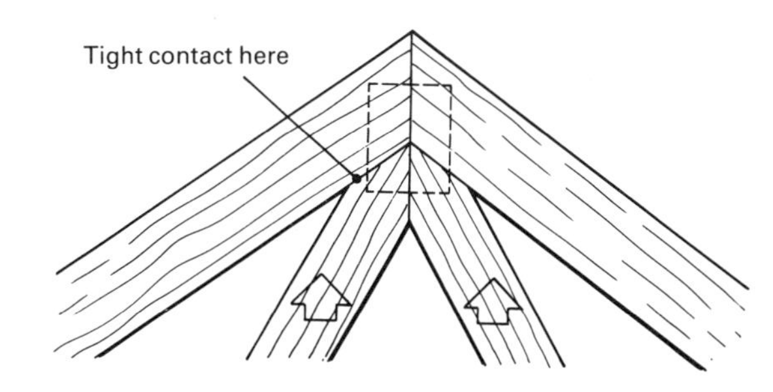

Figure 3

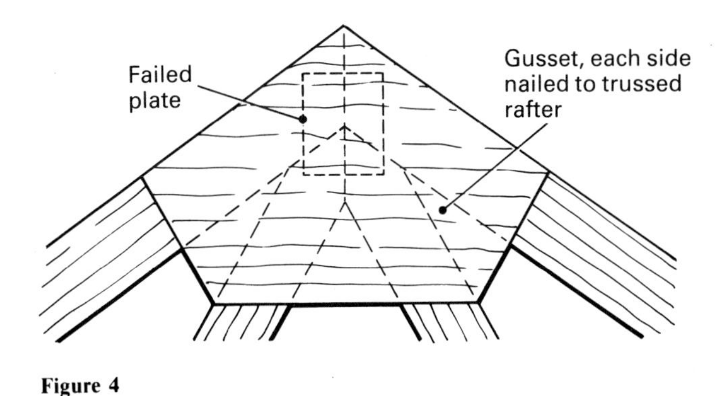

Figure 4

REFERENCES AND FURTHER READING

1 **Defect Action Sheet** DAS 110 'Dual-pitched roofs: trussed rafters — specification of remedial bracing'.

2 **Defect Action Sheet** DAS 43 'Trussed rafter roofs: tank supports — specification'.

PART ELEVEN

Chimneys, Flues and Ventilation

Domestic gas appliances*: air requirements

FAILURES: **Insufficient air for appliance or occupants; over-heating of compartments containing appliances; unacceptable draughts**

DEFECTS: **Vents for combustion air too small; vents for heat dissipation too small; vents badly sited and therefore blocked off by occupants**

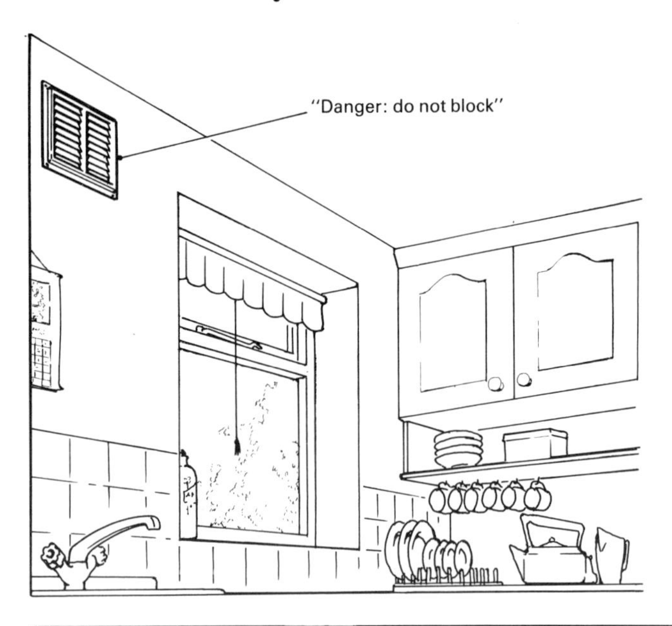

Past BRE surveys have occasionally found 'combustion air' vents with too little clear ventilation area; occupants sometimes block draughty vents.

Provision of adequate air for combustion is a requirement of the Gas Safety (Installation and Use) Regulations 1984.

If ***air supply*** to an ***open flued appliance*** is ***insufficient, poisonous flue gases may be produced and may spill*** into the dwelling; if ***vents*** are ***badly sited, occupants*** may ***block them;*** if ***compartments*** are ***not ventilated*** to dissipate heat losses, ***damage*** may follow.

Figure 1 Open-flued appliances must have adequate air supply

PREVENTION

Principle — Open-flued appliances must be provided with a route by which adequate air for combustion can reach them; both open-flued and room sealed appliances, if in a compartment, must be able to dissipate heat losses.

Practice

Rooms containing only room-sealed appliances

- Do not provide air vents for incoming combustion air (but see below for ventilation of compartments).

Rooms containing only open-flued gas appliances

- Check maximum input ratings of all appliances and provide ventilation to outdoors:
  - — if total is 7 kW or less no special ventilation measures are needed;
  - — if total is greater than 7 kW provide effective ventilation to outdoors at the rate of 450 mm² for every extra kW, Figure 2;
  - — ventilation may be direct (i.e. via a vent in the external wall of the room containing the appliance) or indirect (e.g. via an adjacent room or hall or suspended floor void); if air reaches appliance via 3 or 4 vents in series, increase ventilation area by not less than 50%.

Rooms containing only flueless gas appliances (gas cookers etc)

- Provide an openable window and permanent ventilation direct to outside in accordance with Table 1:
  - — if appliance is a refrigerator, towel rail or airing cupboard heater no special ventilation measures are needed.

Rooms containing both open-flued and flueless appliances

- Establish, using the guidelines above, the largest ventilation requirements from the following:
  - — the ***total*** ventilation requirements** for any ***open-flued space heating*** appliances (A);
  - — the ***total*** ventilation requirements for any ***flueless space heating*** appliances (B);
  - — the ***individual*** ventilation requirements** of all remaining appliances (C, D, E etc).

  **Note: for open flued appliances the first 7 kW input rating can still be discounted.
- Provide ventilation at a rate not less than the largest of these requirements (A to E etc).
  Note: since flueless appliances are present ventilation must be direct to outside.

*Not exceeding 60 kW rated input.

Compartments containing appliances

- Provide, where a compartment is built to house a room-sealed or open-flued appliance, ventilation via permanent non-adjustable vents at high and low level, Figures 3 to 6, with areas given in Table 2:
  - if ventilation is to a room, hall etc, and appliance is open-flued, the room must be provided with ventilation, to outdoors, via an area of 450 mm² for every 1 kW in excess of 7 kW heat input to appliance, Figure 5.

All rooms

- Specify that vents for combustion air entering a room from outdoors are to be located so that draughts are minimised:
  - e.g. locate close to ceiling and to appliance (but note that air inlet externally must be at least 600 mm from flue outlet);
  - if ventilation is indirect, via another room, vents in internal walls must not be more than 450 mm above floor level (to minimise smoke spread in fire); do not vent via a bathroom, WC, bedroom or kitchen.
- Check, unless only room-sealed appliances are used, whether there is to be a fan (extract fan, fan in open-flued appliance, cooker hood or warm air system) ***anywhere*** in the dwelling:
  - if so, the flue-gas spillage test to BS 5440 Pt 1 must be carried out with the fan running.

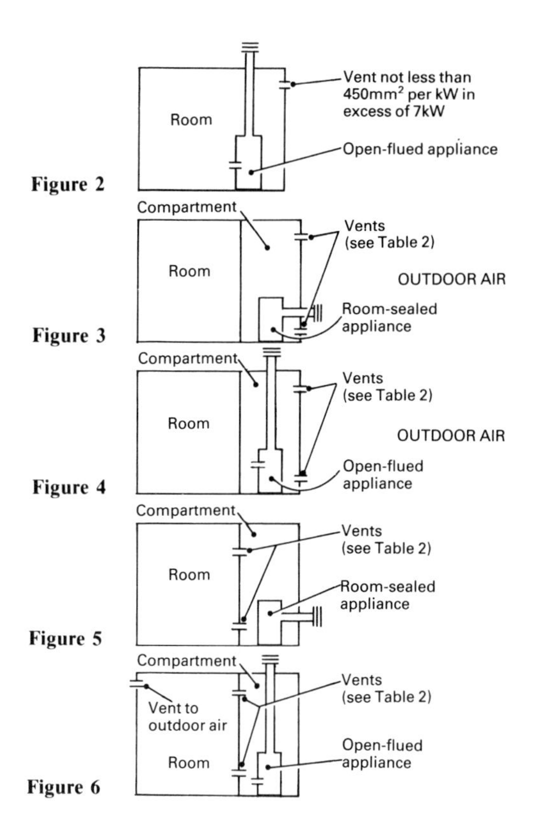

Figure 2

Figure 3

Figure 4

Figure 5

Figure 6

Table 1* Flueless appliances: effective ventilation areas in addition to openable windows

Appliance type	Room volume m³	Minimum effective area mm²
Oven, hotplate grill etc	6 – 9	6500
(Note: no air vent is required if there is a door opening direct to outside)	9 – 11 greater than 11	3500 nil
Instantaneous water heaters up to 12 kW input rating	6 – 11 greater than 11	3500 nil
Storage water heaters, circulators up to 3 kW input rating and wash/boiler machines up to 6 kW input rating	6 – 11 11 – 21 greater than 21	9500 3500 nil
Note: installation of the appliances above is not recommended in rooms of less than 6 m³		
Drying cabinets up to 2 kW input rating	any size	9500
Fixed space heater in a room, input rating 50 W per m³ of heated space but not exceeding 3 kW total input rating	any size	9500
Fixed space heater in a room, input rating as above but total exceeding 3 kW	any size	9500 + 3500 for every kW (or part) in excess of 3 kW

Table 2* Air vent areas for compartments: mm²/kW heat input

Position of opening	Open flued: Air from room	Open flued: Air direct from outside	Room sealed: Air from room	Room sealed: Air direct from outside
High level	900	450	900	450
Low level	1800	900	900	450

Type of appliance

*Adapted from BS 5440, which expresses areas in cm² units

Domestic draughtproofing: balancing ventilation needs against heat losses

FAILURE: **Boilers, fires not operating properly; airborne contaminants not clearing; condensation problems occurring.**

DEFECT: **Air flow through a particular room is not sufficient to vent that room; air flow is not sufficient for the safe operation of the heating appliance used in that room.**

Action has been taken over recent years by house occupiers, community action groups such as Neighbourhood Energy Action and others to conserve heat in a room or dwelling by fixing simple draughtproofing around external doors and windows. Some house occupiers may seal room vents, fit replacement windows, or install secondary glazing further to reduce unwanted cold draughts and heating costs. These draughtproofing measures can improve the living conditions for the occupier provided there is still an adequate supply of fresh air to the room and its occupants. If ***full draughtproofing*** measures are taken ***without consideration*** for the ***ventilation requirements*** of that particular room, ***fuel-burning appliances*** may ***not function properly, air-borne contaminants*** may ***not clear, condensation*** and subsequent ***mould*** growth may become a ***problem.***

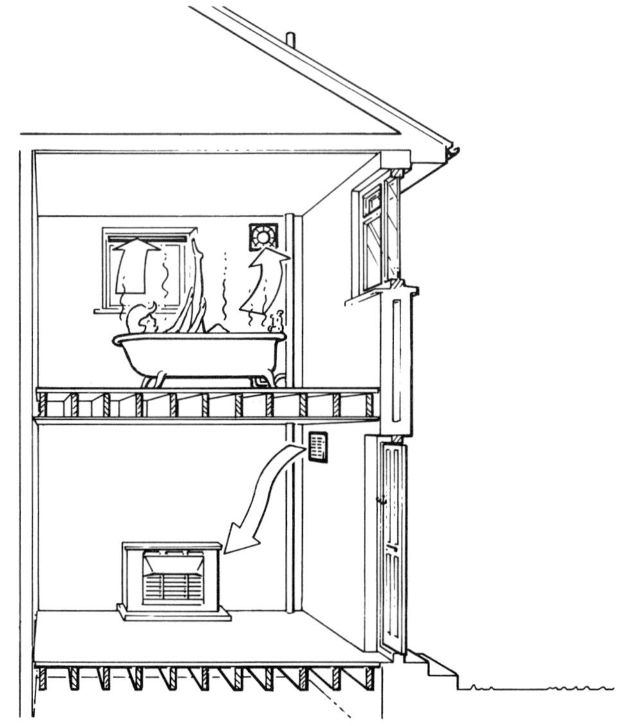

Figure 1 Maintain essential air flows

PREVENTION

Principle — Draughtproofing of a dwelling should not be carried out in isolation; consideration should also be given to providing controllable ventilation. The air requirements of fuel burning appliances must be satisfied. Excess water vapour will need to be removed by ventilation to prevent condensation.

Practice

- Ensure each room has provision for the necessary air changes:
  - — Provision for natural ventilation must be made in all habitable rooms of dwellings; ventilation opening should be at least 1/20th of the floor area of room or space[1]: NOTE: The 1/20th requirement may be reduced to 1/30th under the Building Standards (Scotland) Regulations;
  - — fine control of part of the above ventilation provision, or provision of additional controlled ventilation to give up to 4000 mm^2 free area of trickle ventilation in each room or space is normally adequate for achieving control of the air-borne contaminants and moisture without introducing unacceptable draughts, Figure 2;
  - — existing permanent ventilation in a bedroom and in a living room may then safely be sealed provided no fuel-burning heat appliance is used in that room (see Table 1 for air requirements). Draughtproofing of these rooms can proceed after checking for condensation problems and mould growth;

    NOTE: Security against forced entry of any trickle venting should be checked. Purpose-provided ventilators are more suitable than partially open windows;
- Check rooms for condensation problems or mould growth[2]:
  - — a room with a serious condensation problem or mould growth should not be draughtproofed. Identification of the cause of the mould should be ascertained and remedial measures taken[3];
  - — a living room or bedroom can be draughtproofed if the mould growth is light and easily removed and also if permanent ventilation is provided. If permanent ventilation is not provided, at least 2 m of window

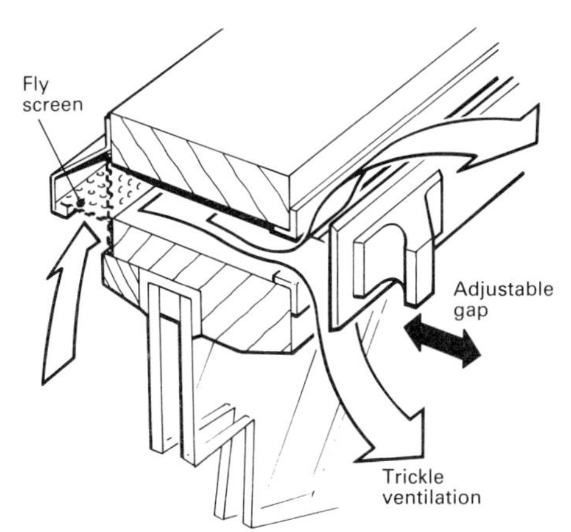

Figure 2 Example of an adjustable slot ventilator fitted into window head

opening perimeter should not be draughtproofed;
- mould growth in kitchens or bathrooms will usually be remedied only if moisture extraction at source is used. Other rooms, not affected in the dwelling, can be draughtproofed.

- Check that a means of increasing ventilation in the kitchen and bathroom during or immediately after the generation of water vapour is available to help prevent condensation problems:
  - windows in kitchens and bathrooms should not be draughtproofed unless extractor fans are installed;
  - internal doors from the kitchen and bathroom may be draughtproofed, to prevent spread of moisture to the rest of the dwelling. NOTE: If the kitchen/bathroom is made airtight, the extractor fan may not have an adequate air supply — it may stall or create a large negative pressure — which in RADON GAS[4] prone areas of the UK is unacceptable.
- Check with the Gas Area Office before installing an extractor fan in any room (including kitchens) containing an open flued gas fired appliance ie an appliance which draws air from the room for the combustion process and exhausts into the outside air:
  - extractor fans are **not permitted** in any room (including kitchens) containing open flued appliances burning solid fuel or oil.
- Check that the air flow for fuel burning space heating appliances is available according to Table 1 and that any original permanent ventilation provided has not been blocked or covered:
  - a room containing an open-flued gas fire (item 3, Table 1) should have 2.5 m of window opening perimeter left without draughtproofing to ensure that adventitious air is available otherwise provide permanent ventilator of open area greater than 3500 mm².
- Ensure that a room containing a flueless space heater (ie one which draws air from the room and exhausts into the room (items 4 and Note, Table 1) is not draughtproofed unless 9500 mm² of permanent ventilation is provided. NOTE: Flueless space heaters should not be permanently fixed in a bedroom or bathroom.
- Check that the air flow for flueless domestic gas appliances is available according to DAS 91[5]:
  - draughtproofing of the room containing the appliance is not acceptable;
  - other rooms in the dwelling can be draughtproofed;
  - installation of an appliance listed in Reference 5 is not permitted in rooms with volume less than 6 m³.
- Check that any secondary glazing does not interfere with the opening of windows or permanent vents.
- Advise the occupier of the dwelling on the need for regular maintenance of all combustion appliances.
- Ensure that new construction has 4000 mm² of controllable trickle ventilation per room or space:
  - rooms or spaces, without trickle venting, incorporating minor construction works, should not be draughtproofed until that construction has substantially dried out.

Table 1 Air supply requirements for fuel burning space heating appliances

Type of appliance	Requirements for permanent opening to the outside air in the room or space containing the appliance as specified in British Standards
1 Balanced-flue heating appliance	None — air supply provided directly from outside
2 Open-flued gas appliance including gas fire with back boiler but excluding room gas fire	Permanent opening required: (i) for a decorative (solid fuel effect) appliance, an area of 1800 mm² for each kW of rated input over 2kW (ii) for any other appliance, an area of 450 mm² for every kW of input rating over 7 kW
3 Room gas fire, open-flued	No requirement for permanent openings; it is assumed that there is a minimum adventitious area of 3500 mm²
4 Flueless gas space heater (fixed)	Permanent opening of at least 9500 mm² and an openable window required. Appliance must NOT be fixed in a bedroom or bathroom
5 Open solid fuel fire	Permanent opening of at least 5500 mm² or 50% of the throat opening, whichever is the greater
6 Other solid fuel flued appliance	Permanent opening with total area equal to at least the combined areas of the primary and secondary air inlets to the appliance
7 Oil burning flued appliance	Permanent opening of at least 550 mm² per kW of appliance rated output

NOTE: In the case of flueless space heating appliances (LPG, paraffin), no requirement for permanent opening but adequate ventilation air is essential. Ventilation as in 4 above is recommended.

REFERENCES AND FURTHER READING

1 **The Building Regulations (England and Wales)** 1985: Approved Document F
2 **Defect Action Sheet** DAS 16 Walls and ceiling: remedying recurrent mould growth
3 **BRE Digest** 297 Surface condensation and mould growth in traditionally-built dwellings
4 **Department of the Environment** The Householders' Guide to RADON
5 **Defect Action Sheet** DAS 91 Domestic gas appliances: air requirements.
BRE Digest 306 Domestic draughtproofing ventilation considerations
Department of the Environment Energy Efficient Renovation of Houses — A Design Guide
The Building Standards (Scotland) Regulations Part K
The Building Regulations (England and Wales) 1985 Approved Document J
BRE Digest 319 Domestic draughtproofing : materials costs and benefits.

Balanced flue terminals: location and guarding

FAILURES: **Terminals damaged by traffic, passers-by, external doors, windows; injury to passers-by in restricted spaces; flue gases unable to disperse freely; plastics, painted and similar materials heat damaged**

DEFECTS: **Terminals not guarded where necessary; terminals badly sited in relation to other features; surfaces vulnerable to heat damage not shielded**

Figure 1 '.... both terminal and door may be damaged ....'

BRE surveys of housing under construction have sometimes found that balanced flue terminals are badly sited.

If an ***unguarded terminal*** intrudes on an access way it ***may injure*** passers-by, especially in the dark, ***or*** it may ***be damaged;*** if a ***terminal*** is ***too close to other features,*** flue ***gases*** may ***not disperse*** freely, or susceptible ***materials*** may be ***damaged by heat,*** if a ***terminal*** is sited ***within a door swing,*** both ***terminal*** and ***door*** may be ***damaged.***

Balanced flue terminals: location and guarding

DAS 92
December 1986

PREVENTION

Principle — Terminals must be sited so that flue gases can disperse freely and be guarded or shielded where they might otherwise be damaged or cause damage or injury.

Practice

- Consider, when deciding boiler type and location, consequent position of its balanced flue terminal.
- Specify that a guard is to be fitted to any terminals which are less than 2 m above ground paving, balcony or other level to which people normally have access:
  — many terminals have sharp edges and corners.
- Check that a guarded terminal (below the 2 m level) will not obstruct passage of people, equipment etc:
  — where space is restricted, access with – say – a wheelbarrow or dustbin may be impeded, while even in wider access ways the additional space taken up by a guard may limit the use that can be made of the access.
- Check for possible conflict between terminal and door or casement swings.
- Ensure that the location of terminals meets the minimum separation distances in Table 1:
  — distances that are common to both 'natural' and 'fanned' draught appliances are illustrated in Figures 2 to 7.
- Check that terminals whether for natural or fanned draught appliances, will be located not closer than 850 mm below a plastics gutter, eaves or verge soffit and not closer than 450 mm below a painted soffit:
  — otherwise specify an aluminium alloy sheet shield, not less than 750 mm long, fitted to the underside of the gutter or soffit above the terminal.

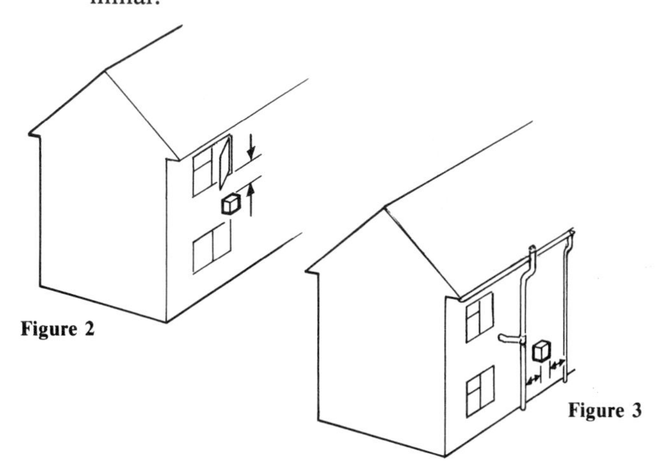

Figure 2

Figure 3

Table 1

Terminal location	Minimum separation distance (mm)	
Directly or partly below any door, openable window or ventilator, Figure 2:	300	
Near vertical drain pipes, soil & vent pipes; Figure 3:	75*	
Above ground, balcony or similar level; Figure 4:	300	
Facing a vertical wall or other surface; Figure 5:	600	
Vertically above or below another terminal (same wall); Figure 6:	1500	
Horizontally relative to another terminal (same wall); Figure 7:	300	
	Natural draught	**Fanned draught**
Below gutters, drain pipes, soil & vent pipes:	300*	75*
Below eaves:	300*	200*
Below balconies:	600	200
Facing another terminal:	600	1200
Near salient or re-entrant corners; Figure 8:	600	300

* Heat-sensitive materials (eg plastics, painted surfaces) may need protection at these minima

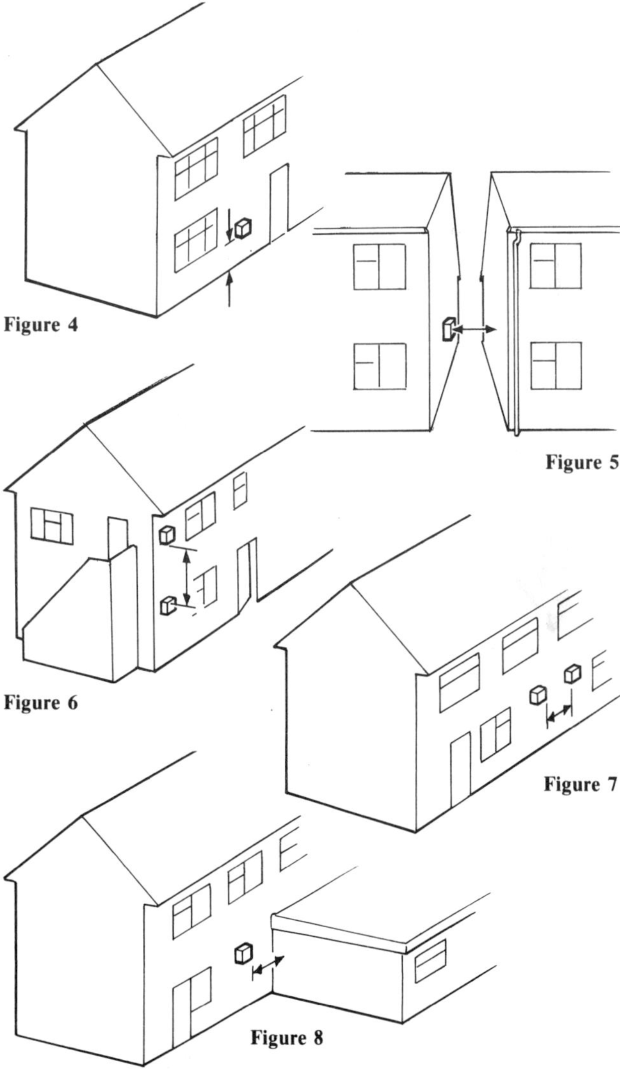

Figure 4

Figure 5

Figure 6

Figure 7

Figure 8

Chimney stacks: taking out of service

FAILURES: **Damp penetration and staining of chimney breasts; water vapour conducted from rooms to roof void; increased condensation in rooms**

DEFECTS: **Flues not vented, exposed disused stacks not weatherproofed; flues vented but open to roof space; room ventilation inadequate following sealing of flue**

Figure 1 Redundant stacks?

In rehabilitation work, chimneys formerly serving open fires are usually taken out of service; sometimes inadequate consideration is given to ventilation.

Disused stacks taken down to below roof level do not need ventilation to keep them dry unless they are on an external wall. But, if ***disused stacks*** that are ***on an external wall*** or remain ***exposed above roof level*** are ***not protected*** from ingress of rain and flues ***not ventilated,*** internal ***damp staining*** is likely. If ***flues,*** whether on external or internal walls, are ***ventilated so that water vapour is conducted from rooms to roof void*** copious ***condensation in the roof void*** may occur. If ***internal ventilation*** is ***inadequate*** when a room is no longer ventilated by a flue, ***condensation in living areas*** may result.

Chimney stacks: taking out of service

DAS 93
February 1987

PREVENTION

Principle — Exposed disused chimneys must be kept dry by weather protection and ventilation.

Practice

Chimneys on internal walls

- Specify that stack is to be taken down to below roof level and capped; do not ventilate flues, Figure 2.
  - — if the internal wall is a separating wall specify measures to maintain fire stopping below roof covering[1].

Chimneys on external walls, eaves elevation

- Specify that stack is to be taken down to below roof level, capped, and flues ventilated to external air at head and foot, Figure 3.

Chimneys on external walls, gable elevation

- Specify that flues are to be ventilated to external air at head and foot, either capping the stack and venting via air-bricks, or using proprietary venting hoods, Figure 4;
  - — if upper part of stack is unsound it can be taken down to just above roof level, capped, and flues vented to external air at head and foot.

- Do not ventilate any disused flue so that water vapour from rooms will be ducted to roof voids.

- Consider what additional means of ventilating living areas needs now to be provided to reduce the risks of condensation.

Caution: a disused flue in a stack containing others that remain in service should not be capped: there is a risk that flue gases may leak into it through the withes and be drawn down into the accommodation.

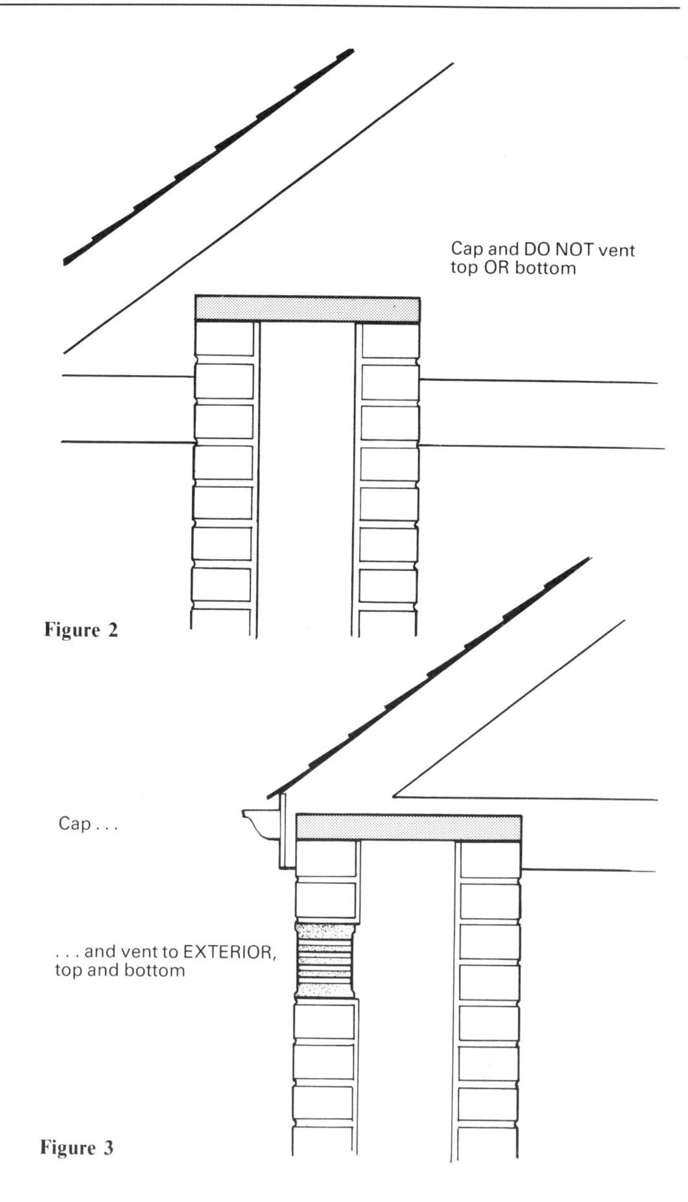

Figure 2

Figure 3

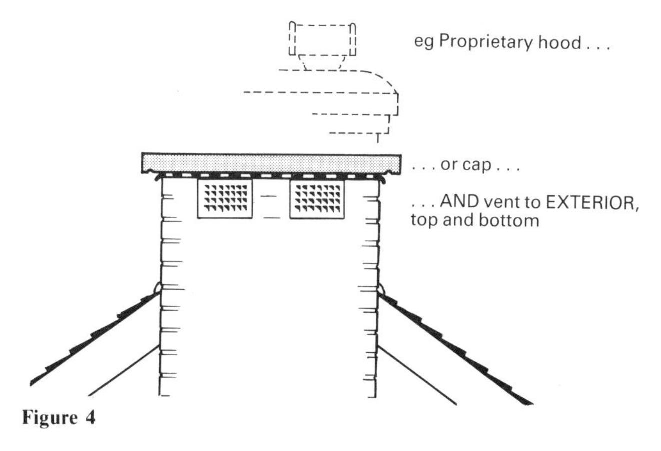

Figure 4

REFERENCES

1 *Defect Action Sheet* DAS 8 'Pitched roofs: separating wall/roof junction preventing fire spread between dwellings'.

Single wall flue pipes for class II appliances — location, fixing and shielding

FAILURES: Leakage of flue gases; ignition of combustible materials

DEFECTS: Flue pipes not sleeved; flue pipes not adequately supported; jointing disturbed; flue pipes too close to combustible materials; flue pipes not fully connected to discharge to external air

Figure 1 An unsupported flue

BRE surveys found instances of flue pipes touching sarking felt, combustible insulation in roofs, timber wedges and chipboard floors; pipes were seen both totally unsupported, Figure 1, in roof spaces, or supported by timber wedges between pipe and combustible construction materials. The Building Regulation requirement for separation between a flue pipe and combustible construction was commonly not met.

If ***flue pipes*** are ***not*** properly ***supported, not*** correctly ***spaced from combustible materials*** and ***not sleeved*** where required, there is a risk that ***joints will open*** and leak flue gases (eg into the roof space) and that combustible ***materials*** may be ***subjected to heat.***

Single wall flue pipes for class II appliances — location, fixing and shielding

DAS 60
September 1984

PREVENTION

Principle — Flue pipes must neither leak nor ignite construction materials.

Practice

- Specify that flue pipes are to be supported at not more than 1800 mm intervals, Figure 2:
  — instruct site staff to check that supports are robust and well fixed, that no combustible packings are used between flue pipe and fixing, that sockets are not installed inverted and that installation is completed so that the pipe discharges to external air.
- Specify that there is to be a support immediately under every socket, Figure 2 — including those at changes of direction, Figure 3.
- Specify that flue pipes, where they penetrate a combustible floor, wall, roof, ceiling, or partition are to be provided with non-combustible sleeves with a minimum clear air space of 25 millimeters between flue pipe and sleeve, Figure 4:
  — additional design measures will be needed to maintain the performance; eg weathertightness, fire resistance — of elements so penetrated.
- Specify that pipes must not be jointed within the floor depth.
- Design casings for flue pipes using only non-combustible materials for both frame and linings.
- Specify a minimum distance of 25 mm between flue pipe and constructed casing, Figure 5.
- Check that nowhere in the completed building design does the flue pipe pass too close to any combustible material, Figure 6:
  — not closer than 25 mm to meet the requirement of BS 5440 Part 1; not closer than 50 mm if adopting the 'deemed-to-satisfy' provisions of the Building Regulations (England and Wales);
  — carry out a three dimensional check, eg of proximity of flues in roof spaces to truss bracing or other combustible parts such as combustible insulation, sarking, vapour barriers etc.

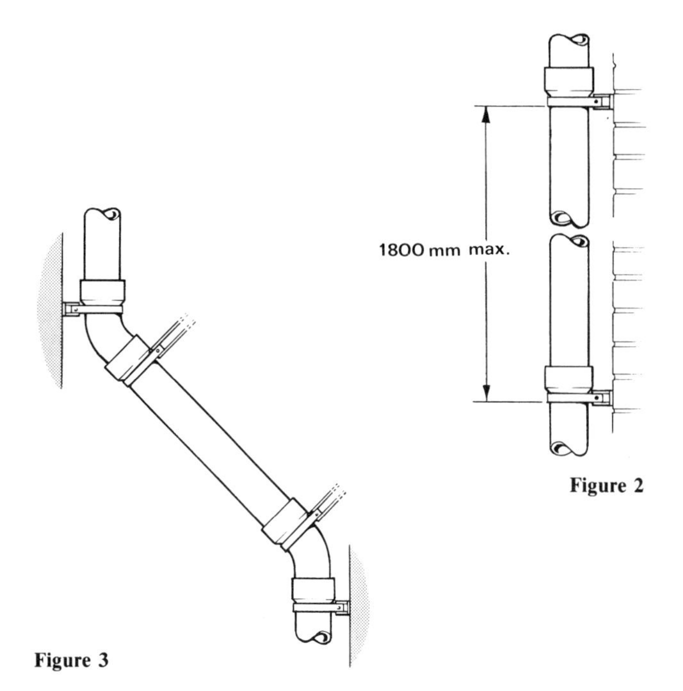

Figure 2

Figure 3

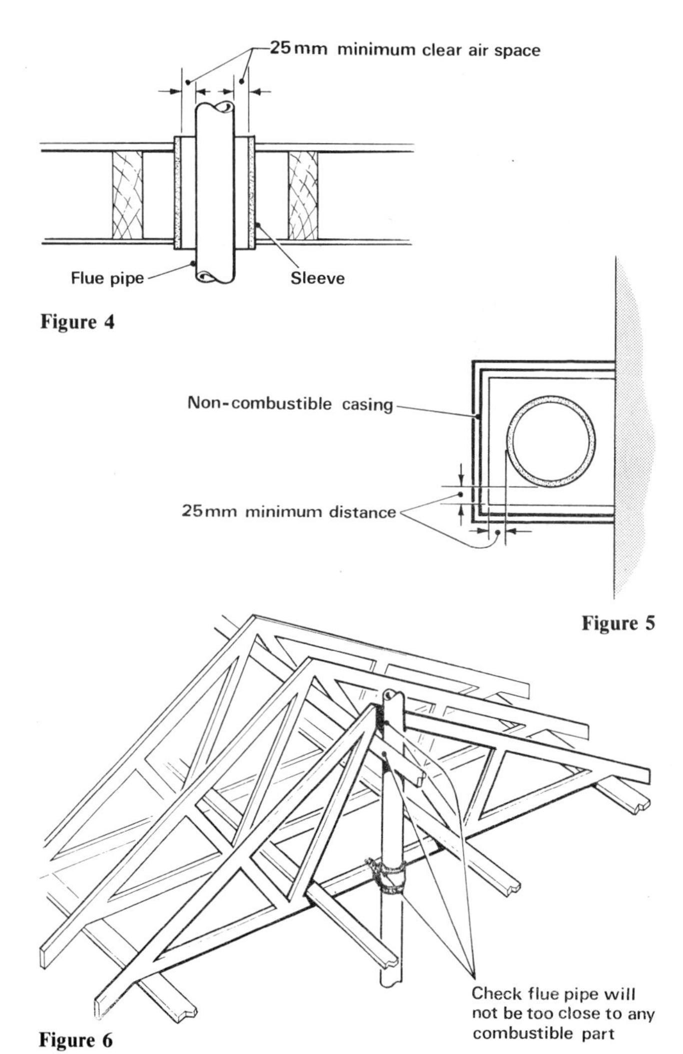

Figure 4

Figure 5

Figure 6

REFERENCES AND FURTHER READING

The Building Regulations 1976 (England and Wales).
The Building Standards (Scotland) Regulations.
British Standard BS 5440 Part 1:1978, 'Code of Practice for flues and air supply for gas appliances of rated input not exceeding 60 kW'.

Domestic chimneys: solid fuel — flue design

FAILURES: **Smoke and gases vent into the dwelling; poor combustion; lack of heat output; condensation leaking from flue into brickwork.**

DEFECTS: **Insufficient draught; 'smoke blow back'; flue temperatures too low.**

Figure 1 Masonry not jointed, flue joints leaking.
Photo by courtesy of the Solid Fuel Advisory Service.

The functions of a domestic chimney are to vent the products of combustion safely to the outside air and induce sufficient air flow through the flue to suit the appliance and its heat output. A chimney will fail to work correctly if the outside air temperature is only marginally different from that inside.

Many problems causing poor chimney performance are due to the design and construction of the flue, Figure 1, as well as to the relation of the flue to the many different appliances which may be connected to it.

The chimney flue requires to have minimum heat loss. A chimney built within a dwelling rather than on an outside wall has the advantage of creating both a warmer house and a warm flue.

If the ***air flow*** (induced by the action of the flue) through an appliance is ***insufficient, combustion*** will be ***incomplete***. If in addition a flue is badly constructed fumes may vent back into the dwelling.

If the ***chimney*** terminates ***below*** the main ridge line, under certain wind conditions there could be turbulence or positive air pressure which may result in '***smoke blow back***' into the dwelling.

If the ***flue temperature*** is ***too low***, ***updraught*** will be ***poor*** and ***condensation*** occurring in the flue may cause ***damage*** to the chimney fabric.

PREVENTION
Principle — The chimney should be designed to induce adequate air flow through the appliance, resulting in complete combustion of the fuel and full ventilation of the flue gases safely to the outside air.

Practice

- Specify a flue 200 to 225 mm diameter (185 mm × 185 mm) for all fireplace openings up to 500 mm × 550 mm, and ensure lining material is suitable for solid fuel[1]. Note that this flue size will also vent other appliances without the need for future structural alterations. Refer to the appropriate Building Regulations, for guidance to flue sizes[2];
  - — for dog or basket grates standing in large openings the cross-sectional area of the flue should be 8:1 ratio for a house, 6:1 ratio for a bungalow (13% to 16% of the flue opening)
  - — a canopy over and envelope around the grate can reduce the effective area of the fire opening, Figure 2.
- Specify that the gap between the refractory liner and the chimney wall is filled with a weak mortar mix to minimise the heat loss from the flue;
  - — in severely exposed locations specify an insulating mortar such as vermiculite or perlite.

NOTE: The liner may require a chimney that is wider than a brick multiple.

- Design a straight flue where possible;
  - — where bends are unavoidable there should be no more than two bends per stack: the angle of the bends should be no more than 45° to the vertical and the traverse should be kept to a minimum, Figure 3.
- Specify that bends are to be formed from proprietary made segments to enable correctly fitting spigot and rebate joints to be made;
  - — bends fabricated on site from straight lengths are unacceptable.
- Specify the chimney throat size;
  - — for an open fire this should be 110 ± 10 mm deep, approximately ¾ width of grate, properly formed to provide smooth non-turbulent transition for the gases from the fire to the flue[3], Figure 4.
- Ensure a permanent (draught-free) supply of air into the room;
  - — for open appliances, 5500 mm² or 50% of the throat opening whichever is greater,
  - — for other appliances, at least equal to combined appliance primary and secondary air inlets,
  - — permanent additional draught-free ventilation can provided by vents near the hearth, or a louvre above an internal door, Figure 5.
- Specify an outlet to the flue terminal not less than the flue diameter to give smooth transition to the outside air;
  - — the flue liner can be continued to form the terminal;
  - — a chimney pot, if fitted, should be parallel not tapered. A tapered pot results in rapid build-up and blockage with coal burning appliances.
- Specify a minimum chimney height of at least 4.5 m measured vertically from top of appliance to terminal for a solid fuel appliance (subject to manufacturer's specific instructions);
  - — terminate flue, where practical, above roof main ridge to avoid turbulence. 'Smoke blow back' may occur if terminated below this level, Figure 6.

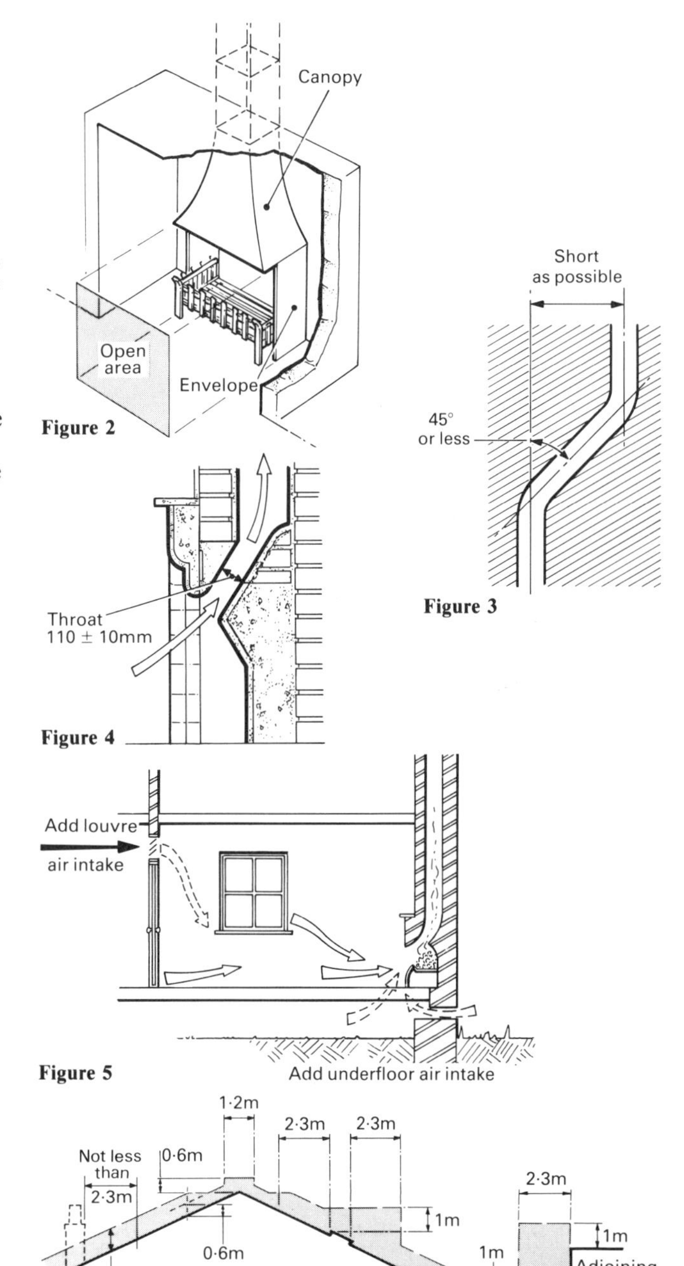

Figure 2

Figure 3

Figure 4

Figure 5

Figure 6 Minimum chimney heights

REFERENCES AND FURTHER READING

1 *British Standard* BS 6461:Part 1:1984. Code of practice for masonry chimneys and flue pipes.

2 *The Building Regulations (England and Wales)* — Approved Document Part J.
The Building Standards (Scotland) Regulations — Part F.
The Building Regulations (Northern Ireland) — Parts L and M.

3 *British Standard* BS 8303:1986. Code of Practice for installation of domestic heating and cooking appliances burning solid mineral fuels.

☐ *The Chimney Guide*. Solid Fuel Advisory Service.
British Standard BS 1251:1970 Specification for open fireplace components.

☐ *Defect Action Sheets* DAS 94 and 95. Masonry Chimneys: DPC's and flashings location and installation.

Domestic chimneys: solid fuel — flue installation

FAILURES: **Smoke and gases vent into the dwelling; poor combustion; lack of heat output; condensation leaking from flue into brickwork.**

DEFECTS: **Insufficient air flow through appliance; blocked flue; holes in flue; masonry and flue liners not jointed properly; flue liner surround not filled; flue terminal unsuitable.**

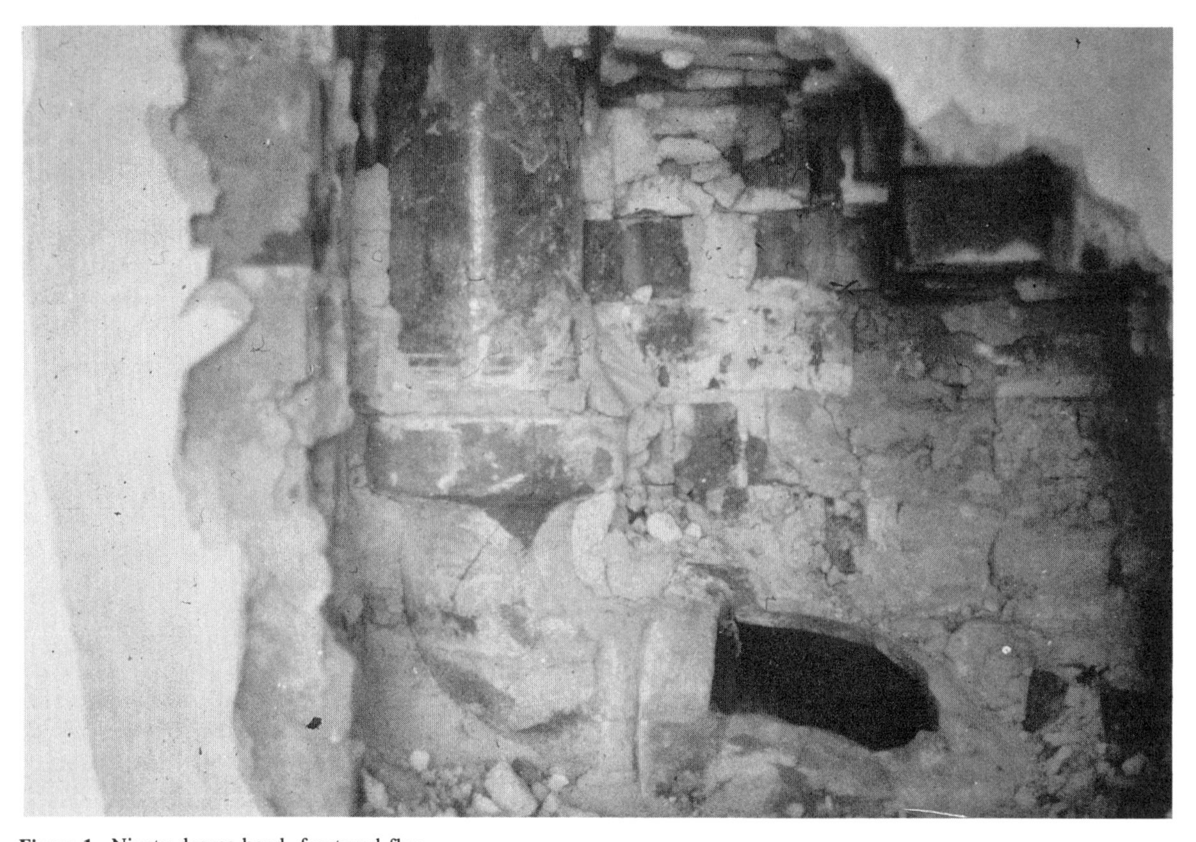

Figure 1 Ninety degree bend, fractured flue.
Photo by courtesy of Solid Fuel Advisory Service.

Problems occur with chimney flues because of construction faults, Figure 1, and the relationship of flues to the different appliances connected to them.

The functions of a domestic chimney are to vent the products of combustion safely to the outside air and induce sufficient air flow to suit the appliance and its heat output.

Domestic chimneys: solid fuel — flue installation

DAS 127
February 1989

If the ***air flow*** through an appliance is ***restricted, combustion*** will be incomplete and ***gases*** may ***vent*** back ***into*** the ***dwelling***. If flue ***bends*** are ***not properly jointed, liners*** placed ***spigot uppermost, liners not packed out*** with the specified mortar, ***flue condensate may leak*** into the brickwork and gases may vent back into the dwelling.

PREVENTION

Principle — The chimney flue should be constructed to minimise heat losses and to give a smooth passage for the flue gases to the outside air. An adequate supply of air through the appliance is essential to give complete combustion and to vent the gases safely.

Practice

- Ensure that the chimney masonry is properly jointed;
  - — unfilled mortar joints are not acceptable.
- Check the flue diameter is that specified;
  - — for an open fire this should be 200 to 225 mm diameter (185 mm × 185 mm) for all fireplace openings up to 500 × 550 mm.
- Check that the refractory lining to be used is as specified.
- Ensure that flue linings are sound and are correctly fitted with sockets or rebates uppermost, Figure 2;
  - — any caulking material is fixed as specified.
- Ensure that any flue offset is set at 45° or less to the vertical;
  - — the offset should be cleared of site debris, Figure 3.
- Check that bends are made properly using spigot and socket or rebate joints;
  - — bends made using cut straight lengths and butted together cannot be properly sealed and are not allowed.
- Ensure that the flue pipe from the appliance does not project into the main flue and cause a restriction to the flue gas flow, Figure 4.
- Check the chimney throat area is as specified;
  - — for an open fire the throat should be 110 ± 10 mm deep and should provide a smooth non-turbulent flow for the flue gases, Figure 5.
- Ensure that the flue is smooth, clean and free from protruding rings of mortar at the lining joints;
  - — a coring-bag is recommended for sectional flues and in-situ lining joints to give a smooth finish;
  - — a coring-ball, 50 mm less that the nominal flue bore can be lowered gently through the flue to the appliance to check the completed flue.
- Check that the flue terminal is of the diameter and at the height specified by the designer;
  - — check with the designer if the flue does not terminate above the roof ridge, have a diameter at least equal to the flue liner, and, if fitted, have a parallel chimney pot.
- Ensure the space between the refractory liner and the chimney wall is infilled with the material specified;
  - — a weak mortar mix is usually specified;
  - — for a more exposed chimney position, an insulating mortar such as vermiculite or expanded perlite and cement may be specified.

REFERENCES AND FURTHER READING

☐ *Defect Action Sheets* DAS 94 and 95. Masonry chimneys: dpcs and flashings — location and installation.

☐ *Defect Action Sheet* DAS 126 Domestic chimneys: solid fuel — flue design.

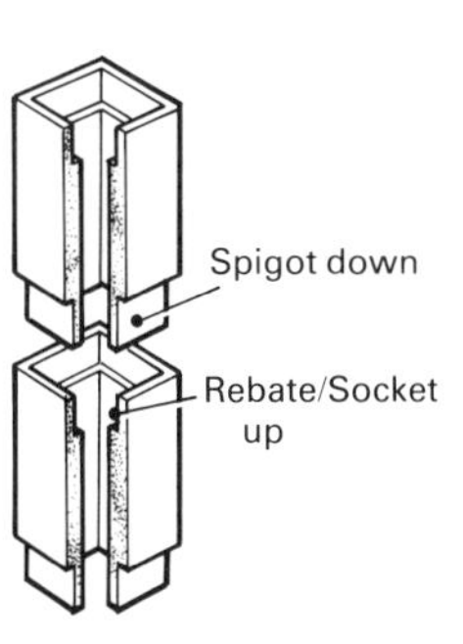

Figure 2

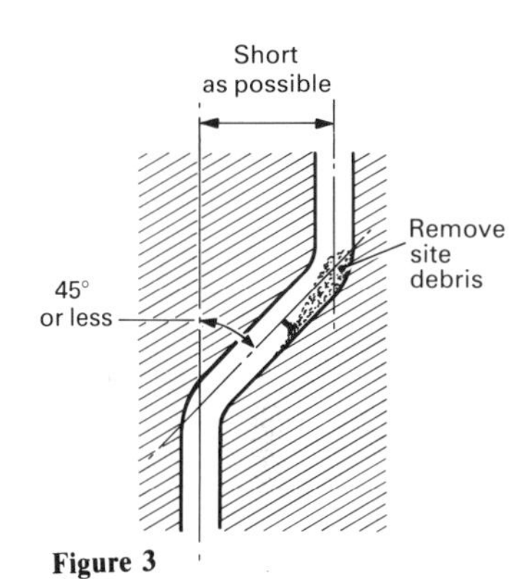

Figure 3

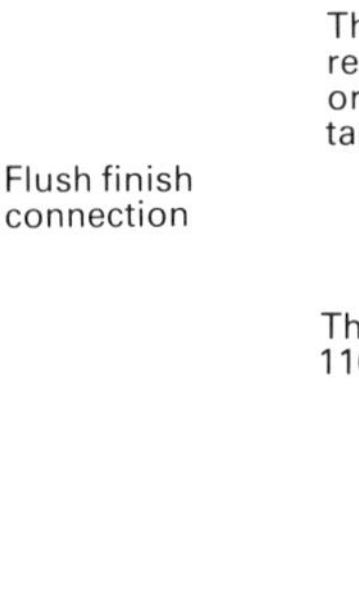

Figure 4

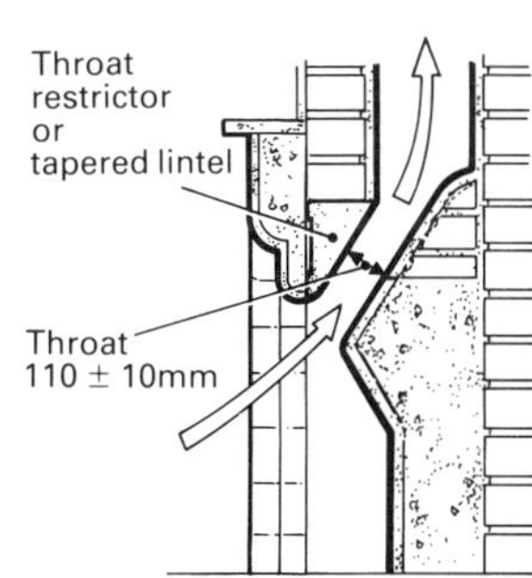

Figure 5

Domestic chimneys: re-building or lining existing chimneys

FAILURE: **Staining of internal plaster and decorations. Staining of external brickwork. Distortion and cracking of stack, cracking and disintegration of renderings. Expansion of brick joints in stack, causing curling and instability.**

DEFECTS: **Chimney not lined and not insulated when built. Fuel or appliance used with the chimney has been changed.**

Figure 1

The flue of a chimney built before February 1966 is likely to be of 225 mm × 225 mm brickwork construction, parged, but not lined.

A chimney used for solid fuel burning may have successfully vented the combustion products (tar acids, ammonia, sulphur compounds and water vapour) safely to the atmosphere. However, the fuel used and the appliance can be changed many times throughout occupation of a dwelling and the flue may become unsuitable.

The Building Regulations give details of the standards which must be met for re-built chimneys and flues. Heat producing appliances are covered under the Building Regulations (England and Wales) 1985: Approved Document J.

When a change is made to the fuel type burned the flue should be checked to ensure that it complies with current Building Regulations.

If ***water vapour*** should ***condense*** continuously in the flue the ***chimney fabric*** may well ***deteriorate. Sulphates*** and ***acids*** will ***attack*** both the ***parging*** and the ***mortar*** joints and ***expansion*** will ***occur.*** Sulphates may be dissolved out of brickwork by the condensate which will accelerate the attack on the mortar **1**. ***Tarry deposits*** and ***salts*** may be carried through the bricks and plaster to ***damage internal decorations.***

PREVENTION

Principle — A chimney should incorporate a flue liner to suit the fuel being used. The flue liner should be insulated throughout its length to ensure that water vapour containing aggressive products of combustion does not condense out during normal operation of the appliance connected to it.

Practice

- Check the condition of the existing flue for blockage, internal damage and leakage by inspection, sweeping the chimney and smoke testing in accordance with BS 6461:
  - — small deposits of mortar in the soot after sweeping indicate a serviceable chimney with no special lining. If a change of fuel is proposed, a flue liner should be considered;
  - — large deposits of mortar after sweeping, or masonry in a fireplace, indicate that remedial work will be needed;
  - — the smoke test will indicate the incidence and location of any leakage from the chimney. Serious leakage must be rectified prior to lining, ie repoint, rebuild.
- Check the size of the flue, Table 1. A liner may be installed (if size permits) to suit the fuel. This liner must be installed and supported in accordance with manufacturer's instructions. The chimney will have to be broken into at the liner joints or selected support points:
  - — flexible metal double skin liners of stainless steel can be used for all fuels but care is needed with wood and peat (see Note 2). Aluminium liners should not be used with solid fuel appliances, Figure 2;
  - — lightweight concrete flues can be cast in-situ using perlite aggregate. The flue is formed by an inflatable tube, removed after casting, Figure 3. Careful site supervision is essential and a period of at least three days should elapse before using the heating appliance at a low setting; five days should be allowed before full normal use or as recommended by installers, (see Note 2);
  - — fibre cement pipes can be inserted in long straight lengths.

Note 1 Other pipes are available, for lining or re-lining but they are generally in shorter lengths and hence will need more support points.
Note 2 Wood and peat are particularly aggressive fuels and give rapid build up of tar and deposits. A chimney fire burns at temperatures exceeding 1000°C and causes collapse of the metal

liner, so the chimney should be swept several times a year. A metal liner should not be regarded as a permanent liner for these fuels. Chimney fires can also split a concrete liner.

- Check the length of the flue:
  - — a gas appliance or fire should have flue lengths not exceeding the values given in Tables 2 and 3 to ensure condensate free operation;
  - — the flue for a solid fuel appliance should be at least 4.5 m in height [2,3].
- Check if the new flue liner can be insulated. This is especially necessary with an external chimney;
  - — gas generates more water vapour than other fuels and this can lead to more chance of condensate problems in a poorly insulated chimney;
  - — insulation materials such as perlite, vermiculite, or mineral or glass wool can be used to fill a void between liner and brickwork. The chimney trays will protect the insulation from rain[4].
- Ensure that all control gear in the chimney is suitable for the fuel to be used, or removed; for example, restriction plates and dampers are not suitable for use with gas.
- Check that the flue serves only one appliance and does not connect into another room except for cleaning purposes. The opening in the other room should be sealed gas-tight.
- Ensure that an unlined brick or block chimney connected to a gas appliance has a debris collecting space below the point of connection with suitable cleaning access.
  - — debris collecting space to be not less than 0.012 m^3, with the base located at least 250 mm below the point of connection.
- Check condition of exposed stack and if in doubt about stability seek a chartered engineer's advice:
  - — a straight stack may need only repointing[5] unless it is breaking up;
  - — a leaning stack should be rebuilt or stabilised. The rebuilt stack or complete chimney should be lined and insulated to meet current standards for new chimney construction;
  - — external rendering of chimneys is not recommended. However, if it is required, the render should be suitable for severe exposure and applied accordingly[6].
- Check extent of any staining within the building:
  - — light staining of plaster may be sealed back by using an adhesive backed metal foil. If the staining is confined to the chimney breast, dry lining with foil backed plaster board fixed to preservative treated timber battens may be used;
  - — serious staining will necessitate the removal of all render and finishes and making good with a 1:3 cement/sand render and finish which should extend at least 300 mm beyond existing staining.
- Check renovated flue:
  - — all access openings should have been made good;
  - — chimney terminal should have been made good;
  - — a coring ball should be lowered to prove flueway;
  - — the smoke test according to BS 6461 should be repeated.

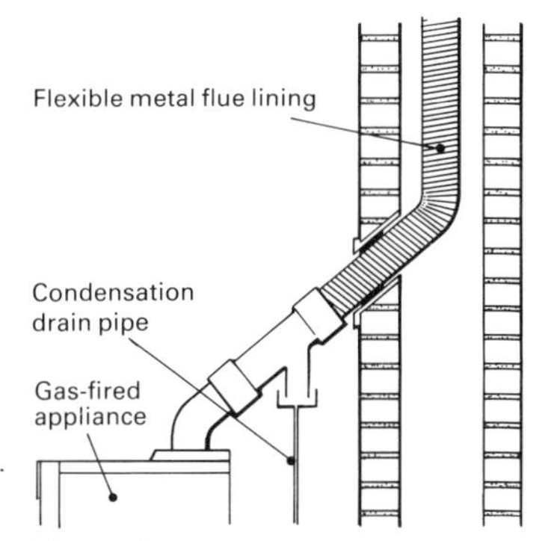

Figure 2

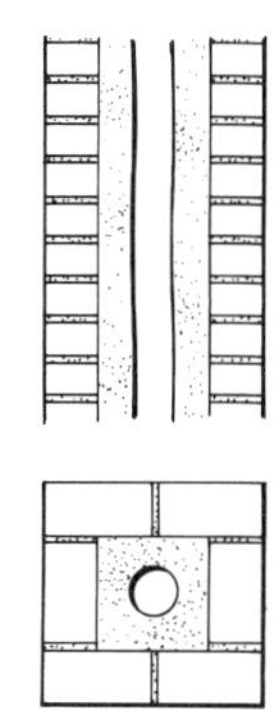

Figure 3

Table 1 Size of flue

Fuel	Type of appliance	Size of flue
Solid fuel	open fire	200 to 225 mm diameter* or 185 mm square
	dog or basket grate in large opening	open area to flue ratio 8:1 for house 6:1 for bungalow
Other solid fuel or oil appliances	output rating up to 30kW	150 mm diameter or square
(flue in chimney)	more than 30kW up to 45kW	175 mm diameter or square
gas (open-flued appliances)	decorative (log effect or other solid fuel fire effect appliances.)	no dimension across axis less than 175 mm
	gas fires	no dimension less than 63 mm. Cross sectional area at least 12000 mm^2. If flue rectangular, the greater dimension not more than 6 times the lesser.
	other appliances	no dimension less than 63 mm. Cross sectional area at least equal to outlet from appliance. If flue rectangular, the greater dimension not more than 5 times the lesser.

* A solid open fire flue is of sufficient size to vent other fuels.

Note: Current designs of condensing boilers are not suitable for connection to open-flues.

Table 2 Maximum lengths for condensate free flues (gas fires only) (extracted from BS 5440: Part 1: 1978)

Flue	Concrete block or brick chimney 13000 to 200000mm^2 flue area	125 mm dia fibre cement or metal flue pipe	
		Single wall	Double wall
Internal	8 m	11 m	23 m
External	6 m	8 m	16 m

Table 3 Maximum lengths for condensate free flues for typical domestic boilers (extracted from BS 5440:Part 1:1978)

Flue	Concrete block or brick chimney Appliance Input rating			Single-walled flue pipes Appliance input rating			Double-walled flue pipes Appliance input rating		
	9kW	17kW	29kW	9kW	17kW	29kW	9kW	17kW	29kW
Internal	6 m	8.5 m	11 m	8.5 m	11.5m	15 m	17 m	23 m	29.5 m
External	3.5 m	6.5 m	9 m	4.5 m	7.5 m	12 m	9.5 m	15.5 m	23.5 m

REFERENCES AND FURTHER READING

1 DAS 128 Brickwork: prevention of sulphate attack.
2 DAS 126 Domestic Chimneys: solid fuel — flue design.
3 DAS 127 Domestic Chimneys: solid fuel — flue installation.
4 DAS 94 Masonry chimneys: DPCs and flashings — location.
5 DAS 72 External masonry walls: repointing.
6 DAS 37 External walls: rendering — resisting rain penetration.
☐ DAS 93 Chimney stacks: taking out of service.

DAS 94
February 1987

Design
CI/SfB 8(21.9)M(A3u)

Masonry chimneys: dpcs and flashings — location

FAILURES: **Water penetration and staining of chimney breasts and ceilings; dampness in roof timbers**

DEFECTS: **Dpcs in chimneys absent or wrongly positioned; roof flashings incorrectly detailed**

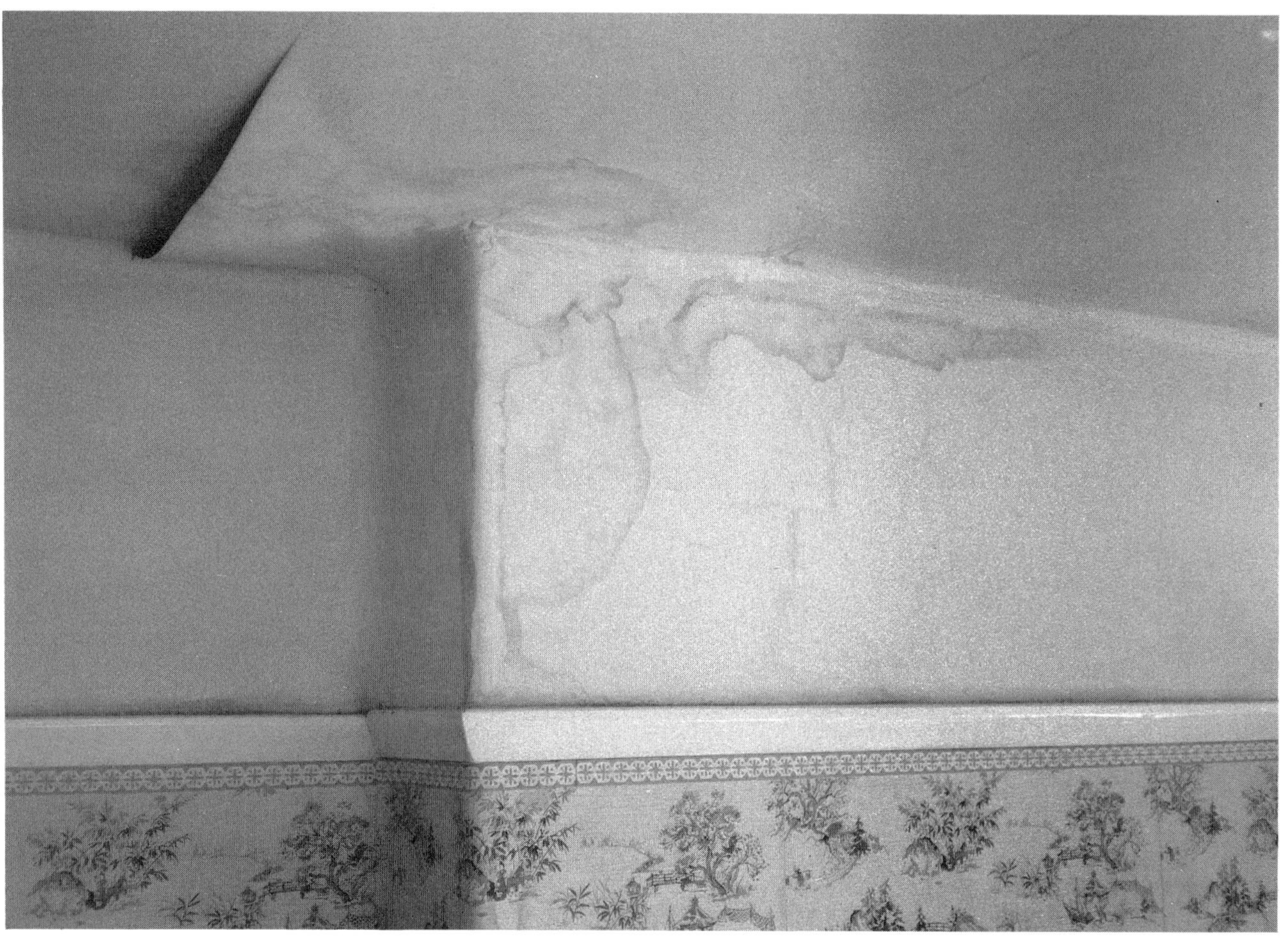

Figure 1

Chimneys are particularly exposed to the weather. Many cases have been seen during BRE investigations, where rain penetration of chimney stacks has resulted in staining of decorations on chimney breasts and the ceiling above — Figure 1.

The advice given is applicable to new construction and also to remedial work that requires rebuilding of a chimney. For capped-off disused chimneys, see DAS 93[1].

If ***damp courses in chimneys*** at the roof junction and beneath the capping ***together with roof flashings,*** are ***not detailed so as to exclude driving rain,*** then ***damp penetration*** and staining on the chimney breast and ceiling ***can result.***

PREVENTION

Principle— Dampness in parts of chimneys exposed to the weather must not penetrate to the interior of the dwelling.

Practice

- Specify a dpc, through the chimney, not less than 150 mm above the highest point of intersection between chimney and roof covering, Figure 2;
  — where the pitch of the roof or size of the stack would produce a large upstand to the flashing, as at A, Figure 2, consider specifying a stepped dpc, Figures 3a and 3b, or two dpcs, Figure 4;
  — two courses of dpc bricks or slates bedded in 1:0 to ¼:3 cement:lime:sand mortar have proved a satisfactory substitute for a sheet dpc where exposure is not severe[2].

- Specify a dpc below the terminal:
  — where the cap is of corbelled brickwork, the dpc should be positioned beneath the corbel, Figure 5;
  — where the cap is a precast coping, the dpc should be positioned directly beneath it, Figure 6.

- Specify that brickwork joints are to be completely filled with mortar and that bucket handle or weathered joints are to be used[3].

- Specify flashings and soakers at roof level, Figure 7;
  — components are side flashings, front apron and back gutter; the need for some or all will depend on whether the stack is on the slope or at the ridge;
  — a cover flashing is required over the upstand on the back gutter;
  — flashings should be chased at least 25 mm into the brickwork;
  — flashings should overlap each other by at least 100 mm.

- Specify for severe exposure, or as an alternative for any exposure, a one-piece horizontal dpc tray through the full thickness of the chimney:
  — tray should be so shaped, jointed and located that water draining from it is directed to outdoors, Figure 8;
  — the tray should be provided with an upstand at least 25 mm high within the flue and coincident with a joint in the flue liner;
  — trays at two levels or a single stepped tray may be needed where a chimney penetrates a steeply pitched roof.

- Specify weight Code No. 4 for lead trays, and that they be well painted on both sides with a bituminous paint, preferably solvent based.

- Check that damp proofing arrangements do not impair the stability of the stack.

REFERENCES AND FURTHER READING

1 *Defect Action Sheet* DAS 93 'Taking chimneys out of service'.
2 Brickwork Domestic Fireplaces and Chimneys: Brick Development Association, February 1981.
3 *Defect Action Sheet* DAS 71 'External masonry walls: repointing specification'.

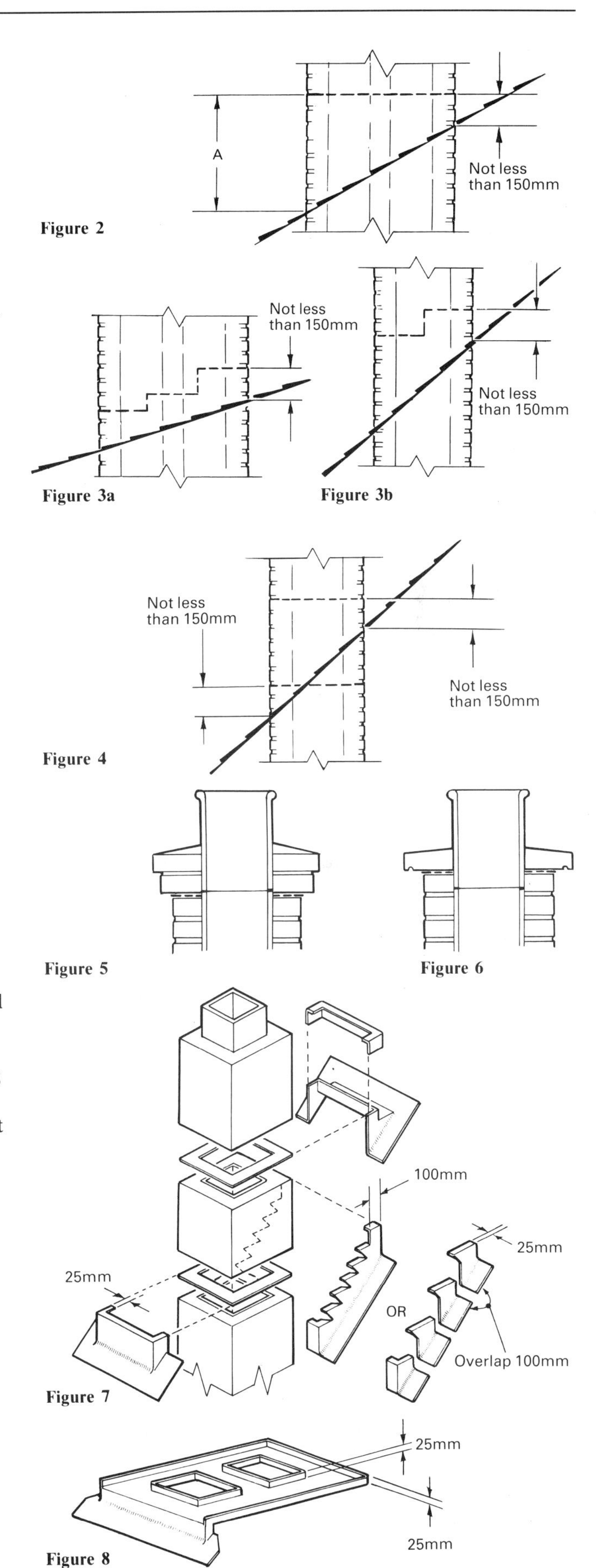

Figure 2

Figure 3a

Figure 3b

Figure 4

Figure 5

Figure 6

Figure 7

Figure 8

Masonry chimneys: dpcs and flashings — installation

FAILURES: **Water penetration; staining of chimney breasts and ceilings, dampness in roof timbers**

DEFECT: **Dpcs omitted or wrongly positioned; roof flashings wrongly installed**

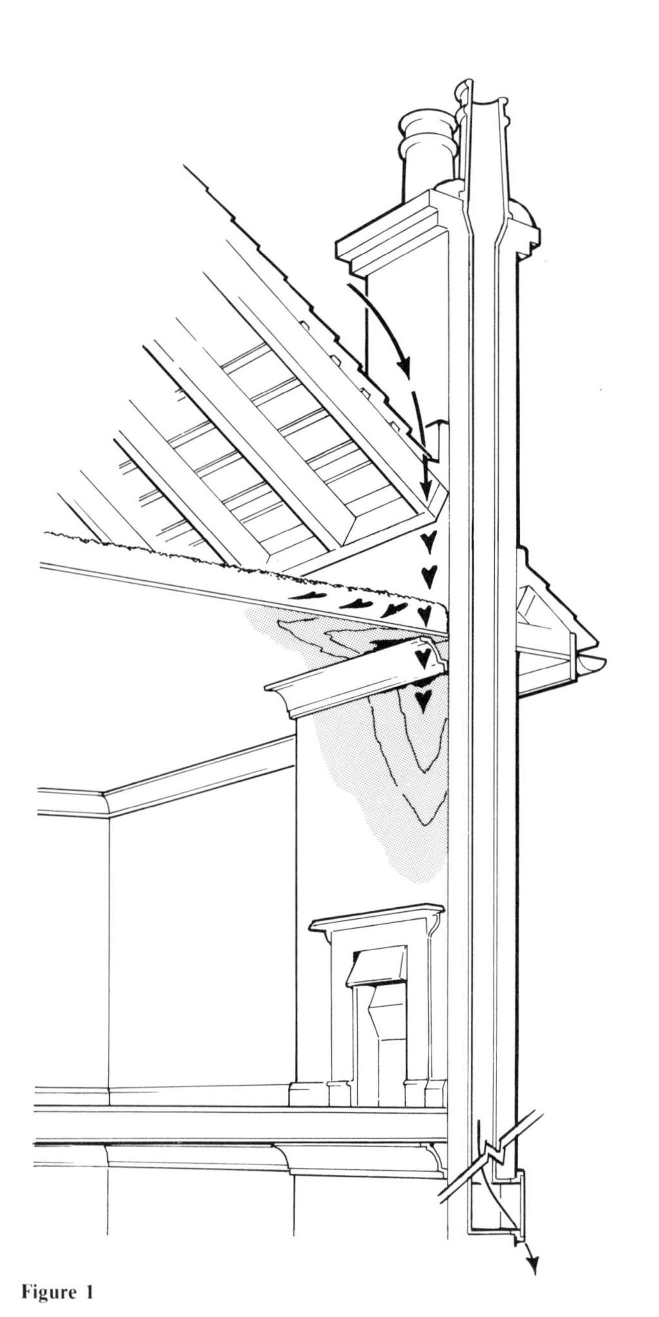

Figure 1

Cases of damage to decorations arising from rain penetration of chimney stacks are frequently seen — Figure 1.

If ***dpcs*** in the chimney at ***roof level*** and ***beneath the cap*** are ***wrongly positioned*** or ***poorly built*** then ***rain penetration and damage to decorations can occur.***

PREVENTION

Principle — Correct construction and positioning of dpcs in chimneys is critical if rain falling on the brickwork is to be kept out of the house.

Practice

- Check where dpcs are specified to be installed, eg beneath the chimney capping, Figure 2a and at roof junction level, Figure 2b, c, d;
  - where the chimney is not too exposed the use of two courses of dpc brick or two courses of slate bedded in a 1:0 to ¼:3 cement:lime:sand mortar may be satisfactory.
- Ensure that the dpc or tray is laid on a bed of mortar, raking out at the front to allow for the tuck under of the front apron, Figure 3.
- Ensure that the front apron is tucked under, not over the dpc or tray.
- Check that trays have upstands of at least 25 mm to inner edges within flues and to back and sides, Figure 4.
- Check that corners in trays form a watertight junction.
- Ensure that lead trays, Code No. 4 (blue marked), are coated both sides with bitumen paint:
  - if a solvent based paint is specified check that correct product is used (eg check that product instructions are to clean brushes in solvent, not in water).
- Ensure that a joint in the flue liner coincides with the level of the tray.
- Ensure that bed and perpend joints are completely filled with mortar and that a bucket handle or weathered joint is used.
- Ensure that flashings and soakers are in the correct position, tucked into the bed joint at least 25 mm and wedged.

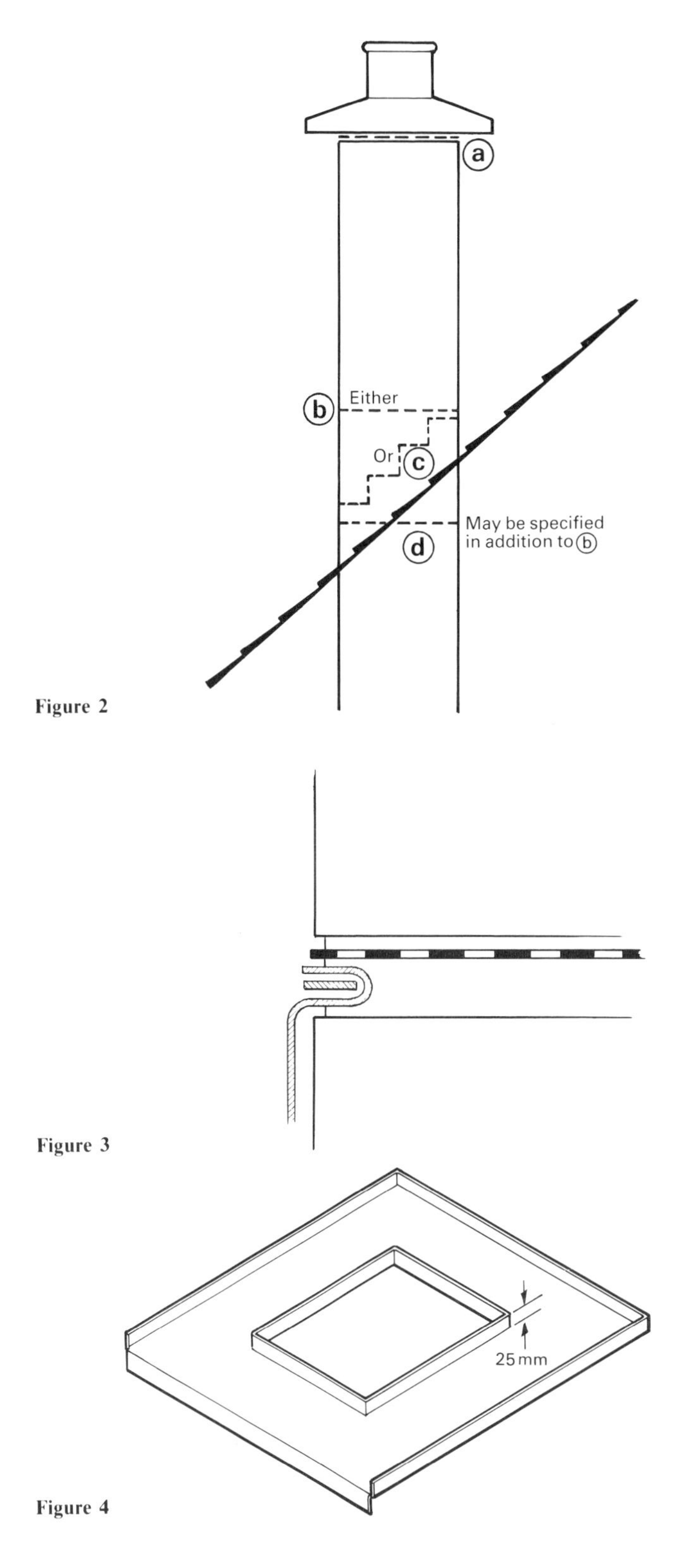

Figure 2

Figure 3

Figure 4

REFERENCES AND FURTHER READING

Defect Action Sheet DAS 71 'External masonry walls: repointing specification'.

PART TWELVE

Windows and Doors

Wood windows and door frames: care on site during storage and installation

FAILURE: **Wet rot following early failure of paint system letting water into unprotected wood**

DEFECTS: **Inadequate protection on site before and after installation; poor preparation prior to installation and painting**

Figure 1 The last rotted window?

In a survey by the AMA, rot in windows and doors was one of the more frequently mentioned problems. A BRE survey of housing under construction observed that inadequate protection for joinery, both before and after installation, and poor preparation before installation and painting, were common.

The need to protect wood windows and door frames from physical damage on site is obvious. However, there are other important requirements. Windows and door frames need to be protected from the weather both to control moisture content and to avoid weathering of primers. If windows and door frames are ***not properly protected***, the advantage of prior 'seasoning' of the timber are lost, timber may become ***too wet to paint satisfactorily*** and may even reach moisture contents at which the risk of rot arises. Paint systems are critically dependent on the quality and soundness of priming coats. If factory ***priming is damaged or weathered***, or contaminated with mortar splashes, early ***failure of the subsequent paint system*** is likely. Most external joinery is now pre-treated with preservative. However, site cutting exposes untreated wood. So, if ***cut ends are not brushed with preservative and primed***, vulnerable ***end-grain*** — usually in permanent contact with potentially wet brickwork for life — ***is unprotected***. All these precautions are necessary even if the joinery has been pre-treated.

PREVENTION

Principle — Maintain both the quality of 'seasoning' achieved during manufacture and the integrity of preservative treatment and factory-applied primers.

Practice

- Store joinery under cover. If outside, stack joinery well clear of ground and preferably on a hard, well-drained surface, covering the top and sides of stack in a way which allows free ventilation. (Figure 2).

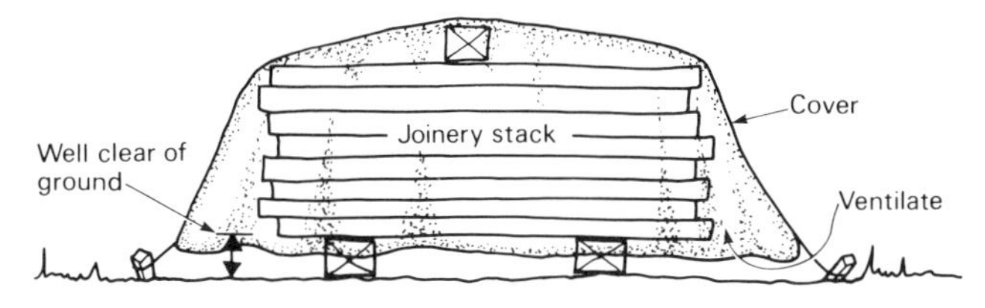

Figure 2

- Smooth sawn-off ends and remove any dust. Treat with at least two flowing brush coats of a suitable[1] preservative, (Figure 3) allowing each to dry before applying the next. Allow final coat to dry before priming. Make sure the primer is compatible with the preservative.

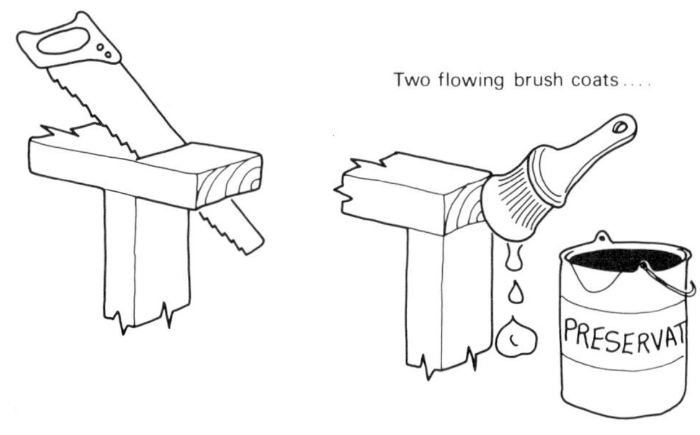

Figure 3

- Inspect factory-applied primer before installation and make good defective areas, paying particular attention to the sides of frames which will be in contact with the building and to exposed end grain.
- If the primer is thin, apply a further coat of primer.
- If the primer is intact but weathered, clean thoroughly before undercoating.
- If the primer has cracked, flaked or chalked severely, or if there is mould growth, remove the primer and apply a fresh coat.
- Clean mortar splashes off primer immediately, paying particular attention to cills and projecting windows.

REFERENCES AND FURTHER READING:

1 *PRL Technical Note* 24, revised 1979 'Preservative treatments for external joinery softwood timber'.

PRL Technical Note 28 (revised 1981) 'Maintenance and repair of window joinery'.

BRE Digest 261 'Painting woodwork'.

PRL Technical Note 12 'Flooring and joinery in new buildings', February 1960, reprinted April 1980.

Code of Practice CP 153:Part 2:1970 'Windows and Roof Lights: Durability and Maintenance'.

Inward-opening external doors: resistance to rain penetration

FAILURES: **Rain penetration leading to wetting and deterioration of internal flooring and floor coverings; flooded mat wells.**

DEFECTS: **Rain-check grooves between door leaf and frame absent or too small; waterbar on cill set too far forward; poorly designed trough cills**

Figure 1 Waterbar set too far forward

Rain penetration past inward-opening external doors is often the subject of complaint (particularly if there is no shelter provided by a porch or canopy).

If rain-check grooves between door leaf and frame are ***absent,*** of ***inadequate cross-section ,*** or ***drain to the indoor side of*** the threshold ***waterbar, rain penetration*** can occur. If, in trying to cure the problem, a ***flexible seal*** is fitted to the ***bottom of the door*** on its ***outdoor side*** it may ***prevent outward drainage*** of water penetrating between door leaf and frame.

Inward-opening external doors: resistance to rain penetration

DAS 67
November 1985

PREVENTION

Principle — Rainwater entering the gap between door leaf and frame at any point around the door must not be able to drain inward.

Practice

Waterbar cills:

- Specify door frames with rain-check grooves at least 6 mm wide and deep:
  — groove should be located next to stop, otherwise hinges will obstruct it.
- Specify doors with the waterbar located on the indoor side of the rain-check grooves so that water running down the grooves drains to the outside, Figure 2a, not the inside, Figure 2b.
- Do not specify a flexible seal, at the foot of an inward opening external door, on its outdoor side; the seal may prevent outward drainage of water entering the gaps between door leaf and frame.

***Trough cills:**

- Specify door frames with rain-check grooves at least 6 mm wide and deep and located next to the stop:
  — note possible need to seal groove at foot, see Figure 3.
- Specify a trough that projects slightly beyond the interior face of the door leaf, Figure 4:
  — otherwise water draining down the edge of the door leaf may miss the trough.
- Specify a trough with stop-ends that will not lead water inward across their tops, Figure 3:
  — the tops of stop-ends should be thin or chamfered, or below the tops of the front and rear faces of the trough, Figure 4.

REFERENCES AND FURTHER READING

Building Research Establishment Current Paper CP 31/78, 'The performance of timber-based doors'.

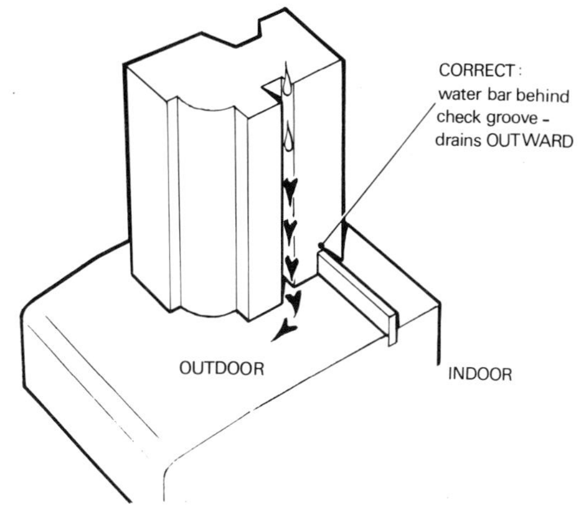

Figure 2a

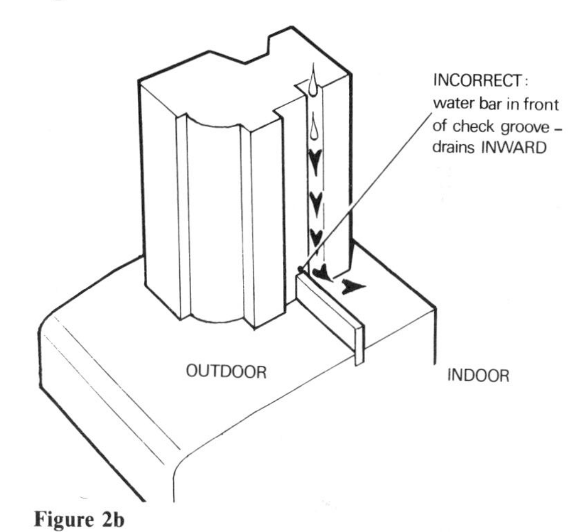

Figure 2b

leads water inward unless thin or chamfered
seal groove
INDOOR
OUTDOOR

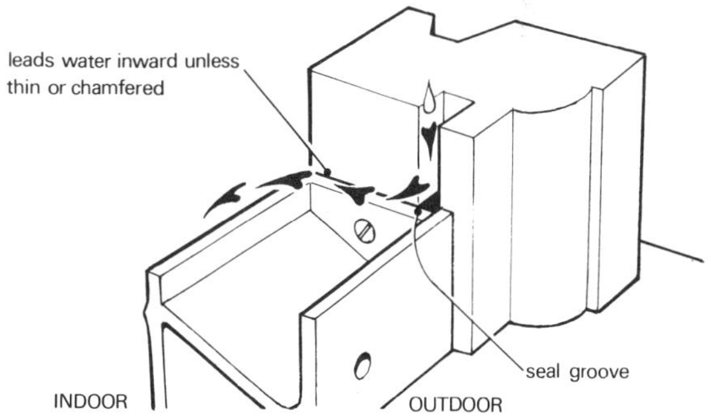

Figure 3

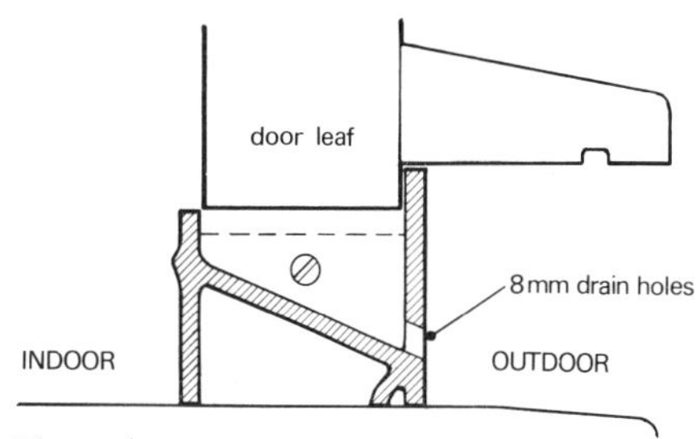

Figure 4

*May impede disabled access

External walls: joints with windows and doors — detailing for sealants

FAILURE: **Rain penetration of joints**

DEFECT: **Incorrect detailing for sealants**

Figure 1

Windows and doors are commonly replaced in rehabilitation work; often jamb dpcs cannot be installed and sealants may form the main defence against rain penetration. In new-build, with jamb dpcs, sealants may be an important additional defence where exposure is severe. In BRE surveys disruption of sealants has been seen even before handover and the use of sealant in the form of a shallow triangular fillet is commonly the cause. Proper design provision for sealants is rare.

If a ***sealant*** has the ***wrong cross-section*** it will ***fail*** when ***movement*** occurs.

PREVENTION

Principle — Joints round windows and doors move: sealants must have adequate section to remain intact and effective.

Practice

- Choose a suitable sealant, taking account of cost, life and compatibility:
  - detailed as shown here, all sealants listed in Table 1 can cope with the amount of movement around domestic windows and doors of average size;
  - some sealants may not be compatible with wood stains or preservatives, or with plastics (check with manufacturers).
- Specify that joint surfaces are to be dust free when sealant is applied.

External walls: joints with windows and doors — detailing for sealants

DAS 68
December 1985

Replacing windows and doors (especially solid walls, no jamb dpcs):

- Design so that there is a groove or rebate to contain a 10 mm x 10 mm section of sealant (Figures 2 a, b, c), specifying back up material (eg foamed polyethylene) to control depth of sealant, Figure 2a, or bond breaker tape where adhesion must be avoided, Figures 2b and c;
  — effective seals are essential where they are the sole defence against water penetration.
- Specify, where a containing groove or rebate is not possible, a ***convex*** sealant bead as in Figure 3;
  — do not permit tooling to a triangular or concave section.

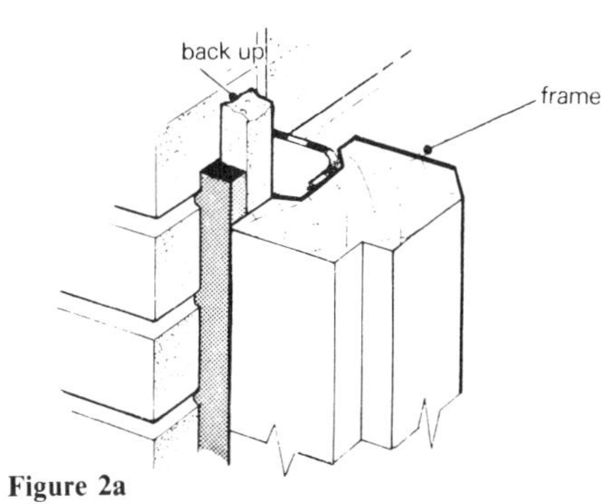

New-build, with jamb dpcs, where exposure may warrant additional defence:

- Specify a convex sealant bead as in Figure 3.

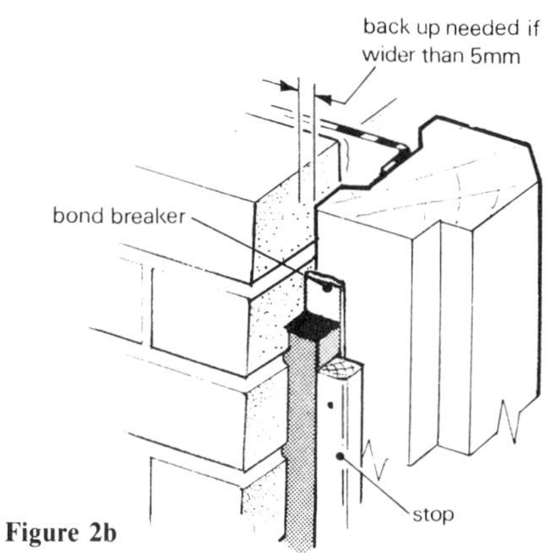

Table 1 Sealants suitable for use around domestic windows and doors

Type	Approx max life (years)	Warnings
Oil-based mastics	10	Painting necessary to achieve maximum life
Butyl rubber	10	
Acrylic (solvent type)	15	Check whether warming needed in cold weather
Polysulphide	20	1-part: slow to cure; 2-part: must be used within time limit after mixing
Polyurethane	20	1-part: slower to cure than 2-part; 2-part must be used within time limit after mixing
Silicone (low or high modulus)	20	

REFERENCES AND FURTHER READING

1 *BRE Digest* 228 Table 2.

2 *BRE Digest* 228 Table 1 and Figure 2.

3 *British Standard* BS 6093:1981: Code of practice for the design of joints and jointing.

The selection and performance of sealants
BRE IP 25/81.

Guide to selection of construction sealants
BS 6213:1982 BSI.

Manual of good practice in sealant application.
Sealant Manufacturers Conference, CIRIA 1976.

Figure 2a

Figure 2b

Figure 2c

Figure 3

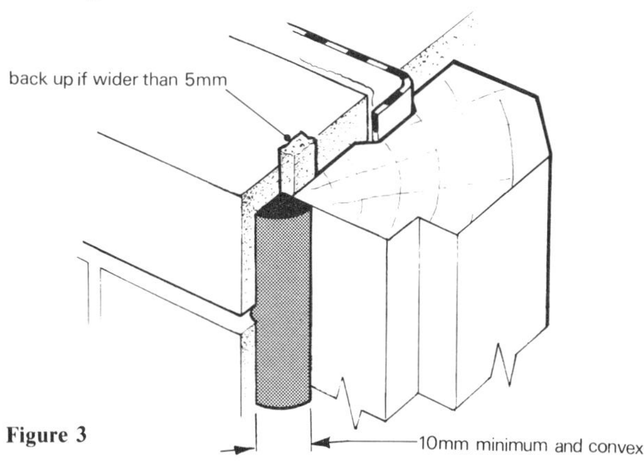

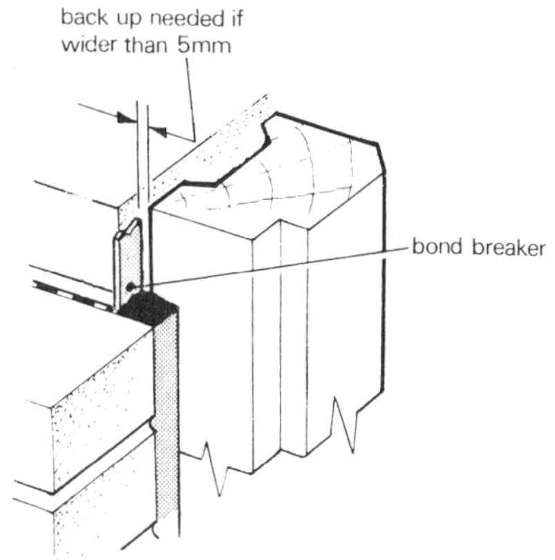

DAS 69
December 1985

Site
CI/SfB 8 (21.9)t4(D6)

External walls: joints with windows and doors — application of sealants

FAILURE: **Rain penetration through joints**

DEFECTS: **Poor bonding, mixing, tooling, joint preparation; sealant too thin; back-up or bond breaker materials not used where needed**

Figure 1 Poor bonding — and slump

Where windows and doors are replaced in rehabilitation work, and installation of a jamb dpc is not practicable, sealants form the main defence against rain penetration. In new-build they may be an important additional defence where exposure is severe. Sealants are usually subjected to movements. Early failure due to poor workmanship can occur. Where there is no joint gap to receive them sealants around windows and doors are often unavoidably applied as corner fillets. They do not work best in this form and if their cross-section shape is wrong they soon fail.

If a ***sealant*** is ***gunned into a joint*** that has ***no back-up*** material to act as a stop ***adhesion*** will be ***poor*** and, since its cross section will be incorrect, ***movements will damage it; rain penetration may follow.*** If a ***sealant,*** applied as a ***corner fillet,*** is tooled to a ***concave*** and feather-edged section, ***it will fail; rain penetration may follow.***

PREVENTION

Principle — Sealants must adhere well to the joint faces and have appropriate cross-section to remain intact.

Practice

- Check that surfaces of joint are sound, clean, dry and free from dust or oil.
- Ensure that appropriate primer, where recommended by manufacturer, is applied evenly to the sides of the joint with complete coverage:
  - — primed surfaces must be allowed to dry before sealant is applied;
  - — in dusty conditions the primed joints should be protected from contamination.
- Check that back-up material (usually foamed plastics strip) is used if joint gaps are open at the back, so that gunned sealant is forced against the sides of the joint, Figure 2.
  - — a minimum sealant depth of 6 mm should be the aim;
  - — bond-breaker tape, if specified, must be applied only to surfaces where adhesion is to be avoided (check design requirements).
- Check that the sealant gun or cartridge nozzle is correct size for joint:
  - — best results are achieved if the cartridge nozzle is cut so that it just fits into the slot, with the cut at about 45° .
- Check that two-part sealants are not mixed until immediately before use:
  - — power-driven mixers should be used at speeds low enough to avoid heat generation or air entrapment.
- Check that sealant is gunned firmly and steadily into joint, forcing sealant against both sides, Figure 3:
  - — if gun speed too fast joint will not be properly filled
  - — seal should be finished neatly by releasing the pressure at, or just before, the end of the run.
- Check that sealants in butt joints are tooled by drawing spatula down the surface, so forcing sealant against joint faces:
  - — spatula should be wetted with water plus a little detergent, detergent must not be applied directly to the sealant surface.
- Check that corner fillets, if they are unavoidable, are formed to a well rounded convex bead not less than 10 mm wide, Figure 4.
  - — back-up material should be used if gap at corner is wider than 5 mm, Figure 4;
  - — do not permit fillets formed to a concave, feather edged, shape, Figure 5.

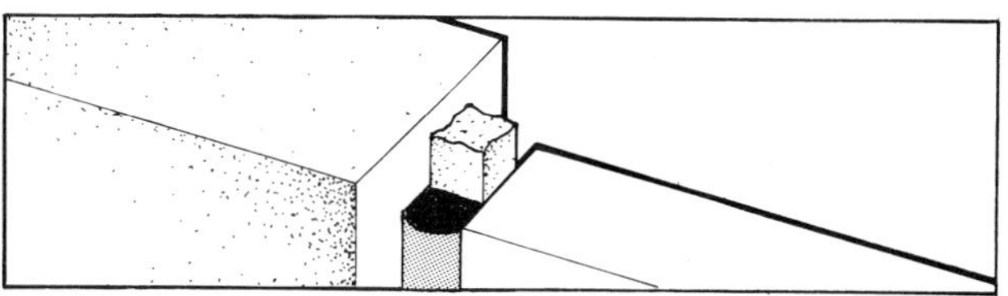

Figure 2

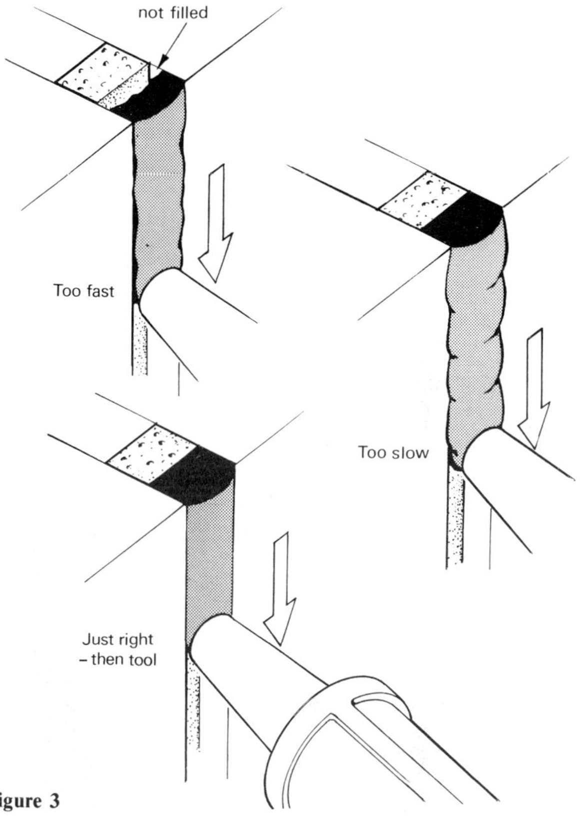

Figure 3

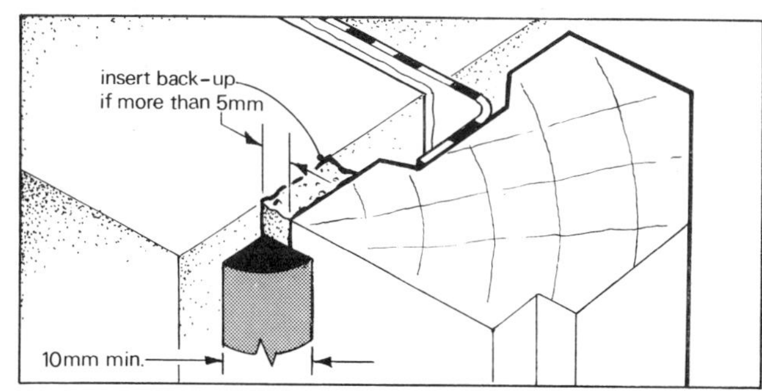

Figure 4

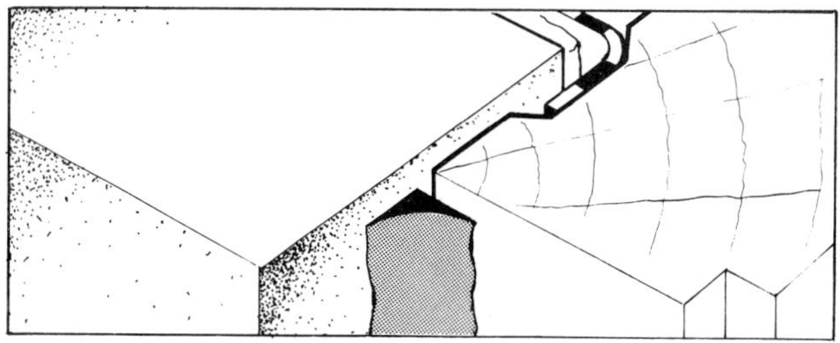

Figure 5

DAS 98

April 1987

Design

CI/SfB 8(31.4)(L3)

Windows: resisting rain penetration at perimeter joints

FAILURE: Rain penetration at window to wall joints

DEFECTS: Incorrect detailing of dpcs and cavity trays; omission of 'stop ends' to cavity trays/pressed steel lintels

Figure 1 Rain penetration at window jamb

A BRE survey of traditional housing in course of construction uncovered many faults likely to lead to rain penetration at the perimeter joints between windows and walls. Internal dampness at window reveals is a frequent subject of enquiry to Building Research Advisory Service. Damp patches, often bordered by white deposits, appear on internal plaster, spoiling decorations, Figure 1. Mould growth and deterioration of materials may follow.

Windows can be located in different positions in relation to the wall thickness. This DAS deals with the circumstance in which windows are located wholly in the plane of the outer leaf*, where the straight jamb joint demands particular care. (For severe exposure conditions a rebated joint is preferable, with the outer leaf mastering the edge of the window jamb.)

If ***vertical dpcs*** at window reveals are ***not fully cloaked*** (overlapped) at their tops, are ***bridged by mortar*** within the cavity, ***do not fully close the gap*** between masonry reveal and window frames, or ***do not shed water outward*** at sill level, ***rain penetration*** can occur.

If ***stop ends***[1] are ***not used*** on cavity trays or lintels acting as cavity trays, rainwater discharge, particularly in cavity filled walls, may wet the inner leaf, producing ***dampness on internal walls.***

*The arrangement is most uncommon in Scotland

PREVENTION

Principle — Water reaching dpcs, cavity trays/lintels should drain outwards.

Practice

DPCs

- Specify jamb dpcs wide enough both to connect adequately to the window frame (or sub-frame) and to project into the cavity by at least 25 mm to avoid mortar bridging. (B, Figure 2.) In practice this will mean a dpc at least 200 mm wide.

- Instruct site staff to ensure that jamb dpcs are the full height of the jamb (D, Figure 2) and lap any sill dpcs in such a way as to drain water out and not in (E, Figure 2). (If no sill dpc is used the jamb dpc should continue to 150 mm below the sill bed.)

- Require any sill dpc to be turned up to form as high an upstand as practicable. (F, Figure 2).

- To accommodate any sealant specified between frame and reveal incorporate a rebate in the frame[2] (G, Figure 2) (sealant applied as a triangular fillet will be short-lived). Note that struck or recessed joints in the reveal will produce 'holidays' in the sealant line. If jamb dpcs are properly detailed, and the work reliably executed, sealants are unnecessary.

- Consider using reveal blocks for neater jamb dpc installation (H, Figure 2).

Cavity trays

- Specify that the cavity tray (or lintel acting as such) should oversail the edge 'C' of the jamp dpc, Figure 2.

- Consider specifying stop ends to cavity trays or lintels acting as cavity trays as recommended by BS 5628:1985; stop ends are essential where cavity fill is to be installed and very desirable elsewhere.

- Instruct site staff to check that wherever stop ends are specified, they are in position;
  - they should be checked for good fit, with no gaps through which leakage might occur.
  - any cleaning off required must not damage or dislodge stop ends.

- Specify weepholes in the outer leaf immediately above the cavity tray;
  - not less than two weepholes over each opening, or open perpends every 4th joint (A, Figure 2);
  - plastics cavity weeps are available for building in as work proceeds.

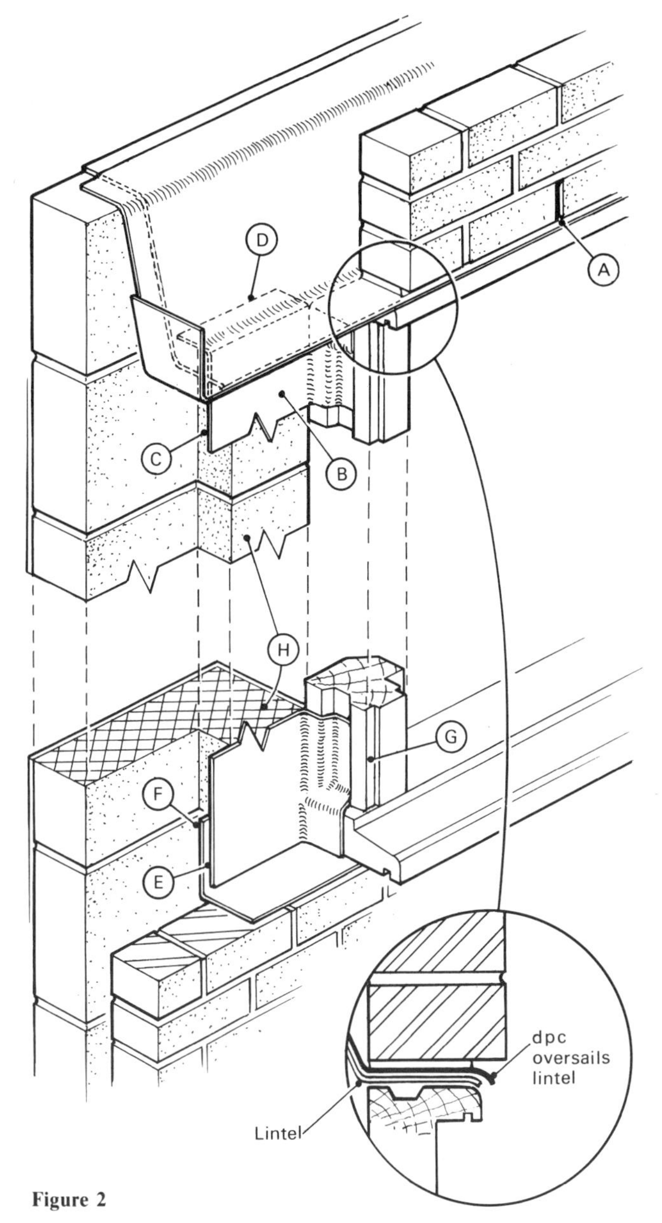

Figure 2

REFERENCES AND FURTHER READING

1 *British Standard* BS 5628 : Part 3 : 1985 'Use of masonry'.

2 *Defect Action Sheet* DAS 68 'external walls: joints with windows and doors — detailing for sealants'

□ *British Standard* BS 6093 : 1981, 'Code of Practice for joints and jointing in building construction'.

Windows, doors and exterior joinery: applying putty, oil-based sealants and solvent-based paints, when weather may be bad

FAILURES: **Loss of adhesion of paint; water-spotting or dust on paint films; poor adhesion of putty and sealants**

DEFECT: **Operating in wet, damp, cold or windy conditions**

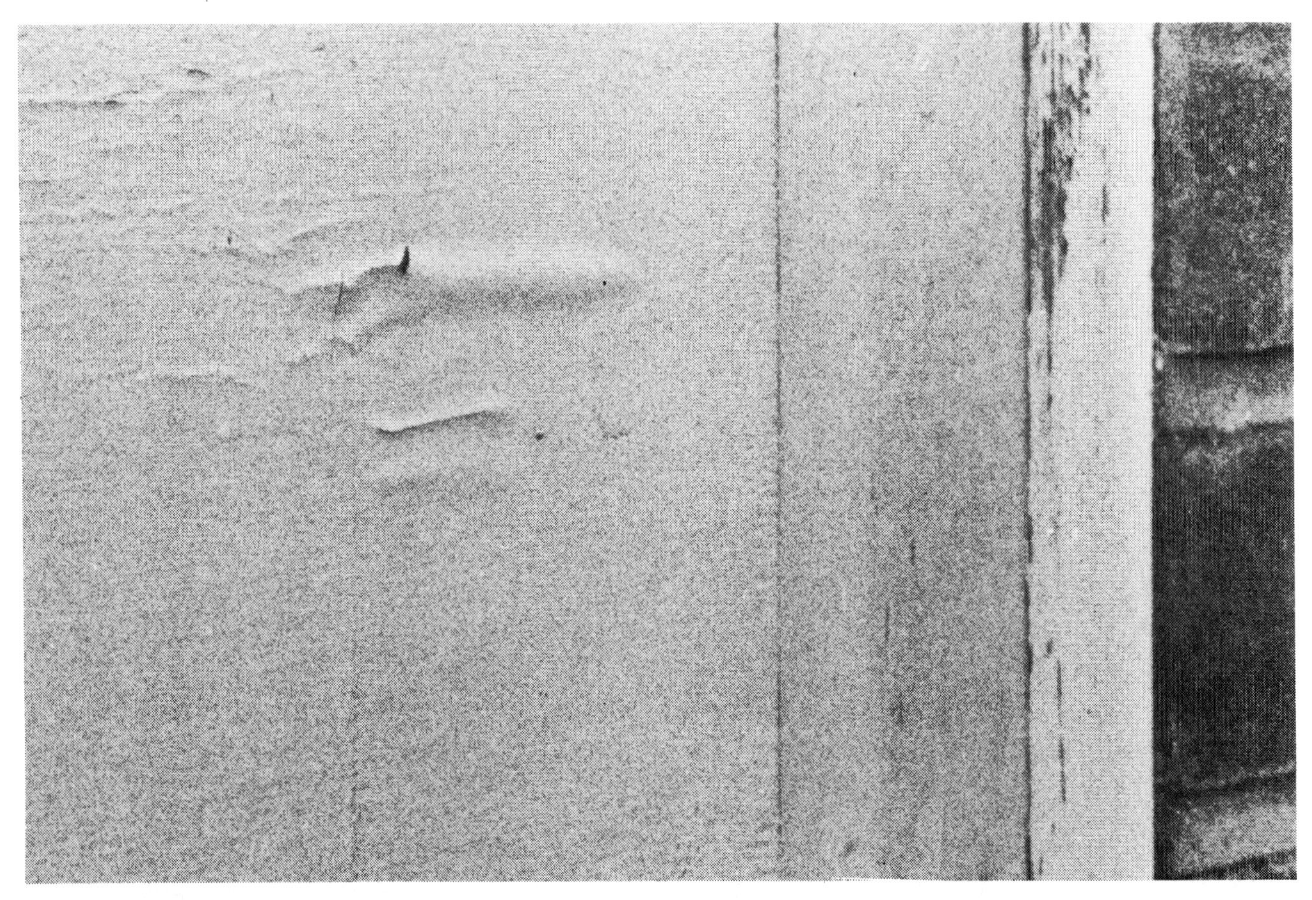

Figure 1 Paint film failure

This DAS applies to putty, oil-based sealants and solvent-based paints. It does not apply to water-based paints which are generally more tolerant of damp conditions during application.

If there is any ***moisture*** or frost ***on surfaces to be painted, puttied or sealed,*** there is a ***risk of*** poor adhesion and premature ***failure.*** If the joinery has too ***high*** a ***moisture content, blistering*** or ***flaking*** of the paint is likely. Rain, falling on fresh paint, can produce a spotted finish and loss of gloss. Wind-blown ***dust*** deposited on surfaces ***can reduce adhesion*** of putty and sealants; ***low temperatures slow*** down ***drying*** so that further dust may adhere to newly-applied paints and sealants.

Windows, doors and exterior joinery: applying putty, oil-based sealants and solvent-based paints, when weather may be bad

DAS 66
August 1985

PREVENTION

Principle — Checks and precautions become critically important if bad weather may occur immediately before, during or immediately after paint, putty or sealant application; in some conditions work should not proceed.

Practice

- Check manufacturers' requirements for conditions for both application and storage of paints or sealants in bad weather.

- Do not paint or apply putty or sealants when there is **any** visible moisture on the surface to be treated, Figure 2.

- Do not paint wood if its moisture content exceeds about 18% (this is about the moisture content it will have in service[2]).
  — measure the moisture content, Figure 3.

- Do not paint when the temperature falls below about 5°C;
  — many paints are very slow to dry in cold weather.

- Do not paint when the wind is strong enough to raise dust (Force 4);
  — note also that high winds may dry paint films too quickly, removing the required 'wet edge'; paint may also be blown on to other surfaces.

- Note that the provision of shelter and heat may enable work to go ahead, but correct conditions must be maintained until, for example, paint dries or sealants cure;
  — when temperatures are low some sealants may need warming to improve workability.

- Remember that pre-recorded forecasts are available — see 'Weather line' numbers in your dialling code booklet. Forecasts specific to contractors' needs can be obtained, for a fee, from your local Weather Centre (see directory).

- For your locality, the Meteorological Office can compute the proportion of working hours in which conditions will be unsuitable for the operations covered by this DAS — a charge is made — England and Wales, telephone 0344 420242, Extension 2278, Scotland, 031 334 9721, Extension 524, Northern Ireland, 0232 228457.
  A list of threshold conditions for many building operations (together with local data for Plymouth) is given in Reference 2.

REFERENCES AND FURTHER READING

1 *British Standard* 6150:1982 Code of Practice for painting of buildings.

2 **Building Research Establishment** *Report* 'Climate and Construction Operations in the Plymouth area' Keeble and Prior. (In preparation.)

Figure 2 Do not apply paint, putty or sealants

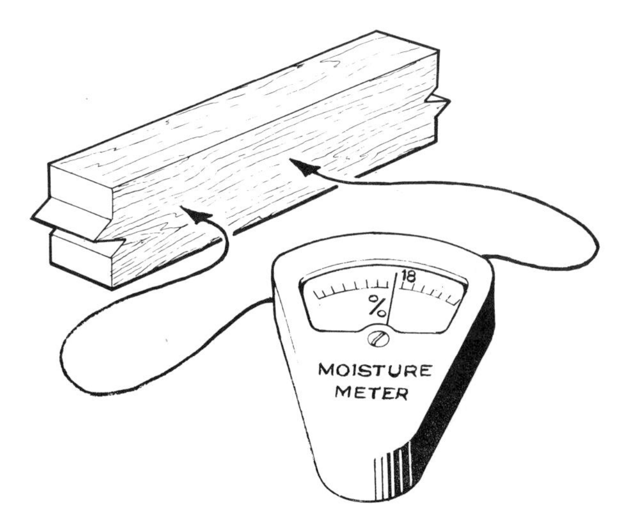

Figure 3 Measure the moisture content

DAS 13

January 1983 Minor amendments February 1985

Design

CI/SfB 8 (31.4)(L6)

Wood windows: arresting decay

FAILURE: **Early signs of decay in wood windows**

DEFECTS: **Design, fabrication and installation faults, which allow excessive penetration of water into joints**

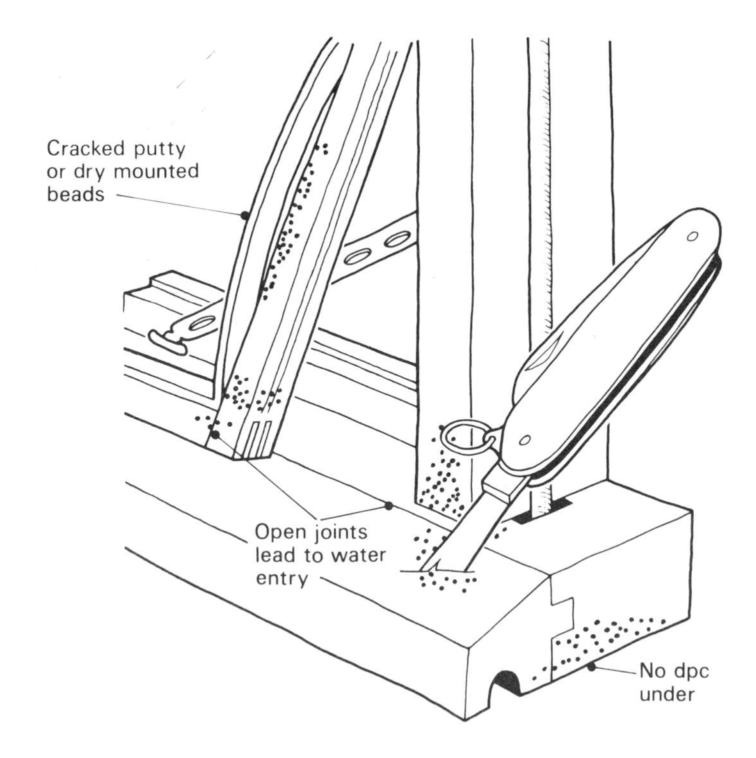

Figure 1 High risk areas in untreated frames

During the last 20 years there has been widespread premature decay in untreated softwood window joinery. Preservative pre-treatment of external joinery greatly reduces the risk of decay but has only been general practice since about 1970 and there is a legacy of untreated softwood windows in service which are at risk of decay.

The parts of the fixed and opening frames most likely to be affected are the bottom rail and the stiles close to the lower joints, particularly where the frame is an opening light of any of the horizontally hinged or pivoted types. The lower ends of mullions and adjacent parts of softwood sills may also be affected. ***If joints gape,*** or if ***glazing putty shrinks*** away from the glass or ***glazing beads are dry-bedded, water can readily enter the wood,*** keeping it sufficiently wet ***to promote decay.*** Both rain, and condensate on the indoor surfaces of glazing, can contribute to the water gaining entry to the lower parts of frames. If decay in window joinery is spotted early enough, it will be possible to prolong the service life of the window. In some cases it may be economic to replace opening lights only, while treating and retaining the remainder of each window.

REMEDIAL WORK

Principle — Treat decay early enough to extend the service life of the joinery and thus to enable replacement to be safely deferred; keep wood dry after treatment.

Practice

- Inspect and establish the extent of decay by probing (Figure 1).

- Use a moisture meter to locate areas where decay, not yet apparent, may occur (moisture contents exceeding about 22 per cent may indicate a risk if it is likely to be persistent).

- Identify frames that are neither structurally unsound (for example, no rot in joints) nor so decayed as to rule out treatment; scrape out surface rot and preservative treat these frames, concentrating on lower joints using preservatives that permit over-painting. Allow time for solvent to evaporate, then flush up with filler — preferably two-pack polyurethane or two-pack epoxy resin types. For preservative treatments consider using:
  brush application (3 full coats, each to the point where no more preservative is absorbed);
  OR pressure injected organic solvent based preservatives (Figure 2) (some member firms of the British Wood Preserving Association can provide this service); treatment may be applied from either indoors or out;
  OR insertion of borax preservative rods in drilled holes near areas of wood affected or at risk, avoiding the tenons (Figure 3); stop the holes with filler as above, sand, and repaint (experience of this technique in service is limited at present but results of trials are encouraging). Treatment may be applied from either indoors or out.

- Reduce water-collecting surfaces (for example, if external glazing beads have inadequate weathering slopes replace them with beads of a better profile); plant weathering fillets on horizontal water-collecting surfaces above wood that is affected or at risk.

- Renew and maintain paint.

- Minimise overheads for scaffolding etc, by integrating the operation with other maintenance work.

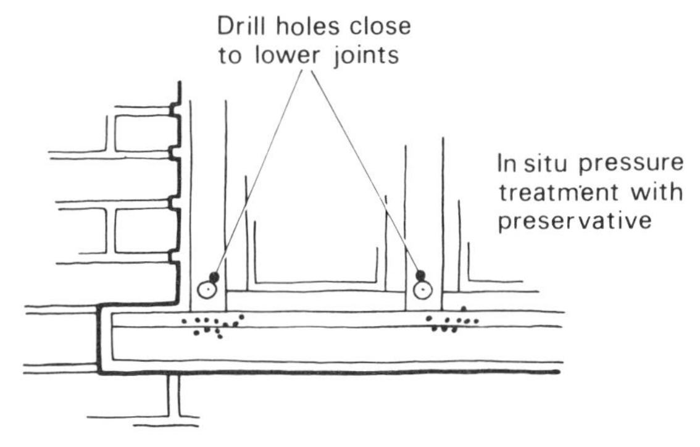

Figure 2

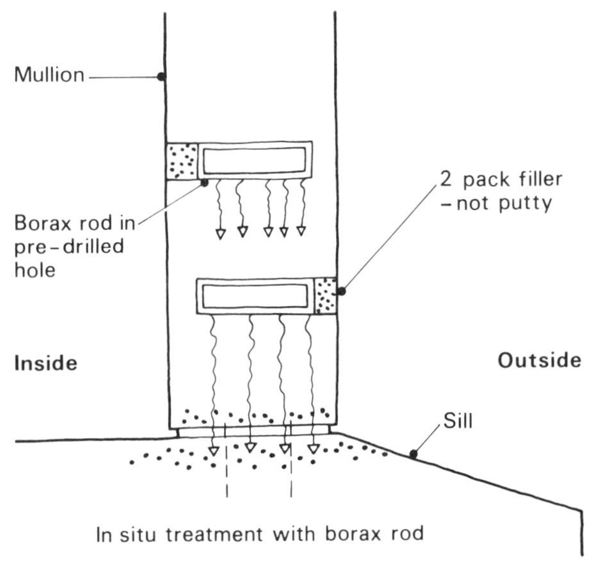

Figure 3

REFERENCES AND FURTHER READING

Henshaw B G. 'In-situ treatment of window joinery'. *Timber Trades Journal* 29 July 1978 pages 19-20.

BRE Information Paper IP21/81 'In-situ treatment for existing window joinery'.

PRL Technical Note 28 'Maintenance and repair of window joinery'.

Lea R G. 'In-situ preservative treatment for joinery in service'. British Wood Preservation Association *News Sheet* No 162, November 1980.

Defect Action Sheet DAS 14 'Wood windows: preventing decay'.

Wood windows: preventing decay

FAILURE: **Severe rot necessitating installation of new frames**

DEFECTS: **Design, fabrication and installation faults enabling persistent penetration of water into joints**

Figure 1 Penetration of water into joints

In recent years there has been an increasing need to repair or replace exterior softwood framed joinery which has failed prematurely through decay. Changes in design, methods of construction and installation into buildings, have all contributed to increased ingress of moisture into the structure of joinery, generally made from non-durable softwood timbers such as Baltic redwood *(pinus sylvestris)*, resulting in decay.

All NHBC and PSA specifications for exterior joinery in buildings demand preservative treatment; preservative treatment is applied as a matter of routine to much other exterior joinery also. But the recommendation to preservative-treat joinery is intended to complement, not to substitute for, good design and fabrication. If the ***design*** of windows ***permits water*** to collect ***on horizontal surfaces*** near joints or glazing rebates (Figure 2), ***if fabrication produces junctions*** between members ***that lead water into the wood*** (Figure 3), or ***if building design does not*** take every opportunity to ***reduce exposure***, the ***service life*** of wood windows ***may be reduced*** despite preservative treatments.

Wood windows: preventing decay

DAS 14
January 1983

PREVENTION

Principle — Minimise exposure to and trapping of water; specify preservative treatment.

Practice

- Design to give protection by construction features (eaves, porches etc) wherever possible.
- Design composite units (eg window/door) with continuous head and sill members.
- Specify frames and beads made from suitably durable timber[1] (the term 'hardwood' is not adequate specification of suitable durability[2]), or specify preservative treatment[3,4] (pressure impregnation achieves better penetration than dip treatments).
- Specify similar treatments for any associated site-fixed joinery such as cheek pieces used with tile hanging.
- Specify windows that have good weathering slopes to all external surfaces (including beads and sills) which could otherwise retain water (Figure 2); sill sections should be formed out of one piece of timber.
- Specify windows that will not allow condensate to be retained on internal frame surfaces or have condensate drainage grooves.
- Specify windows assembled with durable glues[5] (note: sealing end-grain with primer prior to glueing at joints, (Figure 3) can very significantly improve the control of moisture take-up and thus the durability of paint systems and windows. Manufacturers' guidance on suitable combinations of glue and primer should be sought).
- Specify primers to BS 5358:1976 and ensure that primers and preservatives are compatible.
- Specify glazing in accordance with BS 6262:1982.
- Instruct site supervisors to ensure that all site-cut ends, and any parts cut or planed to achieve fit, receive two full brush coats of preservative and are re-primed.
- Instruct site supervisors to ensure appropriate precautions in handling on site[6] to avoid damage to joints and finishes.

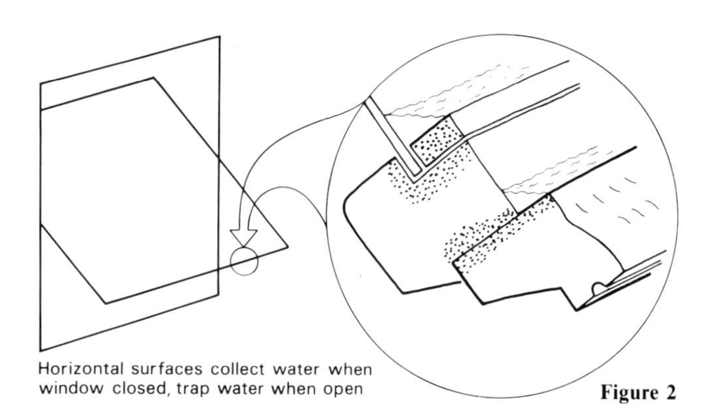

Horizontal surfaces collect water when
window closed, trap water when open

Figure 2

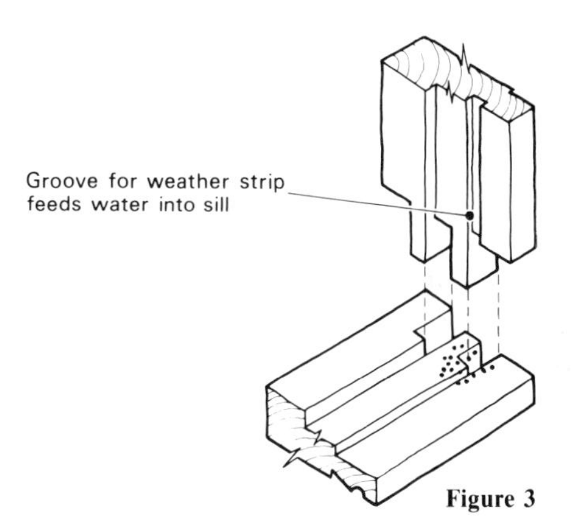

Figure 3

REFERENCES AND FURTHER READING

1 *BRE Report* 'Timber selection by properties — the species for the job: Part 1, windows, doors, cladding and flooring' by C Webster. HMSO 1978 500g £4.75.

2 *BRE Digest* 73 'Prevention of decay in external joinery.'

3 *PRL Technical Note* 24. 'Preservative treatments for external softwood joinery timber' 1967 (revised 1979).

4 *British Standard* 5589:1978 'Code of practice for preservation of timber. Section 2, External woodwork in buildings and out of contact with the ground'.

5 *BRE Digest* 175 'Choice of glues for wood'.

6 *BWF Booklet* 'Care of joinery on site'. London 1978, 2p. Obtainable from British Woodworking Federation.

BRE Information Paper IP10/80. 'Avoiding joinery decay by design' by J K Carey.

Defect Action Sheet DAS 11 'Wood windows and door frames: care on site during storage and installation'.

Defect Action Sheet DAS 13 'Wood windows: arresting decay'.

Windows and doors: reconstituted stone non-structural components; 'plastic' repair using Portland cement mortar – specification

FAILURE: **Shrinkage cracking and loss of adhesion of patch repair**

DEFECTS: **Incorrect specification of materials; patch material too strong for background; inadequate preparation; inadequate key**

Figure 1

Reconstituted stone components of relatively simple section were often used in Local Authority housing for window cills and mullions, window and door surrounds during the interwar and early post war periods. The material is often fairly porous and sometimes contains reinforcement. The material is sometimes subject to local frost damage or to cracking and spalling following corrosion of reinforcement.

There is often a need to carry out individual patch repairs to delay further deterioration when it might be considered uneconomic to use specialist repairers.

Patch repairs in ordinary Portland cement : sand mortar can be relatively durable but should only be used where appearance is not paramount*. Such patching often fails because the repair material is too strong for the background and has been applied in too thick coats. If ***plastic*** (ie mouldable) ***repairs*** made with ordinary Portland cement mortar to spalled or cracked reconstituted stone are ***too strong*** for the ***background*** or are ***applied too thickly or the background*** has been ***incorrectly prepared*** the patch is likely ***to fail.***

REMEDIAL WORK

Principle — Patch repairs must be capable of adhering well to the original material and must be compatible with the strength of the background.

Practice

- Consider whether local patch repairs in ordinary Portland cement : sand mortar are appropriate;
  - — identify area of material to be replaced;
  - — check likely durability of surrounding material;
  - — check for chlorides: their presence in moderate (0.4 to 1.0% chloride with respect to cement content) or high (greater than 1.0% chloride with respect to cement content) quantities is likely to lead to further deterioration due to corrosion of reinforcement. Seek specialist advice.

Note: carbonation tests may not give reliable indications on reconstituted stone.

*Specialist advice should be sought where housing of historic or architectural interest is concerned.

Note: the advice given in this DAS does not apply to deteriorating structural concrete, the assessment and repair of which will need specialist advice.

- Specify that cracked material be entirely removed from the defective area:
  — no feather edges are to be left;
  — best results are obtained by cutting or sawing (or grinding) into the material at the edge of the patch to a depth of at least 15 mm, Figure 2. Note that cold-chiselled cuts can leave cracked edges to the surrounds to the patch which will not be durable;
  — patches at external corners should be specified to be cut back at least 50 mm on both vertical faces, Figure 3, and to a depth of at least 15 mm as shown in Figure 2.

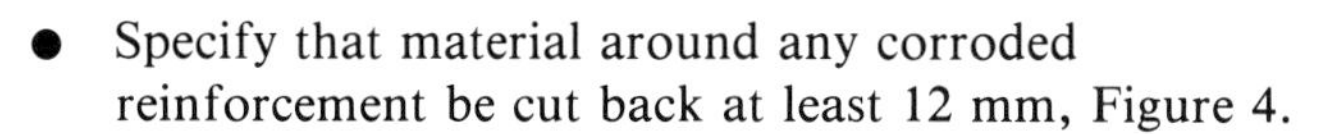

- Specify that material around any corroded reinforcement be cut back at least 12 mm, Figure 4.

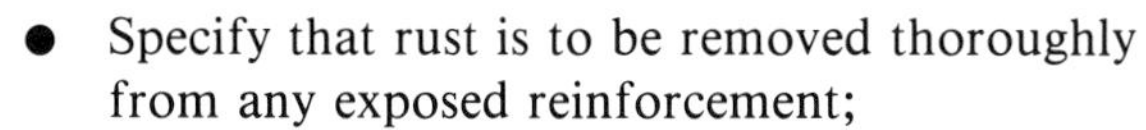

- Specify that rust is to be removed thoroughly from any exposed reinforcement;
  — grit blasting gives best results; needle-gun cleaning may also be effective. Wire brushing and chemical cleaning techniques are not recommended.

- Specify that all surfaces to be patched should be free of dust.

- Instruct site staff that surfaces to be repaired should not be excessively smoothed;
  — a key to smooth surfaces can be improved, where the repair patch is sufficiently large, by applying a 'spatterdash' coat of 1 : 2 cement : clean coarse sand mortar 3 to 5 mm thick. This should be allowed to dry slowly and harden before the next coat is applied;
  — small areas and exposed reinforcement could be coated with a cement grout;
  — non-ferrous wire or dowel may be used to provide a key or to reinforce vulnerable areas.

- Specify an ordinary Portland cement : coarse sand mortar for repair (see BS 5262);
  — the strength of the mix should be at least as weak as, preferably weaker than, the background, the mix being determined by trial on site.

- Specify mortar to be trowelled into place in layers of not more than 10 mm thickness;
  — allow each layer to cure under moist conditions (at least overnight) before rewetting and placing the next.

- Specify that repairs should be protected while curing;
  — repairs must be protected from the effects of freezing or hot dry weather.

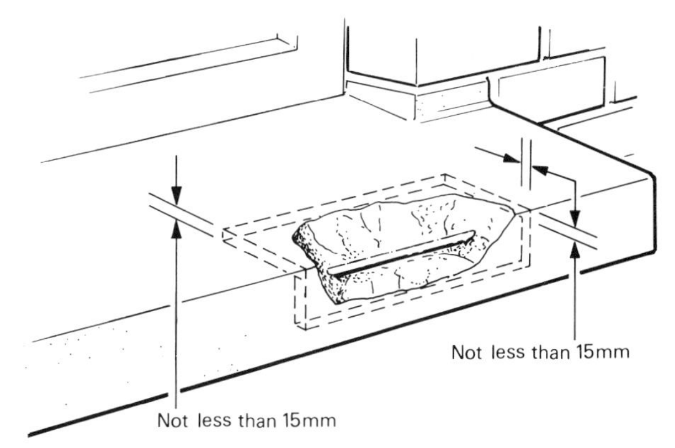

Figure 2

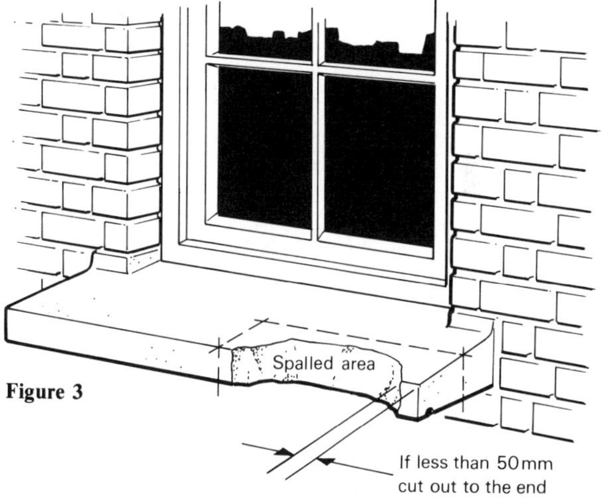

Figure 3

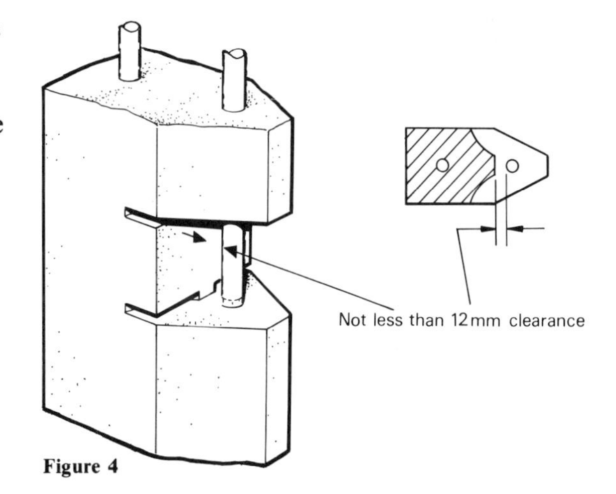

Figure 4

REFERENCES AND FURTHER READING

☐ **English Heritage** *Technical Handbooks* 'Practical Building Conservation': Volume 1 Stone masonry, Volume 3 Mortars, plasters and renders. Gower Technical Press 1988

☐ **BRE** *Digest* 196 'External rendered finishes'

☐ **BSI** *British Standard* BS 5262 'Code of practice: External rendered finishes'

Windows and doors: reconstituted stone non-structural components; 'plastic' repair using Portland cement mortar on site

FAILURE: **Shrinkage cracking and loss of adhesion of mortar patch**

DEFECTS: **Patch material too strong for background; inadequate preparation of material surrounding patch; inadequate key**

Figure 1

Sometimes local repairs are needed to reconstituted stone components. Non-loadbearing components such as window cills and mullions, and window and door surrounds can be patched with ordinary Portland cement mortar provided simple rules are followed.

If the ***mix*** of the patch material is ***too strong for*** the ***background***, if the surface of the ***surrounding material*** is ***not well prepared*** or if there is ***insufficient key***, the ***patch*** is ***likely to fail***.

NOTE: no attempt should be made to repair loadbearing reinforced concrete members such as lintels, balconies or structural frames using the information in this DAS. Refer such cases to a Structural Engineer.

Windows and doors: reconstituted stone non-structural components; 'plastic' repair using Portland cement mortar on site

DAS 123
November 1988

REMEDIAL WORK

Principle — Patch repairs must be capable of adhering well to the original material and must be compatible with the strength of the background.

Practice

- Make sure the areas to be cut away and patched are clearly defined by the specifier.

- Ensure that spalled areas are cut away, as specified, avoiding feather edges to surfaces;
  — saw cutting or grinding to a well defined edge gives best results, Figure 2;
  — square the edges of the cut in line with the edges of the component, Figure 3.

- Ensure that rust is thoroughly removed from any exposed reinforcement.

- Ensure that all surfaces to be patched are free of dust.

- Make sure that either there is sufficient mechanical key or that a 'spatterdash' coat* is applied in accordance with the specification or that all surfaces to be patched, including any reinforcement in the area to be repaired, are coated with ordinary Portland cement grout in accordance with the specification.

- Ensure that surfaces of areas to be patched with ordinary Portland cement mortar are wetted immediately before patching is applied.

- Check that a mix compatible with the background is used;
  — if mix has not been specified use very weak, eg 1:7, ordinary Portland cement : coarse sand mortar.

- Ensure that mortar is trowelled into place using coats of not more than 10 mm thickness;
  — allow each layer to cure under moist conditions before rewetting and placing the next;
  — partial shuttering fixed in place may help to achieve square arrisses, Figure 4.

- Ensure that during curing of the repair, it is protected from frost and from hot, dry weather.

*Spatterdash coat of 1:2 cement : clean coarse sand mortar.

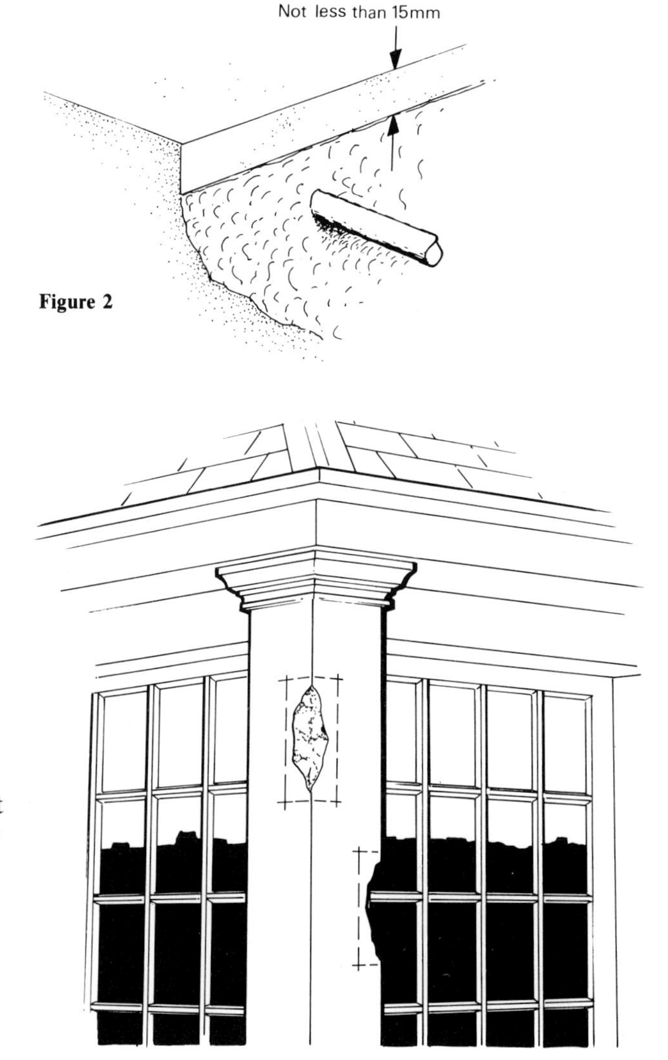

Figure 2

Figure 3

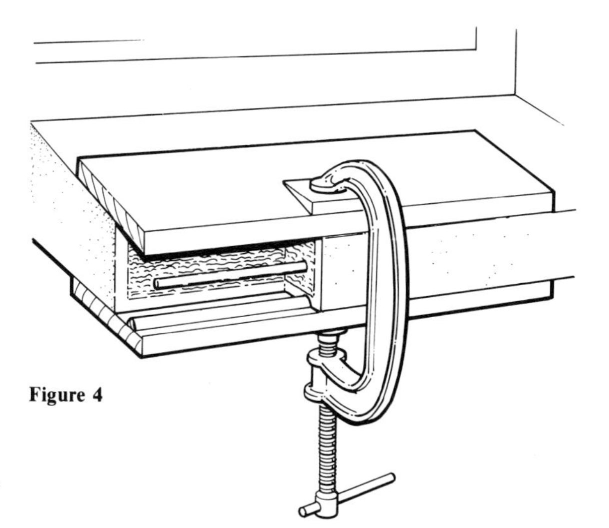
Figure 4

Wood entrance doors: discouraging illegal entry — specification

FAILURE: **Leaves or frames easily forced for illegal entry**

DEFECTS: **Inadequate robustness in door leaves, frames and furniture; inadequate fixing of frames to structure; weak or insufficient hinges; inappropriate locks**

Figure 1

A DOE report of security on Local Authority estates revealed that, among the features contributing significantly to lack of security were doors and locks. Intruders often exploit weaknesses in design, construction and installation of door leaves, frames and furniture to gain entry to a dwelling. Entrance doors to flats are particularly vulnerable and often particularly weak. Locking entrance doors may be ineffective if door leaves, frames or fixings are weak.

If *frames* are *insecurely fixed* to the structure or if hanging and locking *frame sections* are *insufficiently robust*, the *door* could be *vulnerable* to forced entry. If the door *leaf* is of *insufficient* thickness or of *unsuitable construction* it may be easily *broken through*. If *type* or *siting of lock* is *inappropriate* or if *hinges* are *weak* or insufficient the door may be *insecure*.

PREVENTION

Principle — Doors, door frames and furniture must in combination be adequately robust to resist illegal entry.

Practice

- Assess the standard of security needed, eg taking account of the layout of the building and of the estate and local risk: consult the local crime prevention officer and (since requirements may conflict) the local fire officer.
- Specify robust door frames:
  — the depth of rebate (door stop) should be at least 18 mm, either out of solid wood or glued, screwed and pelleted to frame, Figure 2;
  — hanging and locking frame sections for doors with glazed sidelights should be of sufficient section to allow for morticing and rebating, Figure 3.
- Specify that frames are securely fixed to structure:
  — built-in frames should be fixed at 600 mm centres on both jambs of door frame, Figure 4, with galvanised steel cramps set into structure. Fixings at head and cill are advisable.
  — frames fixed into prepared openings should be screwed through the frame into holes drilled into the structure. Fixing into timber slips in mortar joints is not adequate.
- Specify robust door leaves:
  — door leaves should not be less than 44 mm thick;
  — flush doors should be solid core construction;
  — stiles should be at least 119 mm wide;
  — avoid door leaves incorporating weak panelling.
- Specify 1½ pairs of steel broad butt hinges per door, Figure 4:
  — consider, if door is outward opening, specifying 1 pair of hinge bolts set between the hinges;
  — avoid using face mounted hinges externally.
- Specify good locks such as mortice, eg to BS 3621 or rim automatic dead locks:
  — specify the position of the lock on the door so that the jointing of door rails to stile will not be impaired.

Note: although mortices for locks should be no more than 20 mm wide, morticing will nevertheless weaken a door. Consider specifying mortice shields bolted from the inside of the door.

- Specify letter plates to BS 2911, whether or not positioned in the door, no closer to locks than 400 mm and preferably 500 mm;
  — the fitting of internal cover plates would further impede the misuse of letter plates by intruders.
- Consider fitting a door viewer and chain or limiter to front entrance doors.

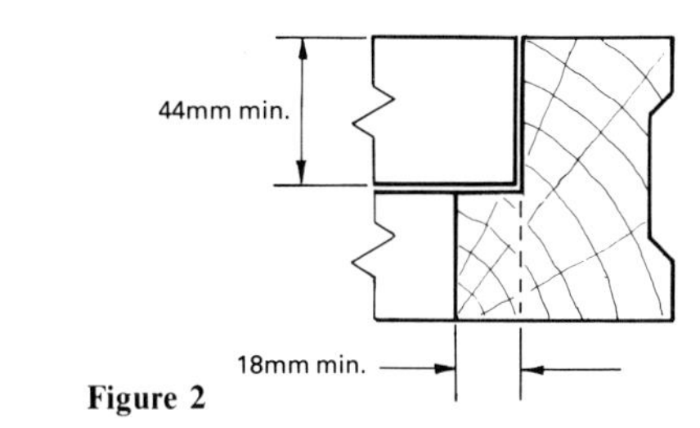

Figure 2

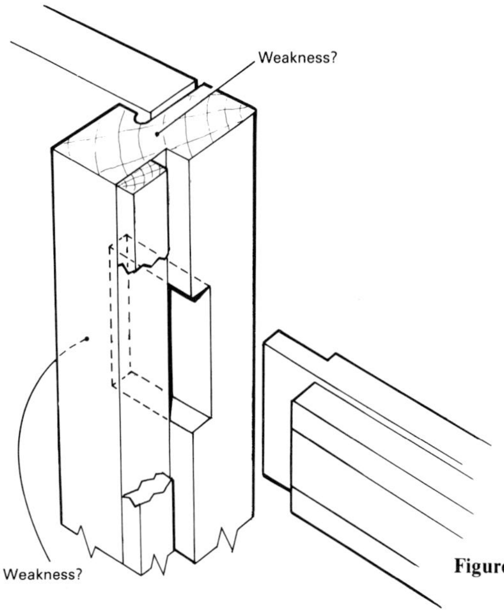

Weakness?

Weakness?

Figure 3

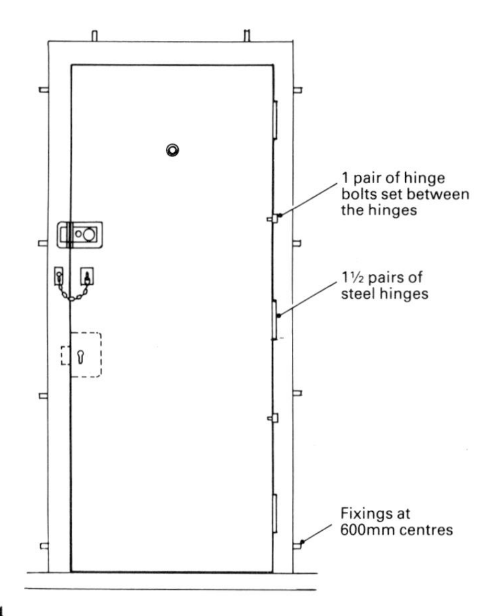

Figure 4

REFERENCES AND FURTHER READING

Defect Action Sheet DAS 88. 'Windows: reducing vulnerability to illegal entry'.
British Standard BS 8220:1986 'Guide for security of buildings against crime. Part 1 Dwellings'.

Windows: discouraging illegal entry

FAILURE: **Windows easily forced for illegal entry**

DEFECTS: **Vulnerable window type or glazing method chosen; inadequate robustness in frames, fixings and furniture; opening lights easily accessible**

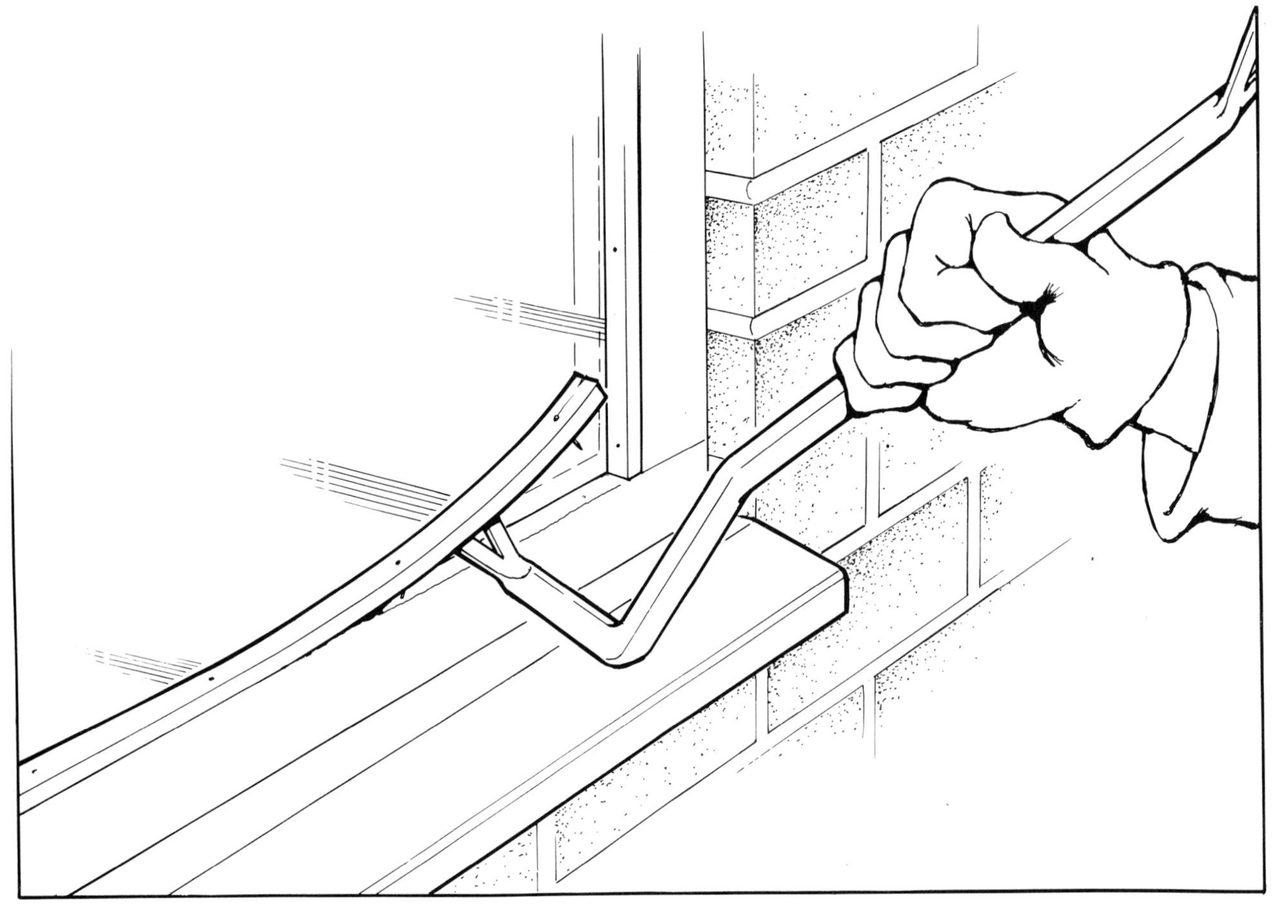

Figure 1 Springing a bead

In recent years the risk of illegal entry and theft at home has increased markedly. The majority of 'break-ins' are made through windows.

Home occupiers are now being encouraged to increase the security of their homes against illegal entry. The aim of security measures is to deter the intruder. ***Inadequate design*** consideration of security can make ***windows vulnerable*** initially and may also make ***subsequent security*** measures ***ineffective.***

PREVENTION

Principle — Windows should be designed to provide adequate resistance to illegal entry.

Practice

- Assess the standard of security needed, eg taking account of the layout of the building and of the estate and local risk: consult the local crime prevention officer.
- Identify windows that are most at risk from illegal entry, such as those where access can easily be gained from outside eg from adjacent roofs, or those concealed from public view.
- Do not specify, for windows identified as at risk, a design where sashes or panes could be removed from outside:
  - louvre windows, Figure 2, and some horizontal sliding windows are vulnerable;
  - flimsy ventilators fitted to apertures in the glass are insecure;
  - 'non-setting' putties are vulnerable; external beads are also vulnerable unless very securely fixed. Note: in some designs of UPVC windows the external glazing beads can be noiselessly sprung out.
- Consider providing secure night ventilation such as trickle ventilation within the frame section or lockable fittings which restrict the opening lights (check that any non-lockable fastenings on adjacent opening windows cannot be reached).
- Specify that windows and sub-frames are securely fixed within the structure:
  - the following number of fixings per window are recommended, Figure 3:
    not more than 1350 mm in height — 2 fixings each side;
    more than 1350 mm in height — 3 fixings each side;
    greater than 900 mm long — 2 fixings, each within 600 mm of corners, along both top and bottom of frames; intermediate fixings at 900 mm centres.
  - specify galvanised steel cramps for 'built-in' windows, Figure 4 and sleeved screws, Figure 5, for units installed into prepared openings;
  - specify that adjacent door and window frames should have common frame members, Figure 6.
- Specify robust fittings; window furniture should not be easily forced or dislodged from outside:
  - avoid hinges or pivots where screws are accessible from outside or specify a non-removable type of screw, Figure 7;
  - specify, where hinge pins might be removed from externally accessible hinges, a type with captive pins.

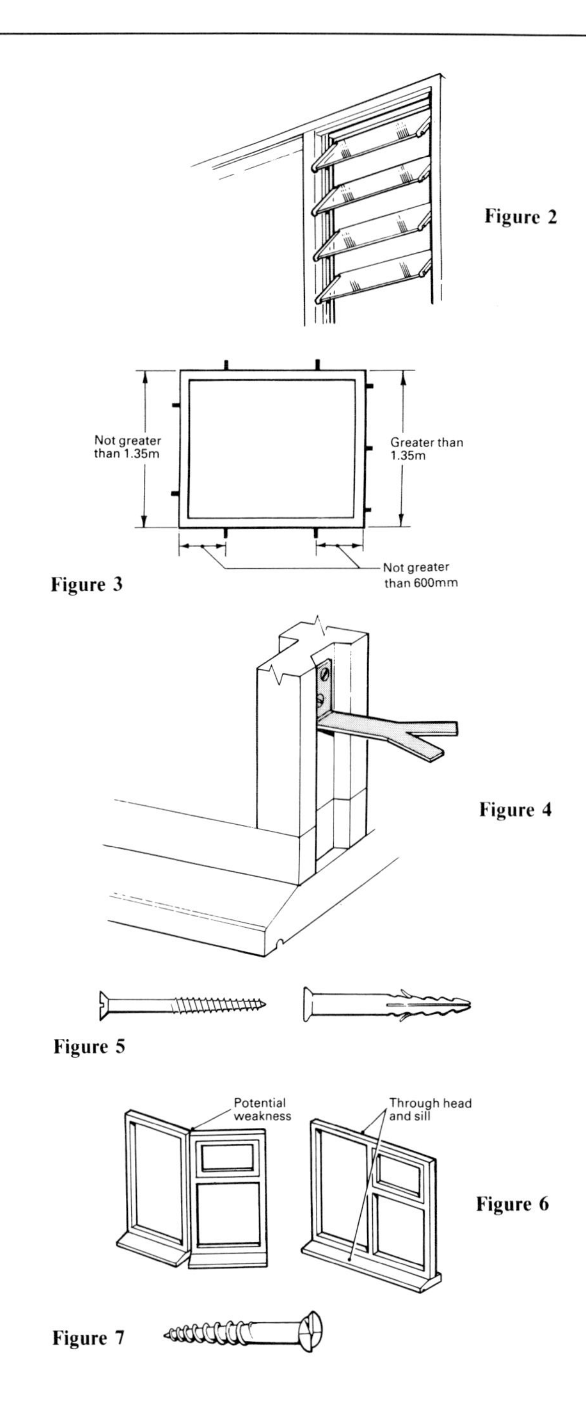

Figure 2

Figure 3

Figure 4

Figure 5

Figure 6

Figure 7

- Make provision for locks for accessible windows:
  - seek specialist advice before specifying locks for aluminium alloy or plastics windows or patio doors;
  - if locks are not to be provided initially specify a window design which will permit effective retrospective fitting.

REFERENCES AND FURTHER READING

Defect Action Sheet DAS 87 'Wood entrance doors: discouraging illegal entry — specification'.

British Standard BS 8220:1986 'Guide for security of buildings against crime. Part 1. Dwellings'.

GLC Development and Materials Bulletin 68, 1973.

PART THIRTEEN

Drainage

Domestic foul drainage systems: avoiding blockages — specification

FAILURE: Blocked drains

DEFECTS: Backfalls, unnecessary bends; incorrect junction design; incorrect gradient or pipesize; long runs serving single wc's; drain damaged where inadequately protected

Flow in a domestic foul drain is intermittent with the discharge normally in a wave form which deposits solids in the pipe to await the next flush. If the system is poorly designed, solids may build up ultimately producing a blockage. Access for rodding is provided to clear blockages arising from abuse of the drain, not as a remedy for its deficient performance.

If drains are designed with ***incorrect gradients, incorrect diameters***, or ***excessive changes of direction***, frequent ***blockages may occur***.

PREVENTION

Principle — Foul drains should carry water and solids to the sewer without risk of blockages.

Practice

- Design soil and vent system within the property to BS 5572: 1978 with a bend at the base of the stack of at least 200 mm radius or preferably two 45° large radius bends.
- Ensure, that pipe sizes are adequate for the service and the number of houses connected:
  - — a minimum nominal diameter of 75 mm where waste water only is conveyed;
  - — a minimum nominal diameter of 100 mm for foul water;
  - — a maximum of 20 houses can be connected to a 100 mm nominal diameter pipe provided the gradient is not flatter than 1:80;
  - — a maximum of 150 houses can be connected to a 150 mm nominal diameter pipe subject to a minimum gradient of 1:150 and a minimum of 5 WCs connected.

  Note: there is no merit in oversizing pipes, it can result in velocities too low to move solids, and thus can create blockages.
- Specify the gradient at which the pipe is to be laid:
  - — a 100 mm nominal diameter pipe, at a gradient not flatter than 1:40, is satisfactory for peak flows of less than 1 litre/second such as occur when the drain serves, for example, only a single sink, wash basin or washing machine.

Figure 1 Excessive bends

  - — a 100 mm nominal diameter pipe, at a gradient not flatter than 1:80, is satisfactory for peak flows of more than 1 litre/second (such as will commonly occur) provided at least 1 WC is connected;
  - — a 150 mm nominal diameter pipe, at a gradient not flatter than 1:150 is satisfactory where at least 5 WCs are connected;
  - — with a good design and a high standard of workmanship flatter gradients have proved satisfactory ie 1:130 for 100 mm nominal diameter pipes and 1:200 for 150 mm nominal diameter pipes;
  - — steep gradients rarely give trouble; rather than laying a pipe of even gradient, a shallow laid drain with a steeper section towards the end of its run may prove more economical to lay.
- Design access points in the drain runs in the following positions:
  - — at or near head of drain run;
  - — at a bend and a change of gradient;
  - — at a change of pipe size;
  - — at a junction unless each run can be cleared from an access point;

Table 1 Maximum spacing in metres of access points

From:	To: Access fitting (1)	Access fitting (2)	Junction or branch	Inspection chamber	Manhole
Start of external drain*	12	12	-	22	45
Rodding eye	22	22	22	45	45
Access fitting (1) Min 150 mm × 100 mm or 150 mm dia.	-	-	12	22	22
Access fitting (2) Min. 225 mm × 100 mm	-	-	22	45	45
Inspection chamber	22	45	22	45	45
Manhole	22	45	45	45	90

*Connection from ground floor appliance or outlet from discharge stack.

Note: If an inspection chamber or manhole is constructed with access fittings the spacing should be as for an access fitting.

Table 2 Minimum dimensions for rodding eyes, access fittings, inspection chambers and manholes

Type of access	Depth to invert (m)	Internal dimensions: Rectangular length width (mm)	Internal dimensions: Circular diameter (mm)	Nominal cover size: Rectangular length width (mm)	Nominal cover size: Circular diameter (mm)
Rodding eye		Preferably same size as drain, but not less than 100 mm diameter		-	-
Access fitting	0.6 or less except where situated in a chamber	150 × 100 225 × 100	150 -	150 × 100 225 × 100	150 -
Inspection chamber	0.6 or less	-	190 mm for drains up to 150 mm dia.	-	190
	1.0 or less	450 × 450	450	450 × 450	450*
Manhole	1.5 or less	1200 × 750	1050	600 × 600	600

* BS 8301 allows clayware and plastics inspection chambers to have a reduced minimum clear opening size of 430 mm in order to provide proper support to the cover and frame.

— in long runs (see Table 1 for maximum recommended spacings).

- Ensure that access point is of the type listed below and at least of the dimensions given in Table 2:
  — rodding eye: a pipe fitting with a removable cover;
  — access fittings: chamber on the pipe with a removable lid;
  — inspection chamber: chamber with an open channel but not working space for a person at drain level;
  — manhole: large chamber with open channel and working space for a person within.

- Check that the position of access points gives working space for rodding, Figure 2.

- Specify removable non-ventilating covers of a pattern to suit the traffic for inspection chambers and manholes:
  — within buildings specify mechanical fixings for covers unless the drain has a watertight access cover.

- Avoid layouts that result in infrequently used appliances:
  — a long branch with a single, little used WC, may block due to drying out of solids.

- Design drain runs to be laid outside buildings wherever practicable;
  — where pipes are laid less than 1.2 m below roads, or less than 0.9 m in fields and gardens, protection should be provided against loads, Figure 3;
  — in all cases, rigid pipes of less than 150 mm nominal diameter laid with less than 0.3 m of cover should be protected by surrounding them in concrete.

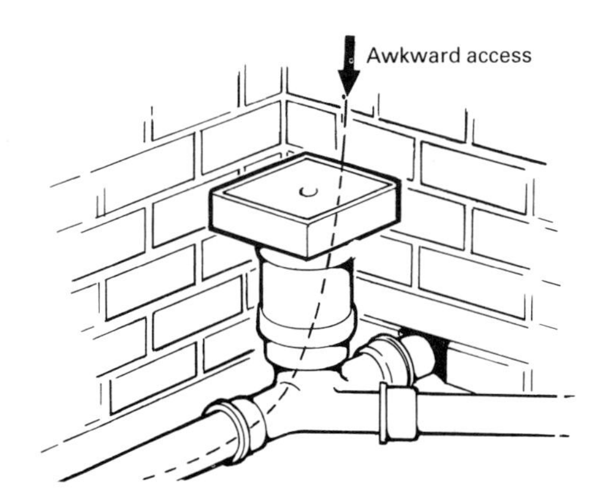

Figure 2

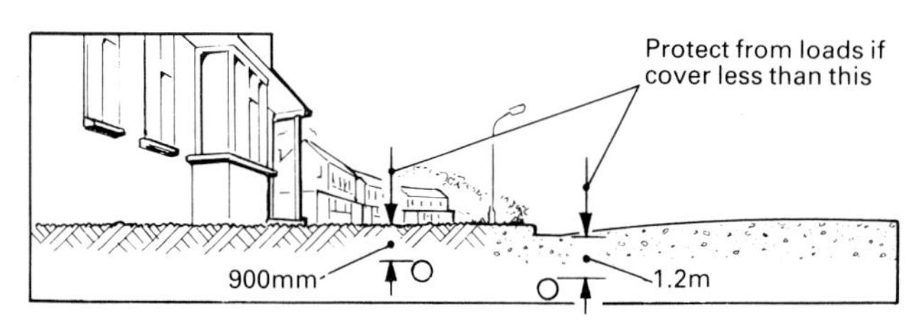

Figure 3

REFERENCES AND FURTHER READING
British Standard BS 5572: 1978. Code of practice for sanitary pipework.
British Standard BS 8301: 1985. Code of practice for building drainage.

Domestic foul drainage systems: avoiding blockages — installation

FAILURE: Blocked drains

DEFECTS: Change of gradient, or backfall, created during laying; incorrect jointing techniques; incorrect connections used; rubble allowed into system; damage from site plant

Figure 1 Mortar blocking inspection chamber

BRE site inspections have shown that rubble is frequently allowed to get into drains so that backing-up of solids on debris occurs when the drains are brought into use. Hardened mortar has been chiselled out with the risk of damaging pipes, and even loose debris may be difficult to clear. Protection of newly laid drains with temporary stoppers is rarely seen on building sites.

If ***drains*** are laid to ***incorrect gradients*** or ***not jointed correctly*** or if they are not ***protected*** from entry of rubble or are ***damaged by site traffic***, ***blockages*** in ***service will result.***

PREVENTION

Principle — Foul drains should be laid such that they convey water and solids without risk of blockages.

Practice

- Ensure that purpose made temporary caps are fitted as early as possible so that builders debris cannot enter pipes, Figure 2:
  — plastic or paper bags are not an acceptable alternative.
- Check that pipes and spigot ends have not been damaged in transit or storage (See DAS 40 and 49).
- Examine factory applied jointing material for soundness and cleanliness immediately before laying.
- Examine jointing components and lubricants for soundness and compatibility with pipes.
- Ensure that the correct adaptor, as recommended by manufacturers, is used where joints have been made between pipes of different materials.
- Check that pipes are accurately laid to the specified gradients and fully supported (See DAS 39).
- Check that access points, Figure 3, are installed in the positions specified:
  — they must be capable of being rodded when required.
- Ensure that site traffic does not dislodge sockets left at slab level:
  — insertion of short length of capped vertical pipework will indicate location to site traffic, Figure 4.
- Test drains for blockages by rodding with a disc or running a ball through the complete layout including branches.

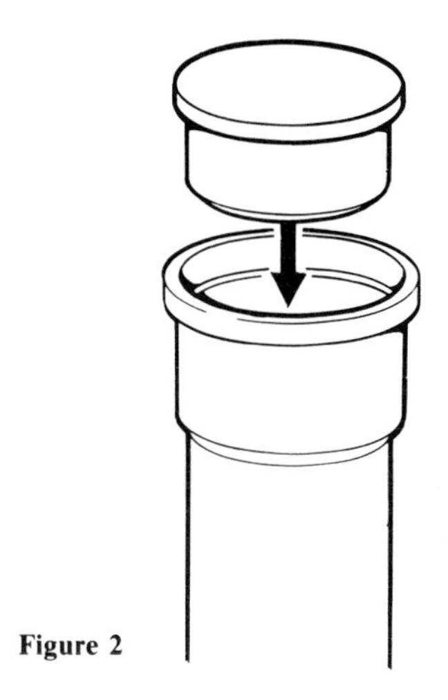

Figure 2

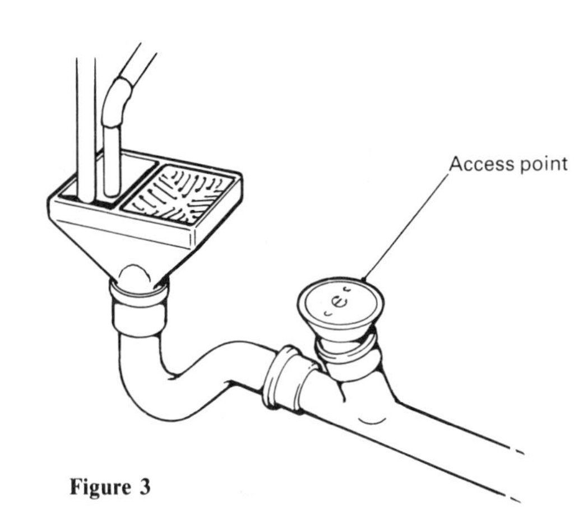

Figure 3

Figure 4

REFERENCES AND FURTHER READING

Defect Action Sheet DAS 39 'Plastic drainage pipes: jointing and backfilling'
Defect Action Sheet DAS 40 'Plastic drainage pipes: storage and handling'
Defect Action Sheet DAS 49 'Clay-ware drainage pipes: storage and handling'.

Plastics drainage pipes: storage and handling

FAILURE: **Leakage from drain**

DEFECTS: **Unprotected outdoor storage for long periods; pipes stacked so as to become distorted; pipes scratched or fractured by rough handling; pipes softened by solvents**

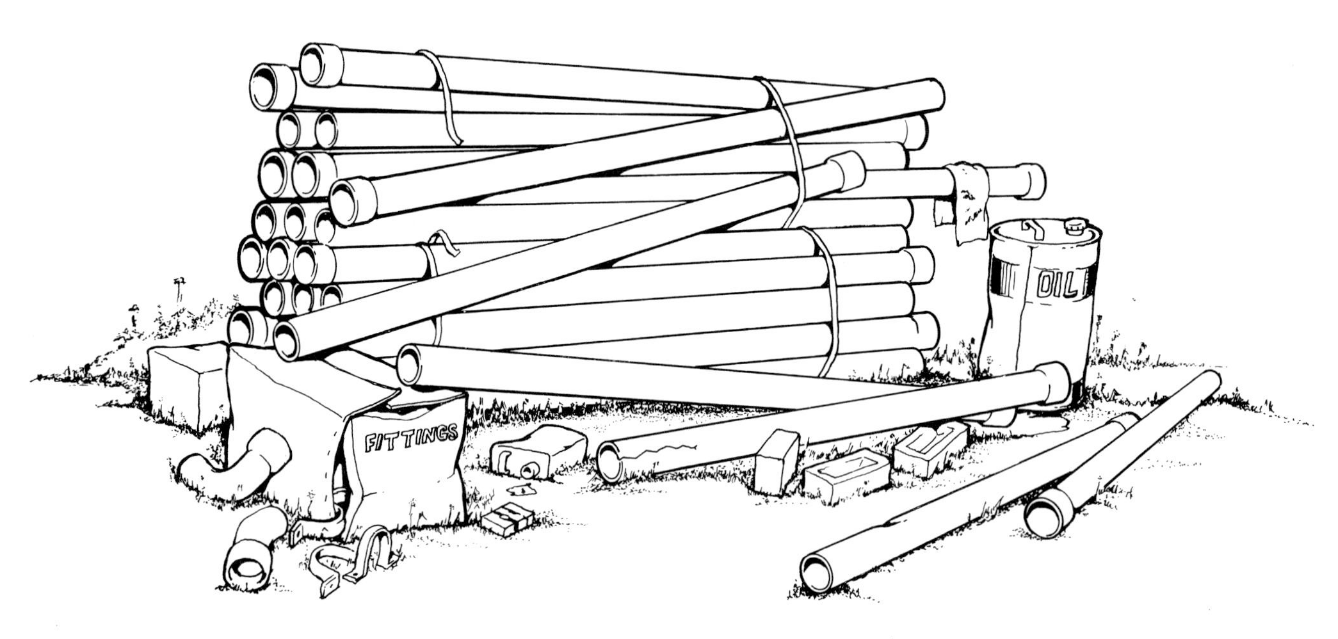

Figure 1 Pipes not stored correctly

In recent years there has been a steady growth in the use of plastics drainage pipes. Handled and laid correctly, they produce a satisfactory drainage system, but their properties differ from those of more traditional materials, and defects caused by bad storage are not always obvious until the drain is tested, or is in service.

If plastics drainage ***pipes*** are ***carelessly transported and unloaded***, or if they are subject to ***point loads or scratching in storage***, they may ***develop cracks***. If ***pipes*** are ***stored*** in ***too high*** a stack they may ***distort*** causing ill-fitting joints. If ***pipes*** are ***stored*** for long periods ***outdoors***, or if temperatures are low, they ***become brittle*** and, if mishandled, may ***fracture***. If ***pipes*** in store are ***contaminated by solvents***, they become more easily ***deformable.***

PREVENTION

Principle — Plastics pipes must be handled and stored so that they are not damaged.

Practice

- Ensure that pipes are unloaded carefully from the lorry; manhandling is best.
  - — take special care not to damage spigots and sockets.
  - — if pipes are mechanically handled they should be supported at two places preferably by protected slings. If pipes are banded, do not hoist by the band: it could break.
  - — put aside for later examination any pipes dropped.
- Check that pipes are stored well away from sands, aggregates, oils, solvents, etc.
- Check that pipes are stored under cover: aged, sun-bleached pipes become brittle.
- Check that pipes are stored in racks, or on a flat surface, or on bearers in stacks not exceeding seven pipes high (Figure 2).
  - — spigot and socket pipes should be stacked with sockets protruding so that the pipes are fully supported along their lengths (Figure 3).
  - — check that pipe ends are protected from damage.
- Ensure that pipes are handled with extra care when temperatures are below freezing: they become increasingly brittle as the temperature drops.

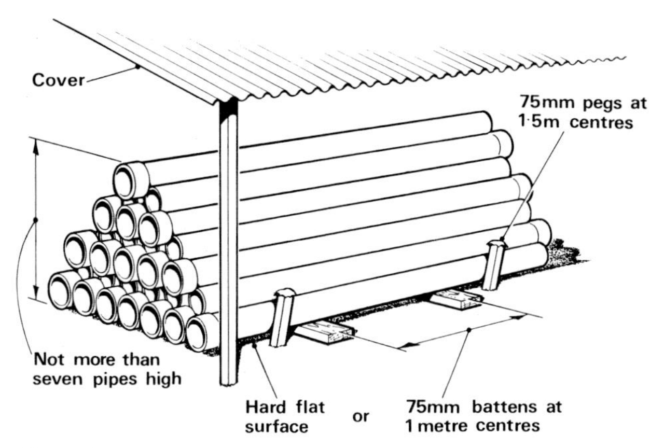

Figure 2

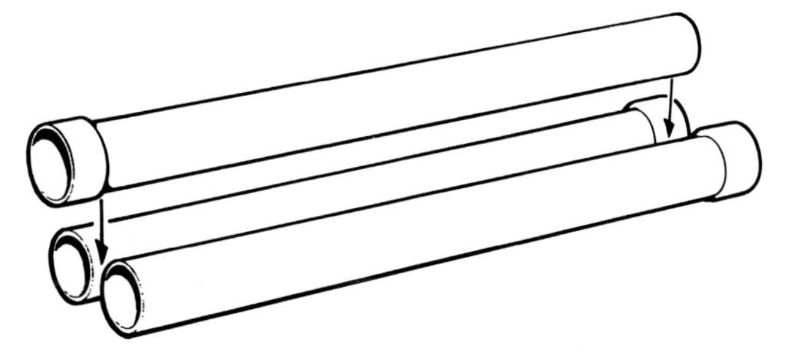

Figure 3

REFERENCES AND FURTHER READING

1 *British Standard* BS5955: Code of Practice for plastics pipework: Part VI.

DAS 39
November 1983 Minor revisions March 1985

Site
CI/SfB 8 (52.7)In(D6)

Plastics drainage pipes: laying, jointing and backfilling

FAILURE: Leakage from drain

DEFECTS: Damaged or distorted pipes; displaced ring seals; unsuitable bedding or fill

Figure 1 Unsuitable backfill

The manufacturers of plastics drainage pipes have observed that the majority of cases of leaking drains proves to be due to bad practice during construction. BRE's experience supports this view.

If ***joints*** are ***dirty*** or made with ***damaged or distorted*** pipes, or if ***ring seals*** are ***displaced*** during jointing, ***joints*** may ***leak*** from the outset. If pipes are not well bedded or if back-fill is stony, leaks may develop in service.

PREVENTION

Principle — Plastics pipes must be accurately aligned, fully supported and have leak-proof joints.

Practice

- Ensure that the trench is cut no wider than necessary:
  — good compaction of sidefill is more readily achieved in a narrow trench and pipes then receive better support against squashing by loads from above (Figure 2).

- Ensure that fill, particularly bedding and sidefill, is properly compacted:
  — pipes can be bedded on trench bottoms provided that the 'as dug' material complies with Appendix A of Reference 1. If it does not, trench must be dug to at least 100mm below invert level and pipe bedded on and surrounded by suitable fill (see Table 1).

- Ensure that granular bedding and back-fill contain no stones which might cause point loads on the pipe:
  — bricks must not be used as temporary supports.

- Check that sockets are cleaned before ring seals are inserted and again before joint is made.

- Check that site cuts are square, evenly chamfered, and free from adhering swarf (Figure 3).

- Check manufacturer's literature for guidance about the need for gaps, at joints, for movement. If no figure is recommended mark pipes, say, 10mm short of socket depth as a guide to assembly (Figure 3).

- Check that only specified lubricants are used:
  — do not permit the use of grease or oil.

- Check (for example using a narrow strip of stiff plastic) that the ring seal has not been displaced during jointing (Figure 4):
  — if in doubt, a jointed pipe that can be readily turned using a strap-wrench, almost certainly has a well-made joint.

- Ensure that all pipe ends are temporarily stoppered to keep out dirt.

- Ensure that after test the trench is filled to the crown of the pipe with granular material before back-filling (Figure 5); pipes in deep trenches in unstable stony soil need to be protected by at least 100mm of granular fill above the crown.

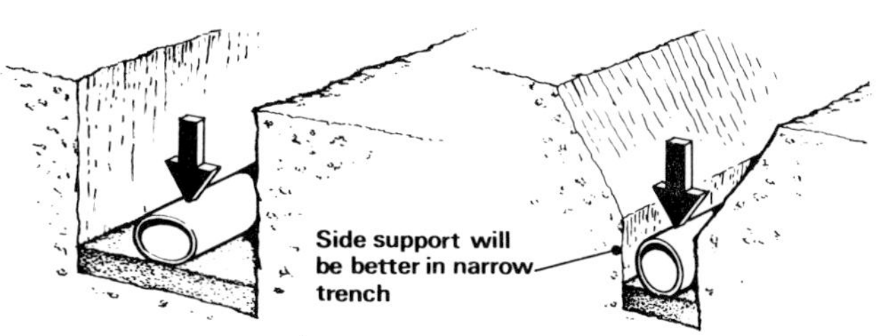

Figure 2

Table 1 Suitable materials for bedding and surrounding pipes.

Nominal pipe size	Material (complying with the requirements of BS882: Part 2)
mm	mm
110	10, nominal single-sized aggregate
160	10 or 14, nominal single-sized aggregate or 14 to 5 graded aggregate

(Taken from BS 5955)

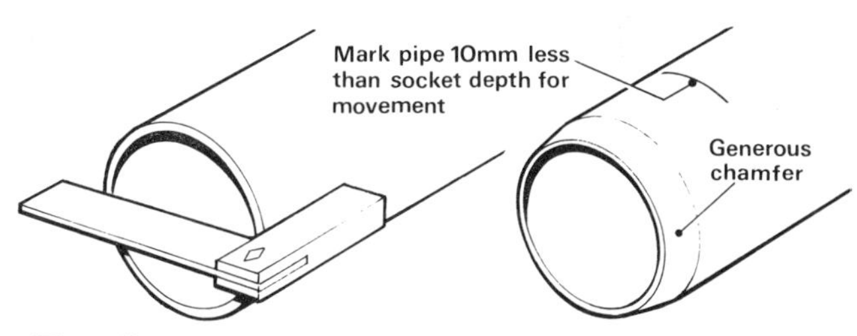

Figure 3

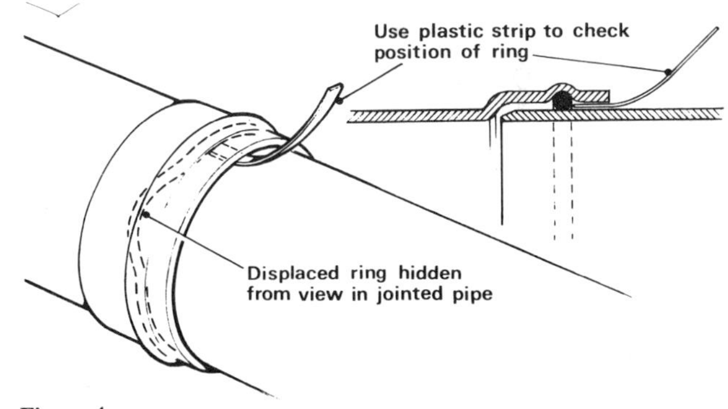

Figure 4

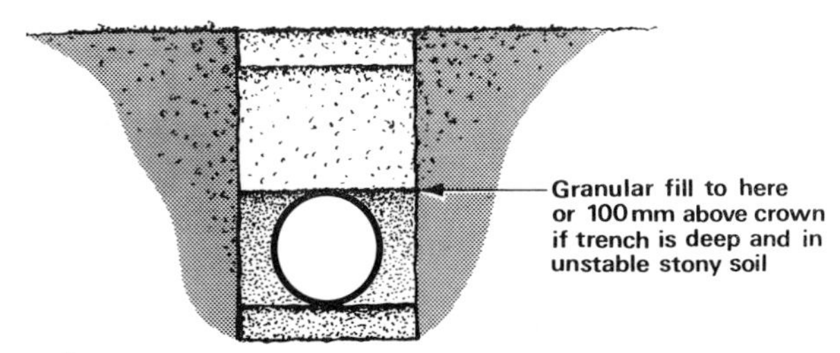

Figure 5

REFERENCES AND FURTHER READING

1 *British Standard* BS5955: Code of Practice for plastics pipework: Part VI.

Clay-ware drainage pipes: storage and handling

FAILURE: Leakage from drain

DEFECTS: Pipes chipped or cracked by rough handling or poor supports in storage; storage badly sited

Figure 1 Pipe stack not pegged

Clay-ware drainage pipes, handled and laid correctly, can produce a satisfactory drainage system. But they are vulnerable to impact damage until laid and protected by backfill. BRE studies show that wastage due to damage more than doubles if pipes are stacked on sloping ground or near roads. Damage, such as chipped spigots and cracks, may go undetected until the drains are tested.

If clay-ware drainage ***pipes*** are ***carelessly transported*** and ***unloaded,*** or if they are ***carelessly stacked*** or stored ***without protection*** from ***impact damage,*** they may be ***cracked*** or ***chipped.***

PREVENTION

Principle — Clay-ware pipes must be handled and stored so that they are not damaged.

Practice

- Ensure that pipe packs are unloaded carefully on delivery:
  — a fork-lift truck is usually best, Figure 2;

  — if pipes are banded, do not hoist by the band, Figure 3.

- Check that storage area is firm and level, where risk of damage by site traffic is small and preferably in a 'site compound'.

- Check that pipes are kept in packs until needed:

  — follow manufacturer's instructions for extracting pipes from wrapped or banded packs, otherwise pack may collapse and pipes be damaged.

- Check that pipes stacked at point of use are pegged to prevent them rolling, Figure 4.

- Check that fittings, couplings, joint rings etc are kept in secure and orderly storage — mislaid parts may lead to improvised connections.

Figure 2

Figure 3

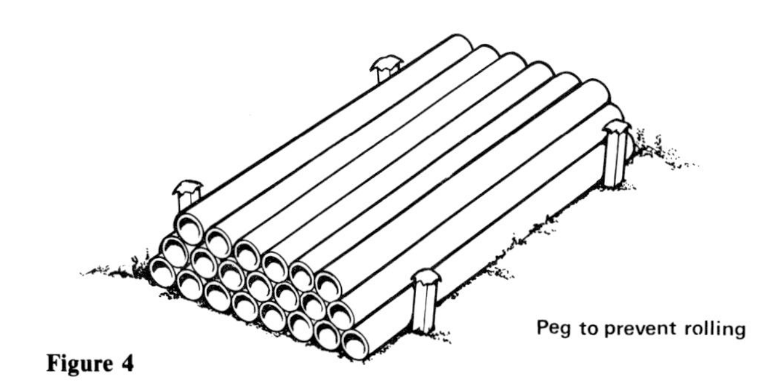

Figure 4

Flexibly jointed clay-ware drainage pipes: jointing and backfilling

FAILURE: **Leakage from drain**

DEFECTS: **Damaged pipes; unsuitable bedding or fill; badly made joints**

Figure 1 Unsuitable bedding

In BRE's experience most cases of leaking drains have proved to be due to bad construction practice. If ***flexible couplings*** or ***ring seals*** are ***dirty*** or if ***pipe ends*** are ***dirty*** or ***damaged, joints*** may ***leak*** from the outset. If ***pipes*** are ***not well bedded,*** if ***bedding*** is ***disturbed*** as pipes are laid, or if ***backfill*** contains, for example, large ***stones, leaks*** may ***develop*** in service.

PREVENTION

Principle — Clay-ware pipes must be accurately aligned, fully supported and have leak-proof joints.

Practice:

- Ensure that pipes are correctly laid on specified bedding:

— pipes may be laid on granular bedding, to specified thickness, which unless otherwise specified should extend the full width of the trench;

— provided it can be hand-trimmed level pipes may be laid directly on the trench bottom unless the ground is compacted sand or gravel, or rock, or very soft; hollows can be levelled using compacted material not greater than 20 mm particle size.

- Ensure that pipes are continuously supported along the barrel and none of the weight is on sockets or couplings:
  - — the bedding under sockets or couplings should be scooped out about 50 mm; if granular bedding used, 50 mm must remain under this scooped recess, Figure 2;
  - — pipes must not be supported on bricks.
- Check that pipes cut on site have clean ends, square to the axis:
  - — follow manufacturer's recommendations for cutting.
- Ensure that jointing surfaces and sealing rings are clean and sound before jointing:
  - — check whether and which lubricants are to be used.
- Ensure that a small gap (see manufacturer's literature) is left between spigot and shoulder of socket, or between pipe ends in a coupling, to allow for length-wise movement.
- Ensure that bedding is not disturbed as pipes are laid:
  - — pipes should not be levered home (eg with a spade);
  - — if a run has to be left temporarily uncompleted, stopper end and lay about 1 m of consolidated bedding beyond the end of the run to prevent settlement of the last pipe.
- Ensure that side and back fill are placed immediately the drain has been tested and approved:
  - — plan work to minimise time between laying and testing.
- Ensure that, before filling trench, pipe ends are temporarily stoppered.
- Ensure that, for side fill, only specified material is used, correctly compacted:
  - — the material should be placed uniformly each side of the pipe and hand-tamped in 100 mm layers until the pipe has 150 mm of compacted cover, Figure 3;
  - — if material from the trench is specified it should be readily compactible and must not contain frozen material, stones larger than 40 mm or clay lumps larger than 100 mm, or site rubbish.
  - — specification may call for granular side fill material to be taken to 100 mm above pipe crown, eg in deep trenches in unstable ground;
  - — over-compaction of pea gravel side fill may lift pipes.
- Ensure that, for backfill, only specified material is used, correctly compacted:
  - — material from the trench must be free from frozen lumps, large stones and site rubbish;
  - — backfill should be placed in not more than 300 mm layers initial thickness, each layer then being well compacted by hand. Do not permit mechanical compaction until two such layers at least have been hand-compacted, Figure 4.

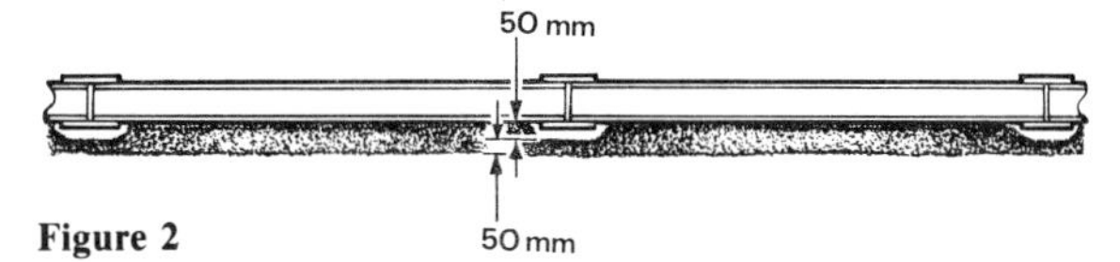

Figure 2

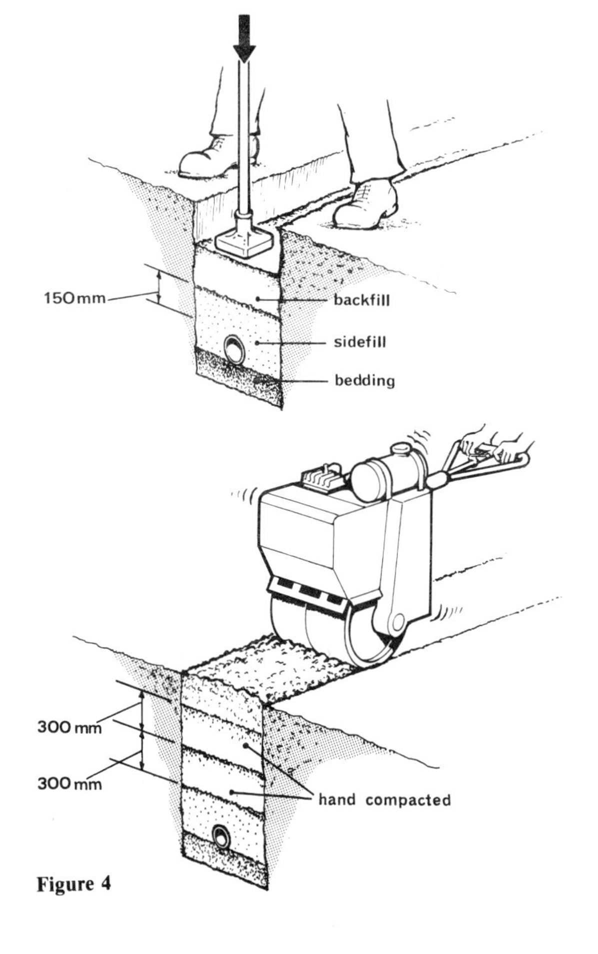

Figure 4

REFERENCES AND FURTHER READING
British Standard *Code of Practice* CP 301 : 1971 : Building drainage.

DAS 41
December 1983

Design
CI/SfB 8 (52.7)In(A3u)

Plastics sanitary pipework: jointing and support — specification

FAILURES: Poor drainage; leaking joints

DEFECTS: Pipe supports missing or ineffective; pipe expansion not provided for

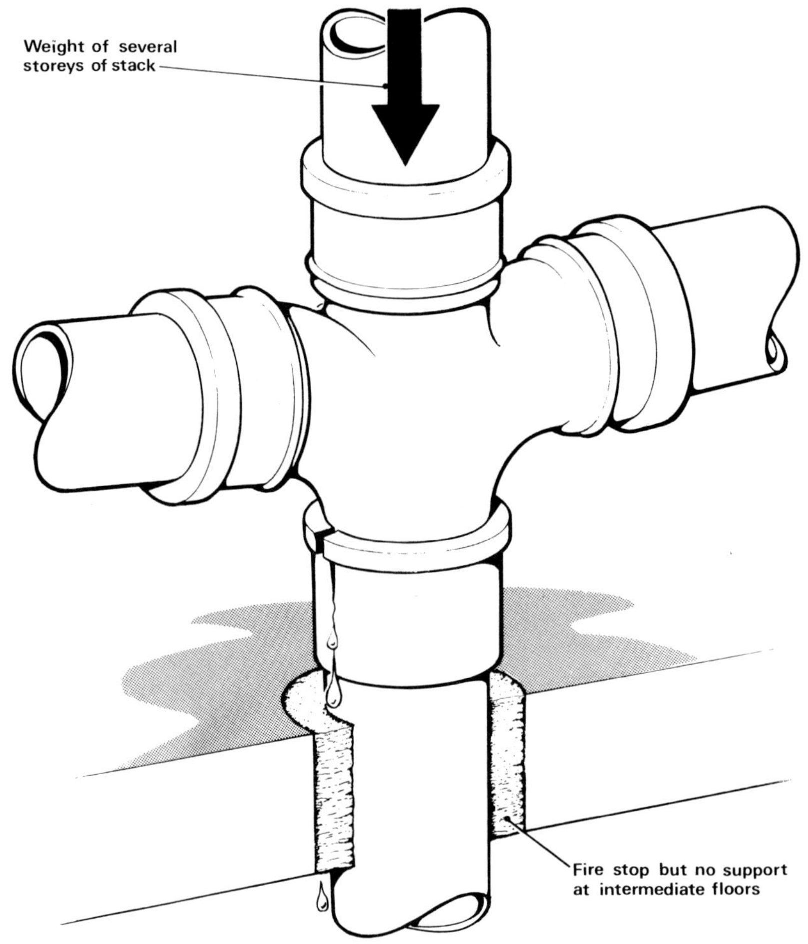

Figure 1 Unsupported stack settles and breaks at the socket collars

Although plastics sanitary pipework has been used for a number of years, there is evidence that designers and plumbers are unfamiliar with some requirements.

It has been known for soil stacks to be unsupported, even over several storeys. Installations sometimes take no account of the thermal movements in plastics soil and waste pipes, produced by the wide range of discharge temperatures.

If every storey height length of ***SVP*** is ***not fully supported*** by fixings below the socket collars, pipes can ***creep downward, opening joints*** and ***changing*** the ***slope of*** soil and waste ***branches***. If ***joints do not have end clearance*** for movement, or if ***waste branches*** are ***not supported*** at correct spacings, pipes can be ***distorted or damaged, drainage*** can be ***impaired***, and ***leaks*** develop.

PREVENTION

Principle — In sanitary pipework, joints must not be disrupted, nor drainage impaired, e.g. by movement in service.

Practice

- Specify that every length of SVP shall be securely supported close below the socket collar (Figure 2).
  — Ducts must be big enough to accommodate support brackets, and to provide access for fixing them.

- Specify that spigots in SVPs shall be fully inserted into sockets, and then withdrawn 6mm to permit expansion.

- Specify supports for waste pipes at intervals not greater than those listed in Tables 1 and 2 (taken from Reference 1).

- Design for thermal movement in runs of waste pipe.
  — 'Push-fit' joints should be assembled with clearance for expansion: check expected movement and relate to number of joints.
  — Solvent-welded joints: provide 'push-fit' couplings at calculated intervals, but not exceeding 1.8m (Figure 3).
  — Sleeve wastes through walls to permit pipe movement.

- Specify the lubricant to be used for 'moving' joints in both SVP and waste pipes (see manufacturers' literature).
  — Lubrication is intended to ease movement in service, not merely to aid assembly.

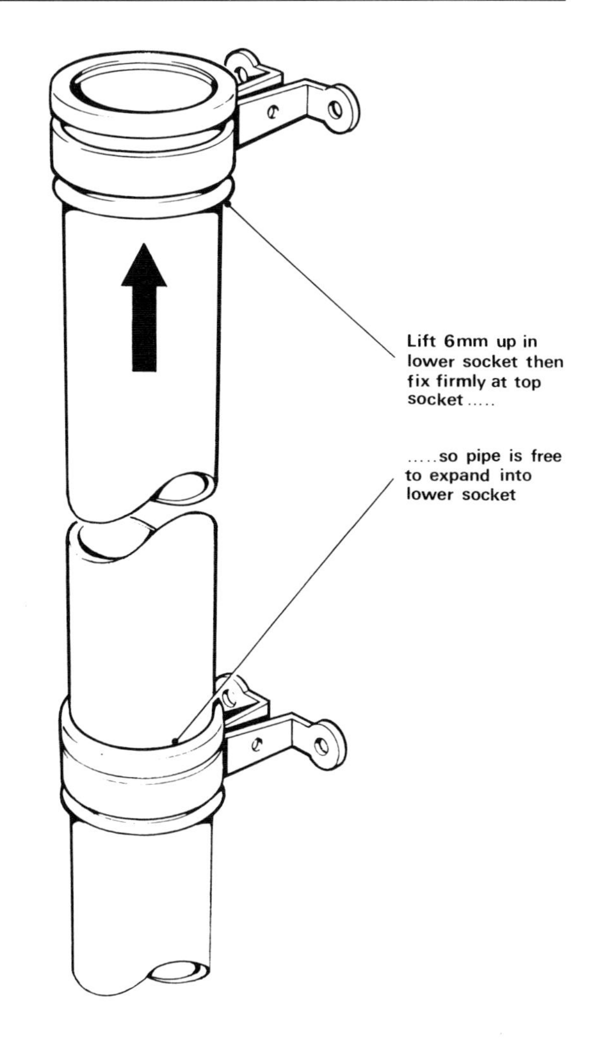

Figure 2

Table 1 Vertical pipes.

Pipe diameter	Support spacing distance
Less than 50mm	1.2m maximum
More than 50mm	1.8m maximum

Table 2 Horizontal pipes

Pipe diameter	Support spacing distance
Less than 40mm	0.5m maximum
Less than 50mm	0.6m maximum
Less than 100mm	0.9m maximum
More than 100mm	1.2m maximum

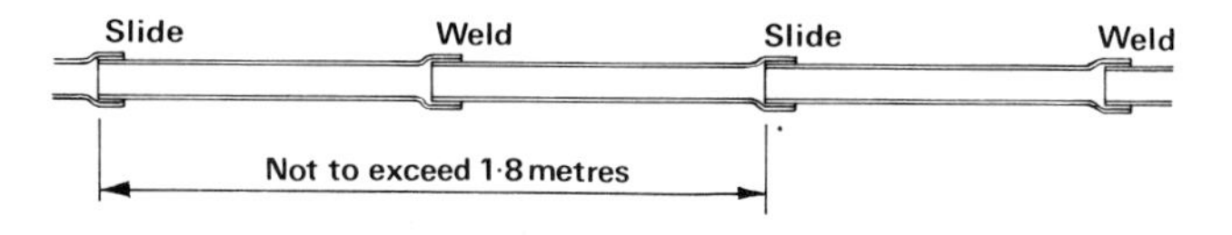

Figure 3

REFERENCES AND FURTHER READING

1 *British Standard* BS5572: 1978: Code of Practice for Sanitary Pipework.

DAS 42
December 1983

Site
CI/SfB 8 (52.7)I6(D6)

Plastics sanitary pipework: jointing and support — installation

FAILURES: Poor drainage; leaking joints

DEFECTS: Pipe not well supported; pipe expansion not allowed for; wrong jointing techniques used

Figure 1 Unsupported pipe distorts, leading to leaking joints

Plastics expand when heated: movement in a 3m length of pipe can be about 5mm in a upvc bath discharge pipe, and nearly twice that amount in a pipe of polypropylene.

If pipework is ***not assembled*** correctly, ***to accommodate*** those ***movements, pipes*** will ***distort, joints*** may ***leak, fixings*** may be ***pulled out***, and ***drainage*** may be ***impaired***.

PREVENTION

Principles — In sanitary pipework, joints must not be disrupted, nor drainage impaired, e.g. by movements in service.

Practice

- Ensure that plastics pipes are supported as specified, or at intervals no greater than those shown in Figure 2.

- Ensure that 'push-fit' joints are provided, with the specified clearance for expansion, and at the intervals specified.
  — where solvent welded joints are used, expansion couplings may be needed: check what has been specified.

- Ensure that 'push-fit' joints are lubricated before assembly (Figure 3).
  — check that specified lubricant only is used (lubrication is intended to ease movement in service, not merely to aid assembly).

- Do not permit plumbers to use 'boss white' to make joints in plastics pipework.

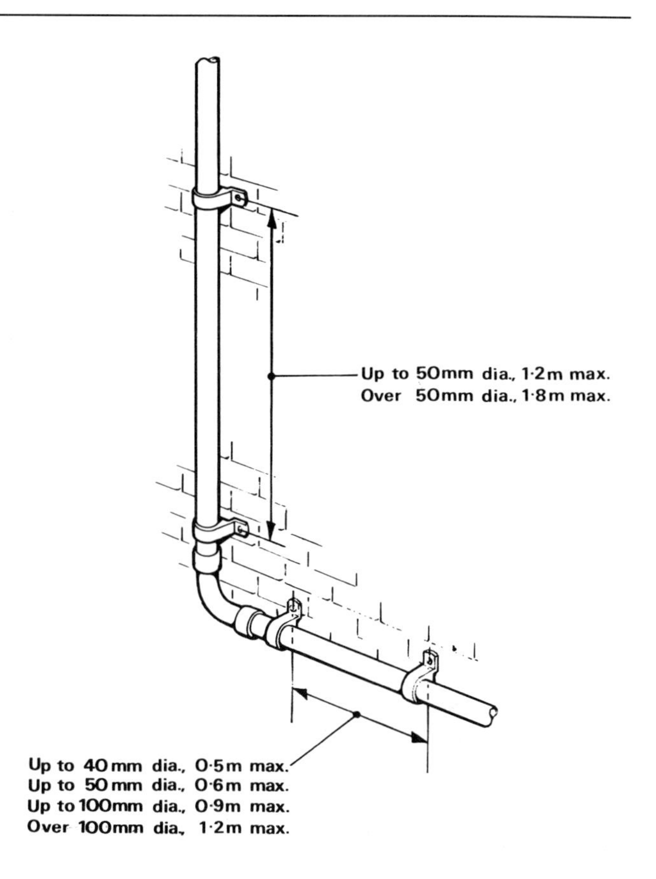

Figure 2

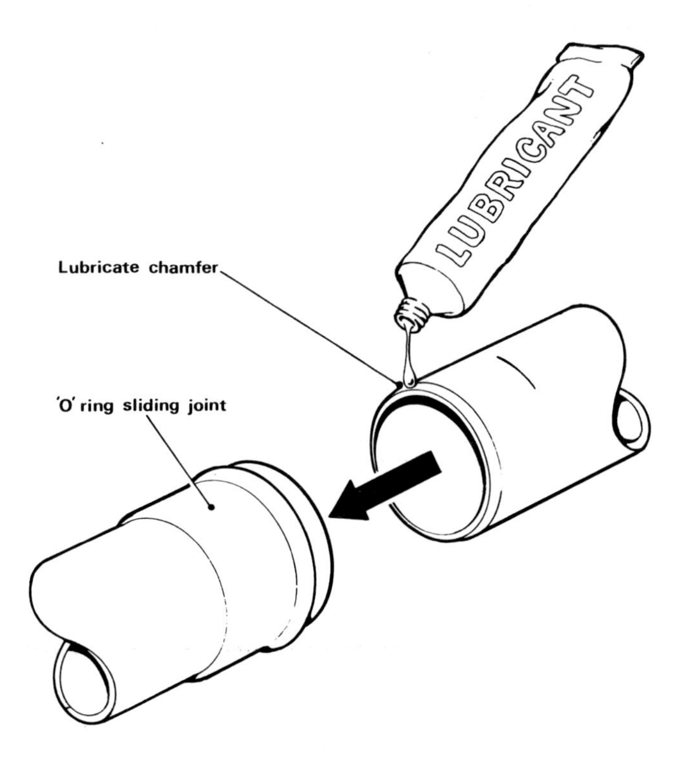

Figure 3

Plastics sanitary pipework: specifying for outdoor use

FAILURE: **Disintegration of exposed UV-sensitive pipework**

DEFECT: **Use of pipes subject to photo-oxidation without suitable protection from solar radiation**

Figure 1 The failed lengths are of unprotected UV-sensitive (polypropylene) pipe

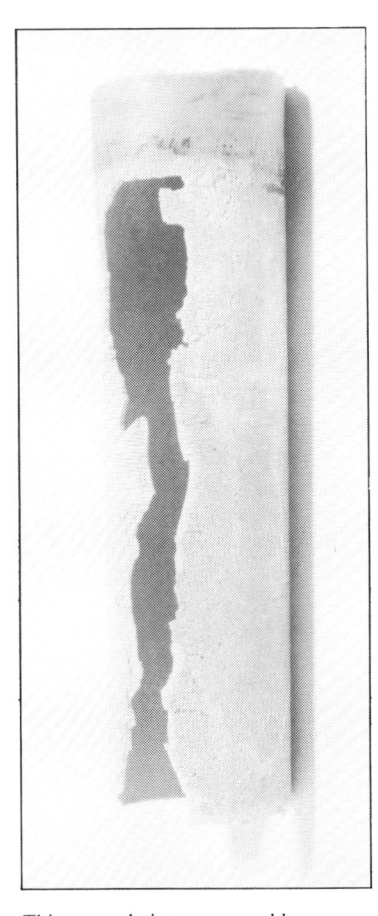

This example is ten years old.
Note the good condition of the spigot.

BRE sometimes receives reports of sanitary pipework which has failed prematurely. Plastics pipes extruded from polypropylene are not suitable for external use unless painted or otherwise protected from sunlight.

If unprotected ***polypropylene*** sanitary ***pipes*** are used where they will be ***exposed to sunlight***, photo-oxidation will result in the pipe ***becoming brittle*** and eventually lead to ***disintegration***.

PREVENTION

Principle — Exposed plastics pipes should be of a material that is adequately durable in sunlight or that can be made so by appropriate protection.

Practice

- Specify that any external sanitary pipework which is to remain unpainted shall be of a material not degraded by sunlight:
  — pipes made from UPVC manufactured to BS 4514 are suitable.

- Specify that externally exposed polypropylene pipe (to BS 5254 or BS 5255) is to be fully protected from sunlight:
  — pipe manufacturers suggest that an exterior masonry paint or a gloss paint (without primer or undercoat) can be applied after degreasing the pipe with white spirit and roughening with fine wire wool.*

- Instruct site staff to ensure that polypropylene pipes are identified so that protection can be given as necessary, Figure 2:
  — such pipes carry, at intervals of not more than 1 m, the manufacturer's identification, the number of the British Standard (BS 5254 or BS 5255) the material code (PP) and the nominal size description, Figure 2.

*Note: weathered but sound polypropylene pipes, already in service, can be painted without prior degreasing or roughening.

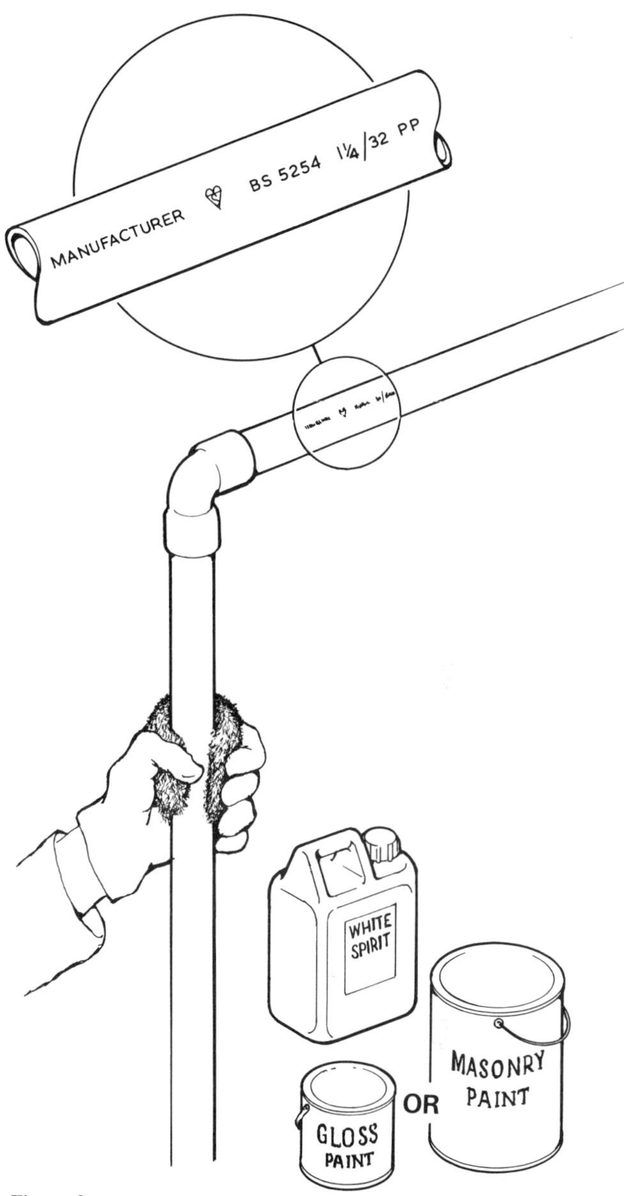

Figure 2

REFERENCES AND FURTHER READING

□ *British Standard* BS 4514: 1983 Unplasticized PVC soil and ventilating pipe, fittings and accessories.

□ *British Standard* BS 5254: 1976 Polypropylene waste pipe and fittings.

□ *British Standard* BS 5255: 1976 Plastics waste pipe and fittings.

Drainage stacks — avoiding roof penetration

FAILURE: Rainwater penetration to inside of roof space

DEFECTS: Flashing incorrectly dressed; poor vent position in relation to tile/slating module; ill-fitted sarking felt

Figure 1

Figure 2

To prevent the loss of water seals in traps, and foul odours entering the dwelling, drainage stacks (excluding 'stub stacks') are usually vented to the outside air by pipes not less than 75 mm in diameter. If the drainage stack is inside the building, the ***drainage stack*** invariably ***passes through*** the ***roof and*** the ***weathertight coverings.*** There is always an increased ***risk*** of ***rain penetration*** to the inside of the roof. An air admittance valve (AAV) will allow air to be drawn into a drainage stack but will not allow air to flow out. Hence the stack may be terminated within a building avoiding any roof penetration and associated problems of damage to underfelt and failure of flashing. The valve must be properly specified and installed.

PREVENTION

Principle — Avoid penetration of the roof surface by the drainage stack.

Practice

- Check that the drainage stack needs to be ventilated. In certain situations, a 'stub stack' may be suitable[1].
- Check that the drainage is not subjected to surcharge:
  — AAV's should not be fitted if surcharging occurs or if intercepting traps are fitted.
- Check the number of dwellings to be connected to the drain:
  — for eleven to twenty dwellings an open vent needs to be provided at the head and mid-point of the system;
  — for five to ten dwellings an open vent needs to be provided at the head of the system only;
  — up to and including four dwellings, 1 to 3 storeys in height, the head of the system may not require open venting, Figure 3.
- Check that not more than two groups of sanitary appliances are fitted to the stack on each floor, Figure 4.
- Check that branch pipework and layout of appliances is in accordance with the Building Regulations and Code of Practice summarised in Table 1[1,2].
- If an AAV is housed within a duct specify that:
  — the duct is adequately ventilated;
  — where there is risk of freezing an insulated cover is fitted;
  — sound insulation be provided to the duct if noise of operation is likely to be a nuisance;
  — access be provided to the stack (for rodding), by fitting the valve using a ring-seal connection.

- If an AAV is housed within a roof space specify that:
  — an insulated cover be fitted to the AAV where there is a possibility of condensation freezing within the valve body (warn site not to discard insulated cover);
  — warn site to check that roof insulation does not interfere with the operation of the valve.
- Check that a separate open vent is provided for septic tanks or cesspools:
  — AAV's must not be used for venting these.
- Specify that the valve shall be fitted in the vertical position and above the flood level of the highest appliance being served, Figure 5.
- Check that the AAV selected is acceptable under Building Control; this requirement is normally satisfied by a current BBA certificate and the proposed installation to be in accordance with the terms of that certificate.

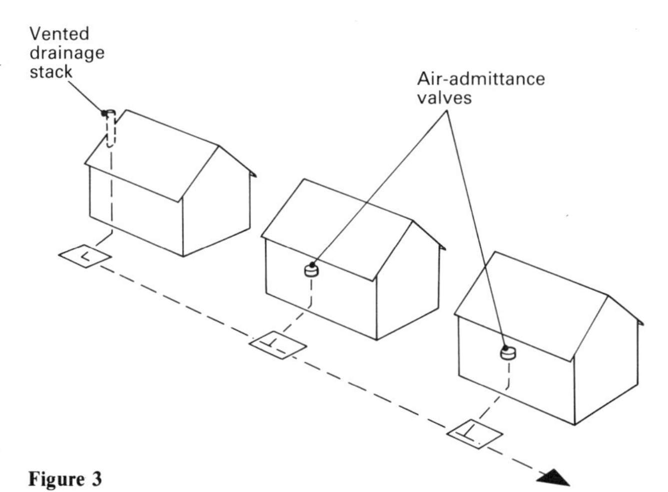

Figure 3

75mm/100mm Valve

A 'group' of sanitary fittings

Max = 10 floors *

Max = 2 groups per floor (A domestic group comprises 1 wc, 1 sink 1 whb, 1 bath/shower, 1 washing machine)

Domestic systems only

> 100mm drain

*Up to 5 storeys lowest branch connection to invert of drain not less than 750 mm. If more than 5 storeys, ground floor appliances should discharge to their own stack, unless discharging direct to a gully or drain.

Figure 4

Table 1 Common branch connections of unvented single appliances to the discharge stack

Appliance	Min trap seal depth (mm)	Min discharge pipe diameter (mm)	Gradient Limits (fall per metre) Min (mm)	Gradient Limits (fall per metre) Max (mm)	Distance to stack Min (m)	Distance to stack Max (m)
WC	50	75* or 100	9	—	—	6
Washbasins (whb)	75	32	20	Refer to design graph see note below	—	1.7
Sinks baths showers	75	40	18	90	—	3.0
Washing and dish washing machines	75	40	18	45	—	3.0
Waste disposal units	75	40	135	—	Refer to manufacturers instructions	

*75 mm can be used for some siphonic WCs.
Note: For pipe arrangements outside the Table limits refer to BS 5572[2] and The Building Regulations[1].

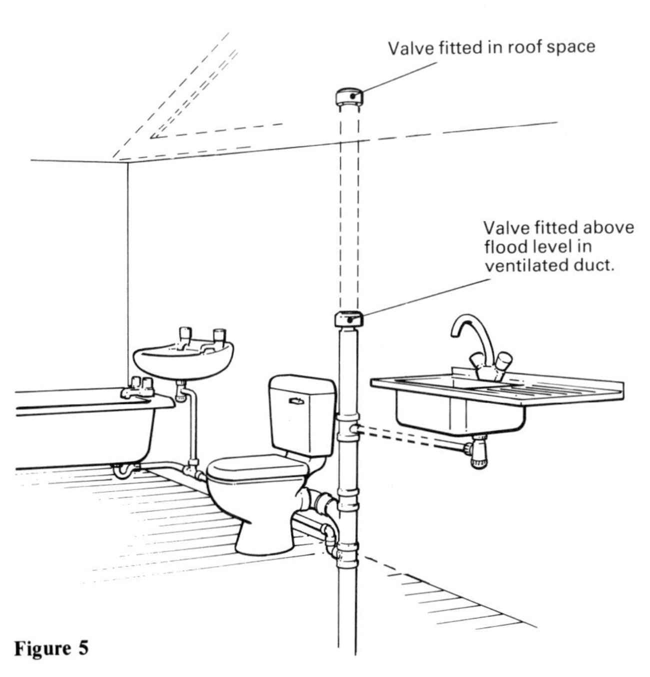

Figure 5

REFERENCES AND FURTHER READING

1 *The Building Regulations* 1985, approved document H1 (1990 edition).
2 *British Standard* BS 5572: 1978 Code of Practice for Sanitary Pipework.
☐ *BRE Digest* 248 — Sanitary pipework: Part 1: Design basis.
☐ *BRE Digest* 249 — Sanitary pipework: Part 2: Design of pipework.
☐ *Defect Action Sheet* DAS 10 (Site) — Pitched roofs: sarking felt underlay — watertightness.

DAS 55
July 1984

Design
CI/SfB 8 (52.5)(A3u)

Roofs: eaves gutters and downpipes — specification

FAILURES: **Rain penetration of external walls; deterioration of walls**

DEFECTS: **Gutters undersized, omitted, poorly supported; downpipes poorly fixed, mispositioned relative to gullies, not accessible for cleaning**

Figure 1 Downpipe or gulley mispositioned: tortuous connection

Gutter and downpipe installations were deficient on many houses in a BRE survey. Downpipes were seen badly sited relative to gullies and, in re-entrant corners, obstructing opening lights — with risk of damage to pipes or windows. Often no gutter was provided to porch roofs when needed.

If ***gutters*** are ***undersized*** (or if downpipes are blocked or displaced), ***water*** spillage can saturate walls and can lead to ***internal dampness*** and to early ***deterioration of walls.***

PREVENTION

Principal — All the water draining from a roof should be conducted without spillage, to its below-ground drainage system.

Practice

- Calculate the effective catchment area *Ae* that will discharge to each gutter:
  - — for a flat roof, *Ae* is the relevant plan area of roof in square metres;
  - — for each face of a pitched roof, *Ae* is the projected plan area plus half the projected elevation area, in square metres, Figure 2;
  - — for a vertical surface draining to a roof, *Ae* is half of that surface area in square metres;
  - — where more than one area drains to a gutter, sum the relevant figures (eg Figure 3).
  (Note that a porch roof, though small, will often have a considerable effective catchment area, Figure 3).
- Calculate the rate of run-off *Q* draining to each gutter:
  - — *Q* is the value of *Ae* divided by 48. (This converts the effective catchment area in square metres into run-off in litres per second, using a recommended[1] rainfall rate of 75 mm/hour; note that this rate is used for all exposure zones: it is a thunderstorm rate.)
- Select from Table 1 a gutter that will accommodate the run-off *Q* when subdivided among a sufficient number of outlets, Figure 4:
  - — gutter outlets can be assumed not to restrict flow at the maximum capacity of the gutter.
- Check that outlet positions will enable direct connection of downpipes to gullies and that downpipes will not obstruct opening lights etc; show downpipe positions on drawings.
- Specify that gutters are to be fixed with a slight fall towards the outlet. A fall of 1 in 350 (about 10 mm per 3 m run) will allow for some possible structural movement.
- Specify how gutters are to be supported:
  - — fascia brackets at not more than 1 m spacing, or closer if recommended by manufacturer;
  - — 'rise and fall' brackets, if used, should be bedded at least two brick courses below top of wall.
- Specify how downpipes are to be supported:
  - — a 30 mm minimum gap is needed between pipe and wall if either needs periodic painting.
- Check accessibility for rodding.

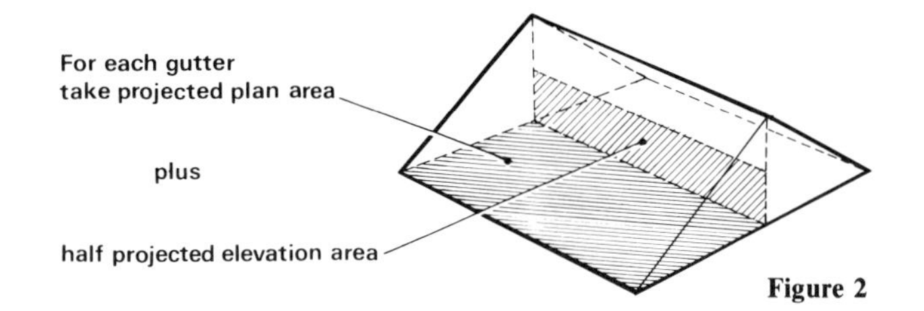

Figure 2

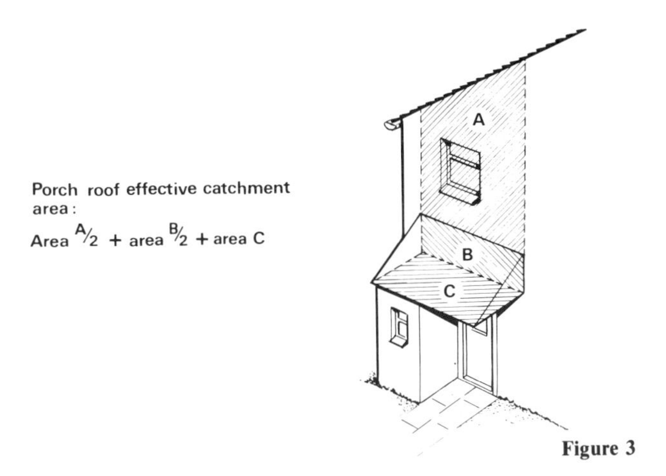

Figure 3

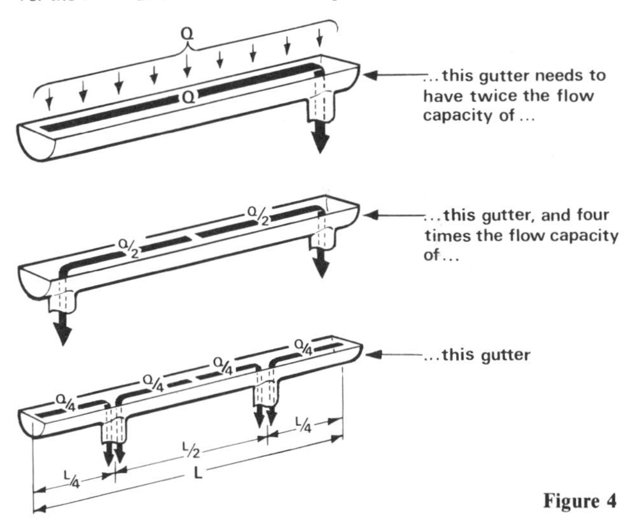

Figure 4

Table 1 Flow capacities of standard eaves gutters (level)

Size of gutter (mm)	Flow capacity (litres/second)	
	True half-round	Nominal half-round
75	0.38	0.27
100	0.78	0.55
115	1.11	0.78
125	1.37	0.96
150	2.16	1.52

REFERENCES AND FURTHER READING

1 *British Standard* BS 6367:1983. Code of Practice for drainage of roofs and paved areas. (Also gives guidance on valley and other gutters.)

DAS 56
July 1984

Site
CI/SfB 8 (52.5)(D6)

Roofs: eaves gutters and downpipes — installation

FAILURES: Rain penetration of external walls; deterioration of walls

DEFECTS: Insecure fixings; gutters installed with backfalls; gutters not installed correctly relative to roof edge; downpipes mispositioned relative to gullies; gutters and downpipes distorted; unnecessary delay before connecting downpipes to gutters

Installation of gutters and downpipes was faulty in various ways on many houses in a BRE survey. Gutters were poorly supported and downpipes were needlessly complicated due to mispositioning. Gutters and particularly downpipes were not fixed as soon as practicable, leading to saturation of walls and to risk of rain penetration and damage, eg by frost.

If *gutters* have *inadequate support* they *sag* and *overspill.* If *plastics rainwater goods* are installed *without allowance* for *thermal movement* they *distort. Water spilling* from *distorted gutters* and *downpipes* or from *unconnected gutter outlets saturate* and may *penetrate* external *walls.*

Figure 1 Walls saturated when connection of downpipes delayed

PREVENTION

Principle — All the water draining from a roof must be conducted, without spillage, to its below ground drainage system.

Practice

- Check that specified types and sizes of gutters and downpipes are used.

- Ensure that gutters are laid to falls:
  — domestic eaves gutters should be laid to a minimum fall of 1 in 350 towards the outlet, about 10 mm in 3 m.

- Ensure that gutters are correctly positioned and aligned:
  — gutters should be fixed with centre line vertically below the edge of the roof covering and close beneath it, Figure 2. Sarking felt (*see DAS* 9[1]) should be dressed into gutter.

- Check that gutters are not twisted or tilted sideways, Figure 3.

- Ensure that outlets are correctly positioned relative to gullies.

- Ensure that gutters and downpipes are well supported with fixings specified. It is important that gutters do not sag:
  — fascia or rafter brackets should be no more than 1 m apart, or closer if recommended by manufacturer;
  — additional support for gutters will be needed at angles and outlets, and intermediate support for downpipes over 2 m long.

- Ensure that joints in gutters and between gutter outlet and downpipe are sealed according to manufacturers' instructions.

- Ensure that gaps for thermal movement are left when jointing plastics gutters and downpipes, Figure 4. Gutters exposed to the sun can reach temperatures well above air temperature:
  — Table 1 gives suggested allowances per socket or gutter joint where no 'gap setting mark' is incorporated in the product. Downpipes should be jointed by 'pushing home' and withdrawing the spigot from the socket by the appropriate allowance.

- Ensure that gutters and downpipes are fixed in position as soon as possible after the roof covering to avoid saturating the constructed work.
  Note: eaves fascias should be fully painted before gutters are installed.

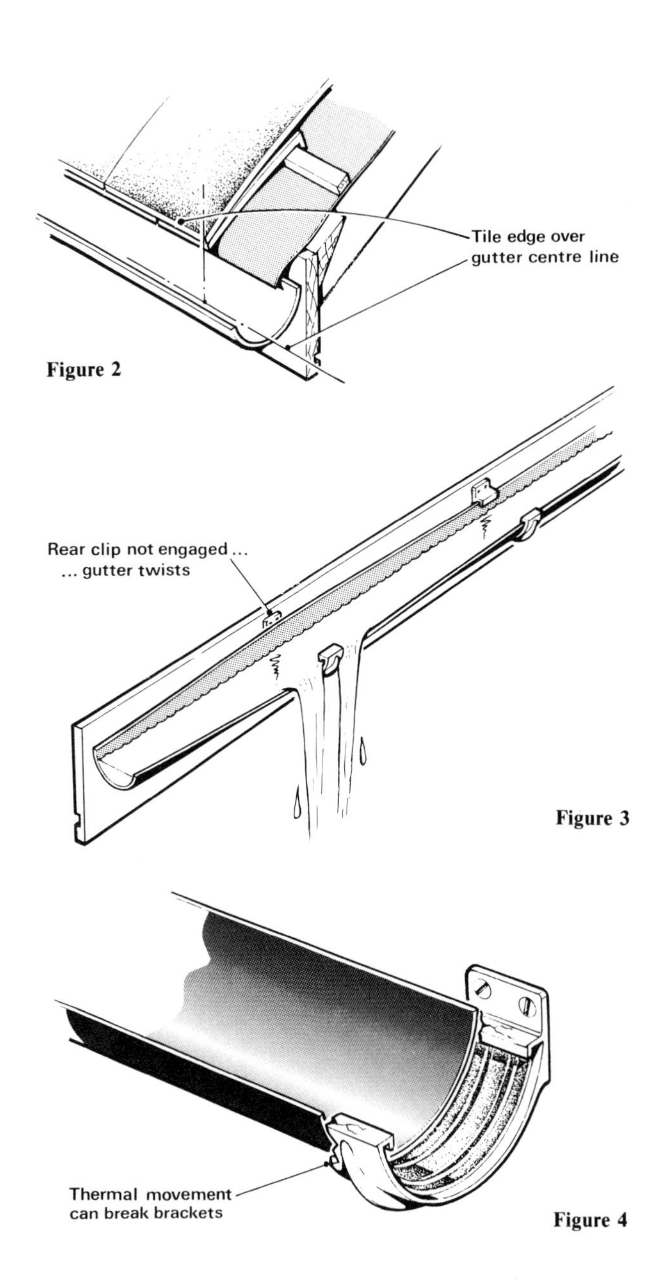

Figure 2

Figure 3

Figure 4

Table 1

Gutter/pipe length	Joint gap at installation	
	Winter	Summer
up to 2 m	7 mm	5 mm
2—4 m	10 mm	7 mm

REFERENCES AND FURTHER READING

1 *Defect Action Sheet* DAS 9 'Pitched roofs: sarking felt underlay — drainage from roof'.
Defect Action Sheet DAS 39 'Plastics drainage pipes: laying, jointing and backfilling'.

PART FOURTEEN

Water Supply and Electrical Services

DAS 141

February 1990

Design

CI/SfB 8 (53.9)(A3u)

Water storage cisterns: warning pipes*

FAILURES: **Collapse of ceiling; loss of strength of chipboard floors; damage to wiring, decorations, furnishings etc. Water damage and staining to outside wall.**

DEFECTS: **Warning pipe of inadequate bore, sagging or with insufficient slope, not protected against freezing. Pipes positioned over doors, windows, balanced flue terminal outlets or where accidental damage is likely. Pipes not vandal resistant. Pipes not suitable for boiling water discharge.**

Figure 1

Cisterns are required by the Model Water Bye-laws (Reference 1) and BS 6700 (Reference 2) to have a warning pipe which indicates water wastage. Cisterns up to 1000 litres capacity require a warning pipe which also serves as the overflow pipe.

It is quite common for float-operated valves to fail after 3 years service and sometimes earlier. Sometimes the failure permits unrestricted flow from the inlet, with costly consequential damage. BRE surveys have found many warning pipes that potentially could not deal even with partial failure of the float-operated valve.

If warning pipe ***bores*** are ***too small,*** or pipes ***sag*** or have inadequate slope — particularly if they are ***not lagged*** or kept ***beneath loft insulation*** to avoid freezing — ***cisterns*** may ***overflow*** causing both ***internal*** and ***external damage.***

* This DAS supersedes DAS 61 Cold water storage cisterns: overflow pipes

PREVENTION

Principle — Warning pipes should be of sufficient bore and slope, to minimise the risk that the cistern will overflow. The pipe material should be rigid and should be suitable for both service and excess temperatures. The pipe should be located to take account of the risk of accidental or deliberate damage.

Practice

- Specify that the warning pipe is of rigid corrosion resistant material and not less than 19 mm diameter.
- Specify that the warning pipe is to be turned down within the cistern and terminate 50 mm below normal water level:
  - — this prevents the entry of cold external air; (Outlet flaps are not recommended: they can freeze shut);
  - — alternatively the outlet can be turned vertically down or terminated in a tee-piece.
- Specify that the float-operated valve is set so that the water level is at least 25 mm below the lowest point of the warning pipe connection, Figure 2.
- Specify that with a vented heating package or hot water supply installation the warning pipe is suitable for working temperatures up to 100°C and its discharge is not directly over a balanced flue terminal.
  - — unvented hot water systems should be in accordance with the Approved Document G3 (1990) of the Building Regulations.
- Specify that the warning pipe is free draining.
- Specify that the warning pipe is adequately supported to avoid sagging.
- Specify that the warning pipe is either lagged or loft insulation is laid over the pipe to prevent freezing, Figure 3.
- Specify that the warning outlet projects sufficiently to overflow clear of the building. The discharge should avoid building features such as windows, doors, balanced flue terminals:
  - — above three storeys internal discharge should be considered;
  - — the outlet should be positioned to minimise vandal or accidental damage.
- Specify that commissioning of the plumbing system should include flushing out pipes to remove grit to ensure correct functioning of any float-operated valve.
- Recommend to the building owner that a regular checking procedure should include renewal of float-operated valve diaphragms and that infrequently operated valves should be serviced i.e. feed and expansion cisterns.

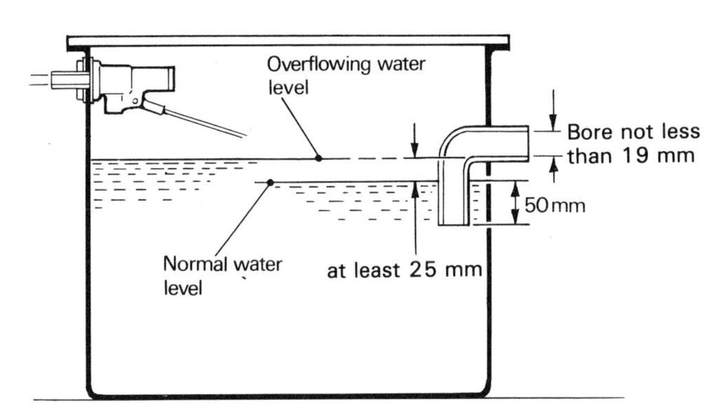

Figure 2

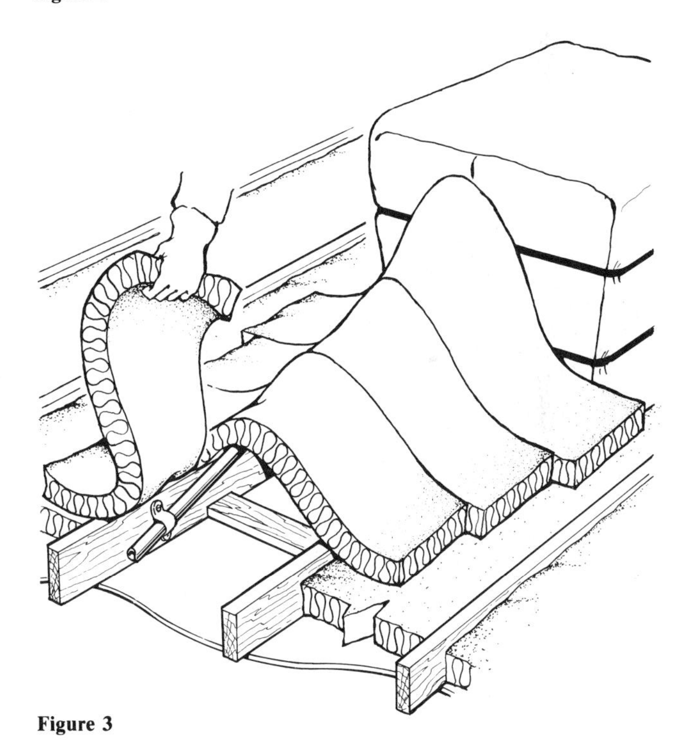

Figure 3

REFERENCES AND FURTHER READING

1 *Model Water Byelaws.* Department of Environment, Scottish Development Department, Welsh Office. London HMSO 1986.

2 *British Standard* BS 6700: 1987: Design, installation testing and maintenance of services supplying water for domestic use with buildings and their curtilages.

Hot and cold water systems — protection against frost

FAILURES: **Damage to building fabric and furnishings following a burst pipe; loss of supply through blocked pipes; risk of explosion of heating system**

DEFECTS: **Lack of, or insufficient insulation on pipes, cisterns and cylinders; pipe entering building at insufficient depth below ground; pipes in contact with outside wall**

Figure 1

Cost of repair to damage caused by a burst is very often out of all proportion to the cost of protection against frost. Figure 1.

If ***system layout*** and insulation is ***not arranged to protect water services, freezing,*** bursts and consequential ***damage*** may follow. At particular risk are pipes above insulation in well insulated lofts and near to ventilated eaves.

Even insulated pipes may freeze if the heating system is inoperative for sufficient time. Insulation only slows down the rate at which heat is lost. Therefore the system layout should not put pipes at avoidable risk of freezing.

Hot and cold water systems — protection against frost

DAS 109
November 1987

PREVENTION

Principle — Hot and cold water services should be positioned and insulated so that risk of freezing is minimised.

Practice

- Specify water services layout so as to eliminate pipe runs in unheated locations:
  - — unheated basements, floor voids, garages and outhouses are all high risk locations;
  - — pipe runs in roof voids should be below over-ceiling insulation;
  - — note that, even in heated areas, some locations may be appreciably colder, eg near permanent vents, windows and external doors or on, or in chases in, external walls.
- Specify that the underground service pipe should have a total depth of cover not less than 750 mm:
  - — local experience in some parts of the country may show that a greater depth is needed;
  - — ensure that underground stop valves are not brought up to a higher level.
- Specify, for dwellings with solid ground floors, either that the entry of the service pipe within the insulated fabric is at least 750 mm from the outside wall or that the service pipe is insulated to indoor standard (Table 1), Figure 2.
- Specify, for dwellings with suspended ground floors, that the service pipe is insulated to outdoor standard (Table 1) for the distance shown in Figure 3).
- Specify for pipes within the insulated fabric which are at risk from freezing (eg pipes unavoidably run close to air vents) insulation to indoor standard (Table 1).
- Specify insulation to surround top and sides of cold water cisterns in roof spaces:
  - — ensure that loft insulation does not cover the ceiling beneath the cistern;
  - — where the cistern is elevated in the roof, insulation is to enclose both cistern and supports to allow warmth from below to reach the cistern base.
- Specify that overflow pipes are to be lagged or laid beneath the loft insulation[1].
- Specify, for pipes run outside the insulated fabric, eg Figure 4, at least the thickness of pipe insulation given in Table 1.
  - — preferably specify an insulant that will not absorb condensate forming on the pipe and so reducing its insulation value (BS 6700: 1987 recommends a vapour barrier round the insulation);
  - — ensure that fixings for pipes and fittings allow enough room for insulation with material of the required thickness;
  - – instruct site staff to ensure insulation is kept dry before fitting, installed with no gaps and continuous over valves (with provision for their operation).
- Specify suitable coverings for the insulation where needed to protect against draughts, mechanical damage, rain, damp atmosphere, subsoil water and vermin.
- Specify heat trace cables for pipes where the consequences of a burst are unacceptable and the risk is high.
- Consider specifying a frost-stat on the heating system.
- Specify drain cocks located so that both hot and cold water services can be properly drained:
  - — where a building may be unoccupied for long periods it is safest to drain the system of water.

Table 1 Minimum thickness of insulating materials (Data taken from BS 6700:1987)

Outside diameter of pipe mm (nominal)	Thermal conductivity of insulating materials not exceeding: 0.035 W/(m.K) Indoor mm	0.035 W/(m.K) Outdoor mm	0.04 W/(m.K) Indoor mm	0.04 W/(m.K) Outdoor mm	0.055 W/(m.K) Indoor mm	0.055 W/(m.K) Outdoor mm	0.07 W/(m.K) Indoor mm	0.07 W/(m.K) Outdoor mm
Up to 15	22	27	32	38	50	63	89	100
Over 15 up to 22	22	27	32	38	50	63	75	100
Over 22 up to 42	22	27	32	38	50	63	75	89
Over 42 up to 54	16	19	25	32	44	50	63	75
Over 54 up to 76.1	13	16	25	25	32	44	50	63
Over 76.1 and flat surfaces	13	16	19	25	25	32	38	50

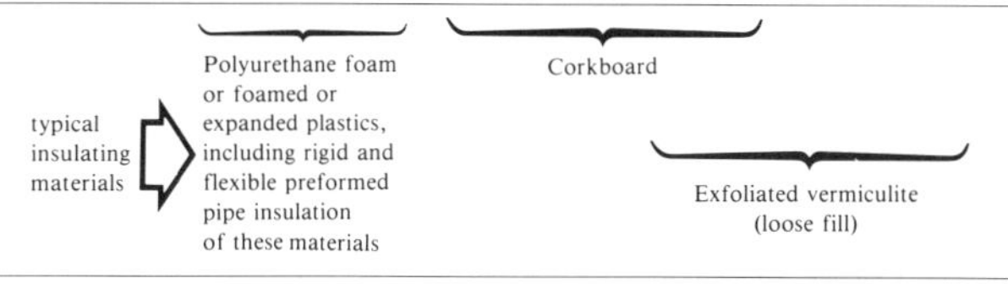

Note: insulation to outdoor standard is recommended for pipes outside the insulated fabric (including loft spaces and under floor void.

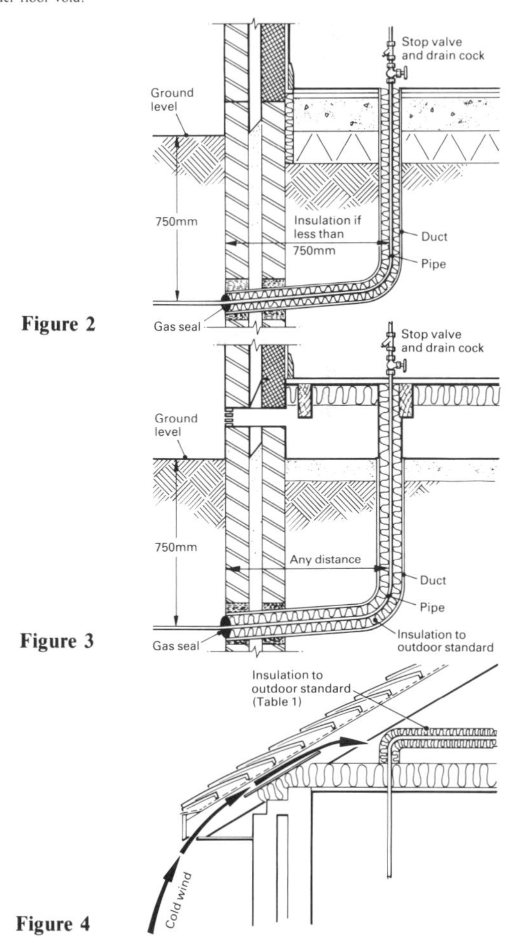

Figure 2

Figure 3

Figure 4

REFERENCES AND FURTHER READING

1 *Defect Action Sheet* DAS 61 Cold water storage cisterns: overflow pipes

☐ *British Standard* BS 6700:1987: Specification for design, installation, testing and maintenance of services supplying water for domestic use within buildings and their curtilages.

☐ *British Standard* BS 5422:1977: Specification for the use of thermal insulating materials.

☐ *British Standard* BS 3958 Parts 1-6 Thermal insulating materials.

DAS 108
September 1987

Design
CI/SfB 8(27.9)Xi(A3u)

Domestic hot water storage systems: electric heating — remedying deficiencies

FAILURES: Insufficient domestic hot water; high electricity bills

DEFECTS: Cylinder too small, poorly insulated; heater operating in inappropriate part of the cylinder; system operating on unfavourable tariff

Electric immersion domestic water heating systems installed some years ago are likely to provide less efficient service than can now be obtained. Old systems do not usually have sufficient hot water storage capacity to make best use of today's tariffs. Water in the lower part of cylinders remains unheated and cylinders are usually poorly insulated. Top entry single boss 3 kw on-peak immersion heaters do not, in general, provide a satisfactory supply of hot water at an acceptable cost. Where electricity is the main fuel and the existing system is outdated, changes will be needed to improve the service and to reduce running costs.

If hot water ***cylinder*** is ***too small*** for the household's use of hot water, and therefore too small to take advantage of cheap rates, ***if*** the ***immersion heater*** heats only the top of the cylinder, or if the ***cylinder*** and pipes attached to the cylinder are ***insufficiently insulated,*** then ***supply*** of hot water will be ***poor,*** and ***electricity bills*** will be ***high.***

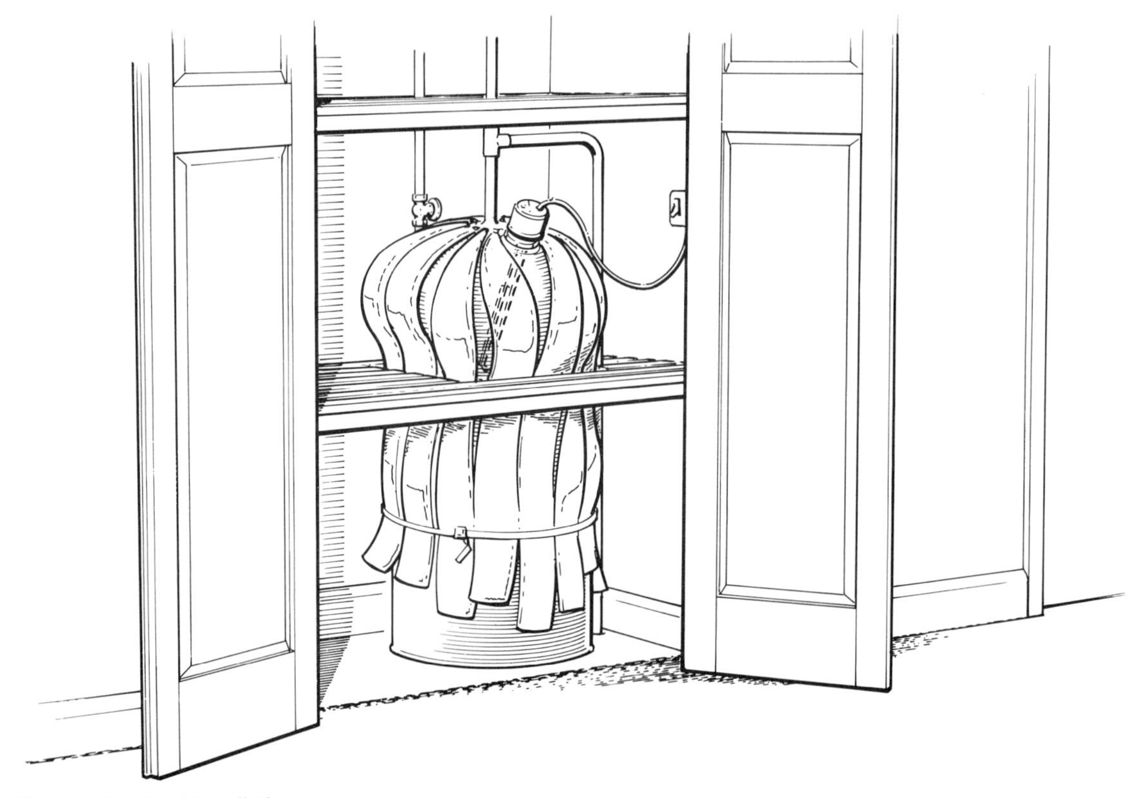

Figure 1 Outdated installation

REMEDIAL WORK

Principle — Hot water systems should be economic to run and thermally efficient.

Practice

- Specify cylinder of adequate capacity;
  - — large cylinders are needed to take full advantage of overnight tariffs;
  - — 210 litres capacity is preferable for a 4 person household, 120 litres is too small.

- Specify cylinder with two bosses and two elements, Figure 2;
  - — where an existing cylinder with one boss is to be retained, replace immersion heater with one with dual element, Figure 3. Note, the thermostats supplied with dual element heaters give closer temperature control.

- Specify cylinder insulation to 'Electricity Board Economy 7' or 'White Meter' specification (a higher standard than given in BS5615);
  - — positive economic benefits can be achieved by insulating cylinders well[1];
  - — factory insulated cylinders are significantly more thermally effective.

- Specify insulation of the first metre of expansion and hot water supply pipe(s) with not less than 15 mm thickness of insulation, Figure 4;
  - — note the need to insulate pipes run in ventilated voids.

- Specify thermostat settings in the range 60-65°C for lower heater element and 50-55°C for upper;
  - — the lower element heats the whole tank at cheap rate; the upper element is in water hotter than its thermostat setting and does not cut in;
  - — the upper element only comes into play when the hot water is almost used up and the next off-peak period has not yet begun.

- Negotiate tariff change:
  - — Local Electricity Boards are willing to advise.

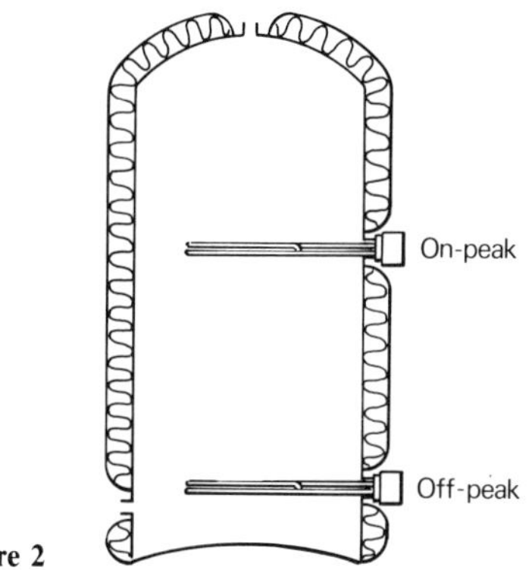

Figure 2

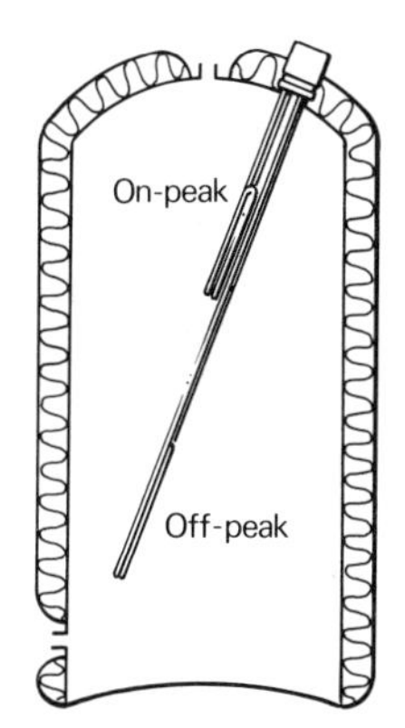

Figure 3

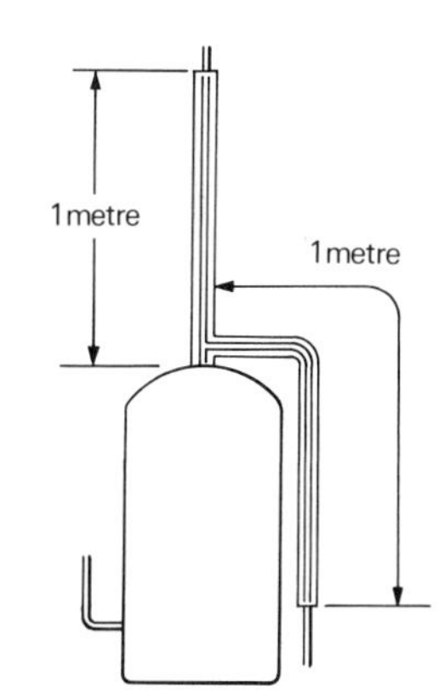

Figure 4

REFERENCES AND FURTHER READING

1 **BRE** Information Paper IP1/87 Investing in energy efficiency: 4 Domestic hot water systems.

☐ **Energy Efficiency Office** Cutting home energy costs. A step-by-step Monergy guide.

☐ **Department of Energy** Domestic Energy Note No. 3 DOE 1978.

Unvented domestic hot water storage systems — safety

FAILURES: Explosion of the cylinder; wastage of expansion water

DEFECTS: System incorrectly installed; lack of maintenance

Figure 1

Traditionally the hot water storage system in the UK has used a cistern to supply water to a hot water cylinder protected in the event of failure of temperature controls by an open vent pipe which relieves any steam generated and protects against rupture of the cylinder by expansion of water.

Changes in the Model Water Byelaws will now permit a hot water storage cylinder in excess of 15 litres capacity to be fed directly from the water main as shown in Figure 2. Statutory safety requirements of unvented hot water storage systems are included in the Building Regulations 1985 — Part G3*.

The cold water pipes to the building may be taken off the supply either before or after a pressure reducing valve A (see Figure 2). A check valve B, is installed downstream of the cold water branch to prevent hot water from flowing back into the cold water pipework.

Certain control devices are needed to guard against excessive pressure in the system. An expansion vessel C, is needed to accommodate within the system the increased volume of heated water. If this vessel fails

*Compliance with the class relaxation No 102 is needed to meet the Scottish Regulation Standard. (In Scotland new regulations are expected to come into force which will be broadly in line with the revised Regulations in England and Wales.)

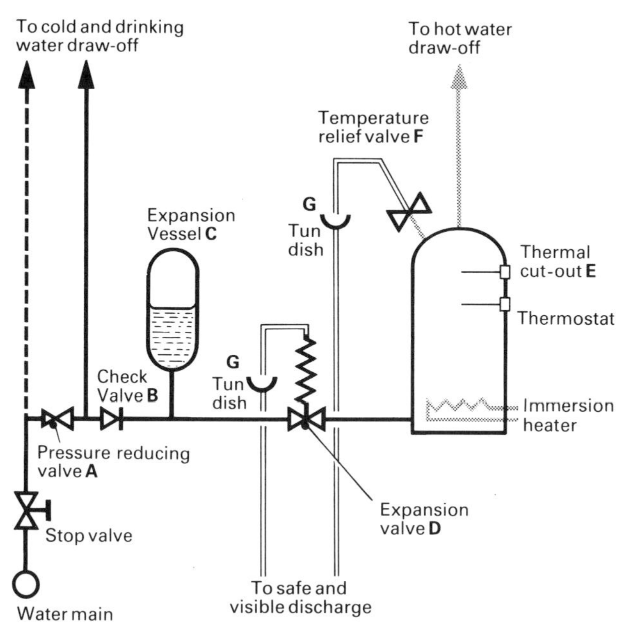

Figure 2

its failure will be indicated by a discharge from the expansion valve D, which also maintains the system working pressure.

To protect against over-heating following failure of the control thermostat, a non self-resetting thermal cut-out E**, acting on each energy supply is fitted. If this should fail explosion protection is still provided by a temperature relief valve F. Both expansion valve and temperature-relief valves discharge via tun dishes to a safe and visible place, eg. over a gully.

If ***over-pressure*** develops in an ***unvented hot water cylinder,*** the system may ***rupture;*** if the water ***temperature exceeds 100°C*** the ***rupture*** will be an ***explosive*** one, Figure 1. Hence a requirement of Building Regulation is that there shall be adequate precautions to prevent the temperature of the stored water at any time exceeding 100°C.

PREVENTION

Principle — The stored water temperature in unvented hot water systems must not exceed 100°C at any time.

Practice

- Specify, in England and Wales, to follow the guidance in approved document G3, an unvented hot water system in the form of a proprietary unit or package that has British Board of Agrément (BBA) certification, or a system that has been designed by a qualified person as appropriate:
  — the installation is to be carried out by an Approved Installer as defined in the BBA certificate.

**Alternatively a low head vented primary circuit may be used.

- Specify that discharges from the discharge pipes (or pipe if coupled together) from the temperature relief valve and from the expansion valve are safely conveyed to where they are visible but will cause no danger to persons using the building:
  — the pipes should be metal;
  — the discharge pipe(s) from each temperature relief valve must be of the same nominal size as the valve and must discharge either directly, or by way of a manifold sized to take the total discharge from the pipes connected to it through an air break over a vertical tun dish. The tun dish must be adjacent and close to the storage cylinder. After the tun dish the discharge pipe and fittings must be at least one size greater and be laid to a continuous fall;
  — the discharge pipes should be located so that the discharge is readily visible.

- Inform the building occupier that maintenance is required on the system; this should be contained in the instruction manual supplied with the system and may recommend that:
  — the expansion and temperature relief valves are manually operated once or twice a year;
  — the pressure in the expansion vessel be checked annually and adjusted if necessary to that specified by the manufacturer;
  — if a discharge is noted from any valve or if any other fault develops then the manufacturers' operating instructions gives details of the action to be taken;
  — any servicing and/or replacement of any components must be carried out by an Approved Installer only.

FURTHER READING

British Standard BS 6700: 1987 Design, installation, testing and maintenance of services supplying water for domestic use within buildings and their curtilages.

BRE Digest 308 Unvented domestic hot water systems.

BRE Report Unvented domestic hot water systems — BR 125 1988.

The Building Regulations (England and Wales) 1985: Approved Document G — Part G3, 1990 edition.

The Building Standards (Scotland) Regulations 1981: Part P (as amended 1986).

Defect Action Sheet DAS 140 Unvented domestic hot water systems — installation and inspection.

Unvented domestic hot water storage systems — installation and inspection

FAILURES: **Failure of heating controls leading to explosion of the cylinder; scalding from steam and hot water; wastage of water**

DEFECT: **System incorrectly installed, commissioned and maintained**

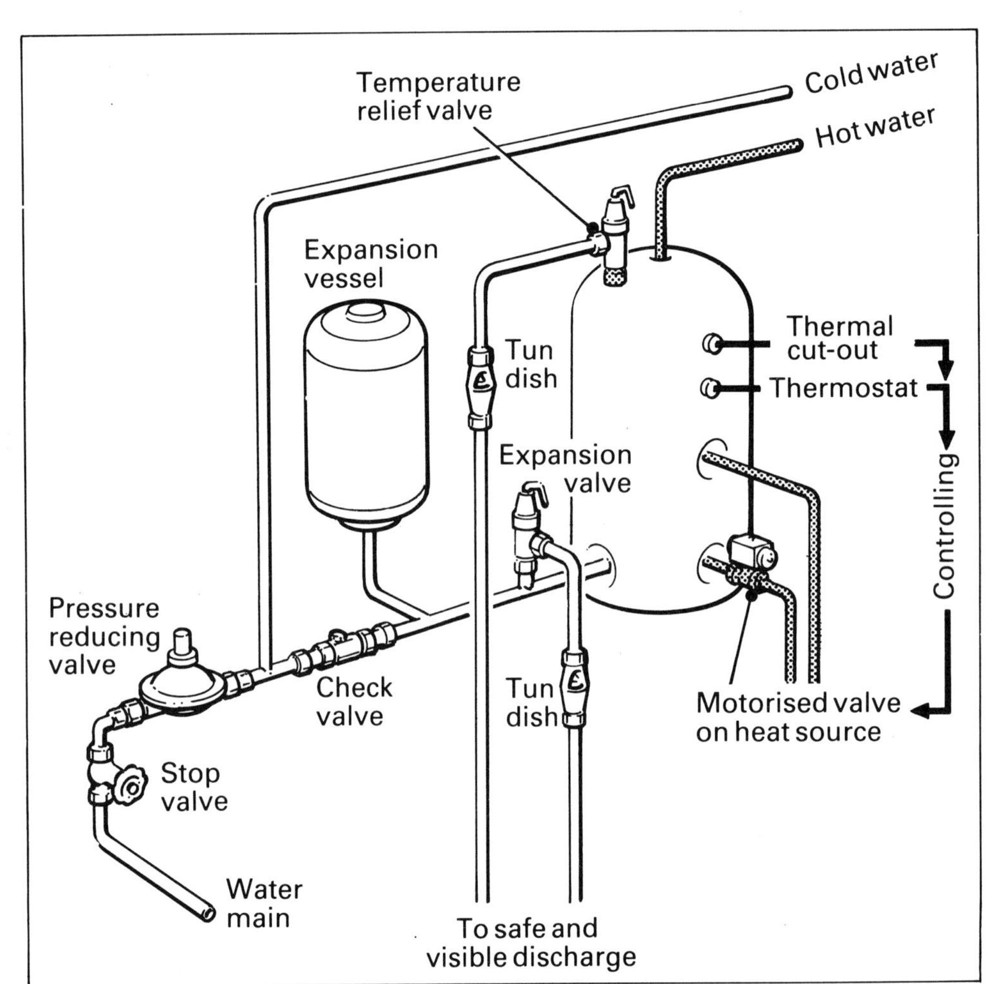

Figure 1
Diagrammatic layout of an unvented hot water system

Changes in Model Water Byelaws now permit the storage in quantity of hot water in an unvented hot water system. The Building Regulations 1985 contain safety requirements for such installations. Valves are generally required to prevent hot water back flowing into the cold water supply pipework and safety devices are required to ensure that the system does not explode.

If an ***unvented domestic hot water system*** is ***incorrectly installed, commissioned*** and ***maintained*** then there is a ***risk*** that the cylinder ***may explode.***

PREVENTION

Principle — The stored water temperature in unvented domestic hot water systems must not exceed 100°C at any time.

Practice

- Use inspection check list on completed installation (Table 1).
- Check the functioning of the hot water system.
  - — water temperature should be maintained at desired temperature;
  - — in normal operation, no water should emerge from discharge pipe(s);
  - — flows from all taps should be adequate but not excessive.

Table 1 Inspection check list

1 BBA approval of operative who installed the system (if self-employed: see 2).

2 BBA approval of firm who installed the system.

3 BBA Certificate for unit or package installed or assurance that the system is designed by a suitably qualified person.

4 The discharge pipe, made of metal and of the same size as the valve discharge connection from each temperature relief valve must discharge either directly, or by way of a manifold sized to take the total discharge from the pipes connected to it through an air break over a vertical tun dish. The tun dish must be adjacent and close to the storage cylinder. After the tun dish, the discharge pipe and fittings must be at least one size greater and be laid to a continuous fall.

5 Discharge from temperature relief valve(s) discharge pipe(s) are located so that it is readily visible (but with no risk to people) eg over a gully.

6 Discharge pipe from expansion valve complies with 4 and 5.

7 Indirect system used with 'units' or 'packages' (see Note 1), must have non-self resetting thermal cut-outs correctly wired to a motorised valve on the primary coil for control of excessive temperature.

8 A temperature relief valve to BS 6283: Part 2 or Part 3 and of appropriate discharge capacity rating is correctly fitted directly to the hot water storage vessel (see Note 1).

9 A non-self resetting thermal cut-out* (to BS 3955 for electrically operated devices or BS 4201 for gas supplied devices) is fitted to every energy supply (see Note 2).

10 A thermostat listed in the BBA Certificate is fitted to every direct heat source (see Note 2).

11 An expansion vessel to BS 6144: and an expansion valve to BS 6283: Part 1 are fitted on the cold feed pipe to the hot water storage cylinder and no valve or other possible obstruction is fitted between this cylinder and the expansion vessel and valve.

12 Check that all operating devices supplied as a kit for fitting on site by the installer have been fitted correctly.

13 Check that all instructions and any other documentation have been left with the system or with a responsible person (eg the householder).

*Alternatively a low head vented primary circuit may be used.

Note 1 A *unit* is a water storage heater factory fitted with safety devices and incorporating any other operating devices. A *package* is a water storage heater factory fitted with safety devices and supplied with a kit containing any other devices to be fitted to the system by the installer.

Note 2 Items 8, 9 and 10 will have been inspected by the manufacturer to comply with BBA Certification and only need to be checked again after maintenance or if malfunction, damage or interference is suspected.

Note 3 All electrical wiring should comply with the 15th Edition of the Regulations for Electrical Installation issued by the Institution of Electrical Engineers; Gas Safety Regulations and Water Authority Bye-Laws must be observed.

Note 4 All systems must comply with the Building Regulations.

FURTHER READING

Defect Action Sheet DAS 139 Unvented domestic hot water storage systems — safety.

DAS 62
February 1985 New edition August 1985

Design
CI/SfB 8 (62)(61.9)(A3u)

Electrical services: avoiding cable overheating

FAILURE: Damage to cable insulation leading to risk of short circuit and fire

DEFECTS: Normal heat dissipation impeded; incorrect choice of cable type or size

Figure 1 Cracking of overheated cable

All electric cables give off heat in use dependent on the load they carry. The heat emitted by any cable operating within its design loading is normally safely dissipated. Overheating due to electrical load alone should not occur if installations complying with IEE Regulations[1] are not misused. But sometimes the normal heat dissipation from a cable is impeded by thermal insulation or because the temperature of its environment is raised by other heat sources. Consequent overheating of the cable may damage its insulation. Cables in power circuits — which may be loaded to full capacity — are more at risk than those in lighting circuits.

In both new and rehabilitation work design must take account of ways in which overheating can occur in otherwise correctly designed and used circuits.

If design does not take into account that ***heat dissipation*** may be ***impeded due to*** the presence of other ***heat sources or*** of ***thermal insulation, cable may overheat — with risk of short circuit or fire.***

PREVENTION

Principle — Cables must not be prevented from dissipating, adequately, the heat generated by their designed loading.

Practice
(Starred items may be more likely to arise in rehab. work)

- Ensure that the location of cables relative to thermal insulation or to sources of heat (such as hot water cisterns or pipes) is taken into account when cable size and type are specified (and check that no subsequent layout changes will alter requirements for cables):
  - — power cables run within, or in contact with, thermal insulation must be suitably derated, Figure 2 (ie increased in size for a given current);
  - — cable exposed to high ambient temperature must be of appropriate type — eg mineral insulated copper clad. (Plastics sheathed cable is normally acceptable in a cupboard containing an insulated hot water cistern, but heat-resisting cable should be used to connect an immersion heater to its isolating switch). IEE Regulations[1] require cable to be derated where ambient temperature exceeds 30°C.

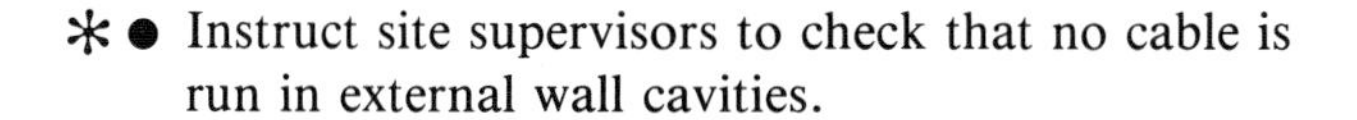

*● Instruct site supervisors to check that no cable is run in external wall cavities.

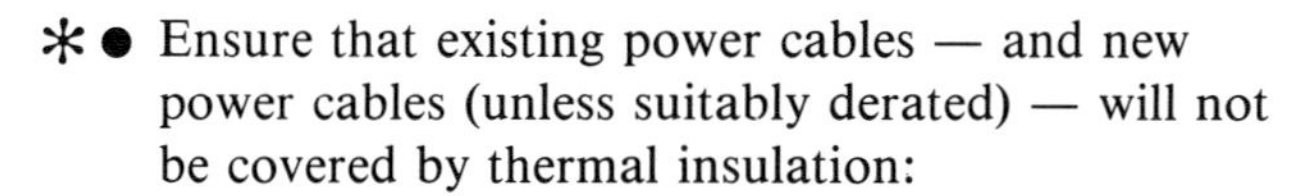

*● Ensure that existing power cables — and new power cables (unless suitably derated) — will not be covered by thermal insulation:
  - — evidence obtained by the Electrical Research Association suggests that cable in locations such as that in Figure 3 should preferably be derated by factor of 0.9;
  - — cable can be kept clear of insulation as in Figure 4 where joist depth permits. (Note: cable must not be installed so that it will be at risk of mechanical damage).

- Check that expanded polystyrene insulation will not be in contact with PVC covered cable : PVC cable sheath will otherwise be attacked.

- Check that no significant cold bridges will occur where insulation must be kept away from fittings that emit heat, Figure 5.

REFERENCE AND FURTHER READING

1 Regulations for Electrical Installations, 15th Edition 1981. The Institution of Electrical Engineers.

Fire hazard from insulating materials, BRE Digest 233, January 1980.

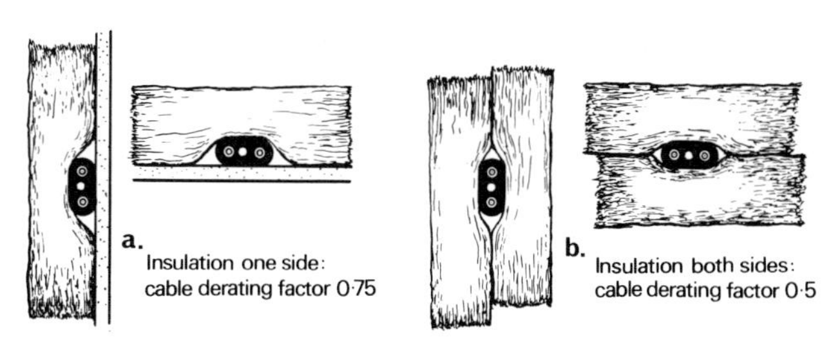

Figure 2

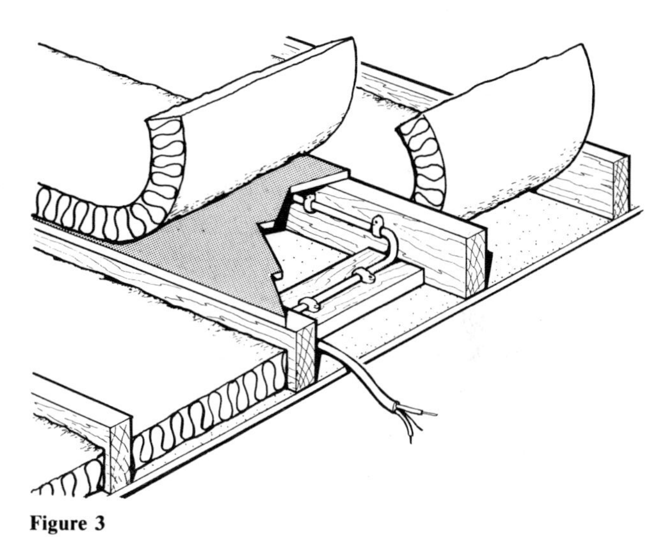

Figure 3

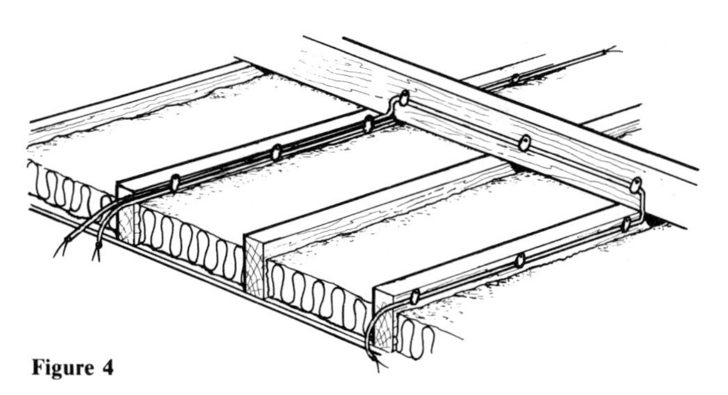

Figure 4

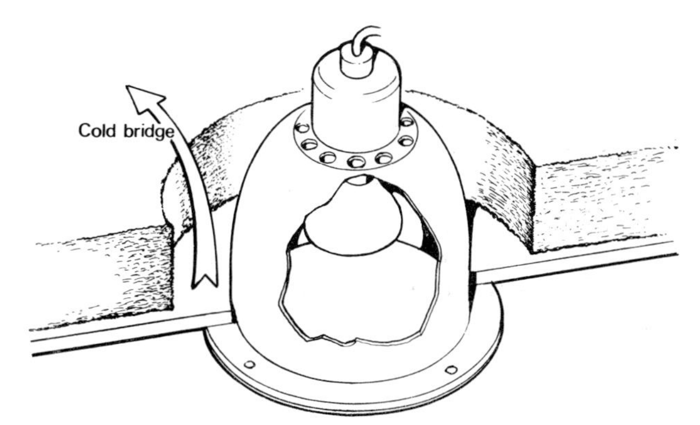

Figure 5

DAS 48
March 1984

Site
CI/SfB 8 (61)(Q2)

Electric cables — stripping plastics sheathing

FAILURES: **Loss of electrical insulation between live and earth; tracking and arcing; possible ignition of combustible materials.**

DEFECT: **Cable sheath improperly stripped when making connections**

Figure 1 Arcing and discharge of molten copper

Instances of arcing have occurred in installations with plastics-covered 'flat twin with earth' wiring cable due to the outer sheathing being removed badly. The defect responsible is probably quite common but, when arcing results, the charring of cable insulation is often blamed wrongly on a bad (overheating) connection. The proportion of fires due to the defect is not known since wiring is damaged in any resulting fire. Every year about 2240 domestic fires have their source of ignition ascribed to fixed installations of electric wire and cable.

If the ***wrong technique*** is used for removing the outer sheathing of plastics-covered flat twin cable, the ***insulation of the live conductor may be imperceptibly cut***. In the presence of even a small quantity of moisture, ***tracking and arcing can then be initiated*** between the live conductor and the bare earth conductor. Once this has occurred a permanent conducting path is formed and ***arcing may continue intermittently*** even in the dry. Molten copper droplets, at very high temperatures, can be violently discharged (Figure 1); ***the heat generated can be a fire risk.***

PREVENTION

Principle — Cable sheathing must not be stripped or trimmed in a manner that may damage the insulation of the conductors.

Practice

- Do not permit sheathing to be cut around the cable as the first step in stripping it (Figure 2a).
- Do not permit the ragged end of the sheathing to be trimmed around the cable after stripping it (Figure 2b).
- Ensure that cable sheathing is stripped safely — eg by making a small cut at the end of the sheathing and tearing the sheathing sideways from the cable (Figure 3a); the sheathing should be cut well clear of the internal wires (Figure 3b).
- Ensure that particular care is taken to strip sheathing correctly where cables run in roof spaces: here especially moisture from condensation, wind-blown snow etc may initiate tracking and arcing in cables close to combustible materials stored in the roof by the householder:
  - — earth sleeving does not remove the need for care: the sleeve ends where the live lead insulation is most likely to have been accidentally cut.

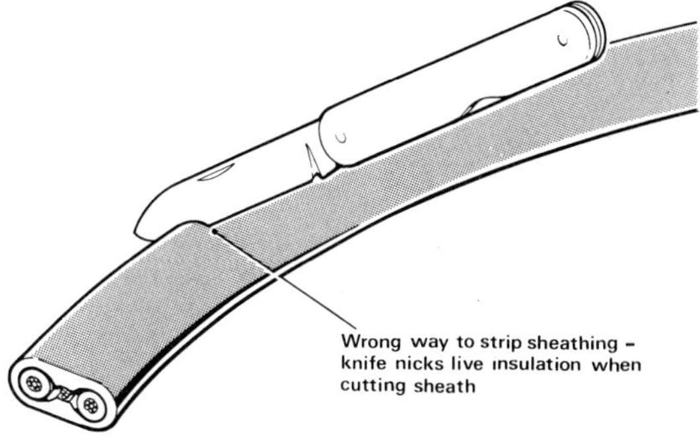

Figure 2a

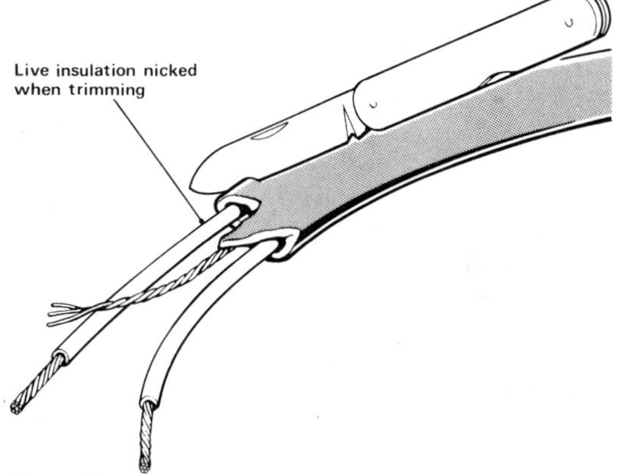

Figure 2b

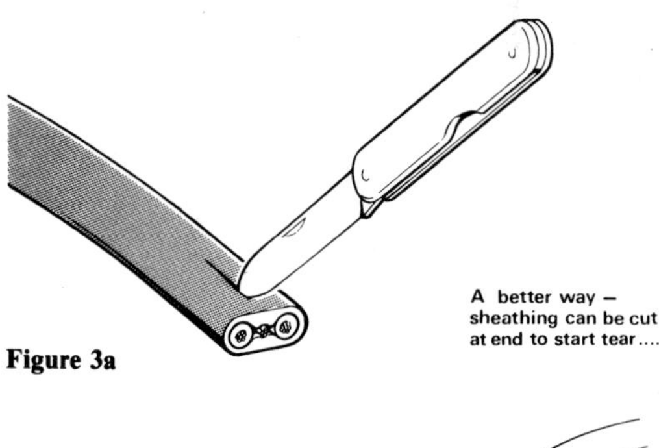

A better way – sheathing can be cut at end to start tear....

Figure 3a

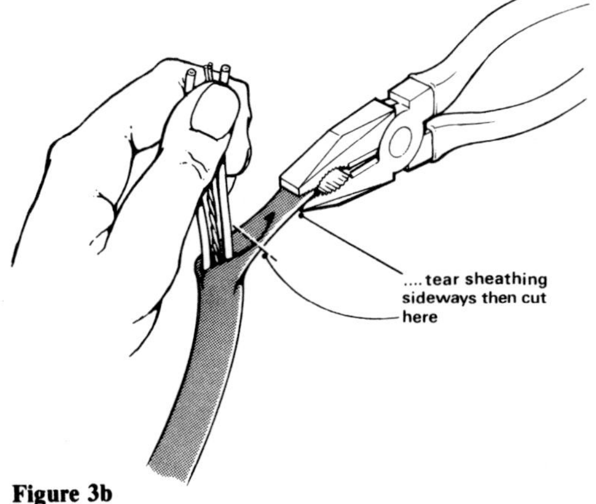

Figure 3b

Subject Index

DAS numbers are shown in square brackets, page numbers refer to the first page of the Defect Action Sheet.